Paul A. Elliott is Lecturer in History in the School
of Humanities, University of Derby. He is also a special lecturer at
Nottingham University. His research interests include historical and
cultural geography, history of science, landscape history, history of
education, and urban and regional history. His most recent book is
The Derby Philosophers (2009).

Enlightenment, Modernity and Science

Geographies of Scientific Culture and Improvement in Georgian England

PAUL A. ELLIOTT

I.B. TAURIS

LONDON · NEW YORK

Published in 2010 by I.B.Tauris & Co Ltd
6 Salem Road, London W2 4BU
175 Fifth Avenue, New York NY 10010
www.ibtauris.com

Distributed in the United States and Canada Exclusively by Palgrave Macmillan
175 Fifth Avenue, New York NY 10010

Tauris Historical Geography: 5

ISBN: 978 1 84885 366 9

A full CIP record for this book is available from the British Library
A full CIP record is available from the Library of Congress

Library of Congress Catalog Card Number: available

Typeset in Stone Serif
Printed and bound in Great Britain by
CPI Antony Rowe, Chippenham

Contents

Figures

Preface

Most of this book was written whilst working as a Research Fellow in the School of Geography at Nottingham University and completed as Lecturer in Modern History in the School of Humanities at Derby University. I am grateful to Robert Mayhew, David Stonestreet, Jayne Ansell and I.B.Tauris for publishing the book, Alan Mauro for formatting and typesetting, and to the anonymous readers who offered criticisms and suggestions. Special thanks are offered to Charles Watkins, Stephen Daniels, Susanne Seymour, Mike Heffernan and other colleagues at Nottingham and to history and humanities colleagues – Ian Barnes, Robert Hudson, Ruth Larsen and Ian Whitehead – and students at Derby for encouragement in a friendly and scholarly environment. Professor John Beckett of the School of History, Nottingham University and the late Neville Hoskins, President of the Thoroton Society encouraged me to explore the history of Nottingham and Nottinghamshire science. I am pleased to acknowledge receipt of a research award from Derby University, which helped with the completion of the book in 2009 and 2010. Over the past twelve years or so, colleagues and friends associated with the Centre for Urban History at Leicester University have also offered much support, particularly Roey Sweet, Richard Rodger and Simon Gunn. The book has also greatly benefited from assistance provided by many library and record office staff especially those of local studies libraries, county record offices and university libraries in Nottinghamshire, Derbyshire, Leicestershire, Norfolk, Yorkshire and Cambridgeshire. On a more personal level, I'd also like to thank my mother Kathleen, my brother Christian, sisters Magdalen and Bernadette and Stuart Eames for their love and encouragement.

As will be evident, amongst numerous academic influences that have inspired the book I would single out in no particular order, Vladimir Jankovic and Jan Golinski for their analyses of Enlightenment science and cultural histories of science, David Livingstone and Charles Withers for their historical-geographical studies of geography and the sciences, Roy Porter's work on Georgian medicine and science, Ian Inkster's studies of the history of science and technology, Desmond King-Hele for his work on Erasmus Darwin, Roey Sweet's books on eighteenth-century urban culture and antiquarianism and Stephen Daniels' analyses of landscape history, Georgian culture and the history of geographical education. As the bibliography and notes demonstrate, a general book of this kind, which wanders through varied fields, relies upon the work of many different scholars. It is a particularly exciting time to study the historical geographies of science and scientific culture, and it is hoped that this book makes some contribution to what is a burgeoning interdisciplinary endeavour.

1 Introduction

Scientific culture was one of the defining characteristics of the English Enlightenment permeating many aspects of Georgian society and culture. In his famous essays on pleasures of the imagination from the *Spectator*, Joseph Addison emphasised the sublime stimulation to the imagination and challenges to the understanding presented by the 'new' natural philosophy.[1] Stimulated by growing British economic and naval power and encounters with lands and peoples new to western eyes, the numerous exotic specimens acquired for private collections, museums and botanical gardens challenged intellectual orthodoxies. At the same time the proliferation of print and ideas and opportunities for travel and tourism by improved communications from superior ships to turnpike roads, bridges and canals fostered philosophical networks and created new audiences for the sciences. Similarly, the expanding cosmos interpreted instrumentally and in terms of natural theology uncovered a multitude of apparently inhabited planets, microscopes revealed countless smaller worlds and miners and engineers exposed mysterious subterranean realms. Stafford has colourfully contrasted the geometrical intellectual simplification of 'Enlightenment epistemologies of schematisation, codification, classification and quantification' with the multifarious complexities and rich geographies of coterminous 'polyphonic, Epicurean, libertine, rococo and romantic epistemologies'.[2] Of course, many Georgians were fully aware of the contrast between a Platonic realm of ideas and divine perfection and everyday reality, and narrowing this chasm was a powerful motivational force. As Withers has emphasised, geographical analysis of the Enlightenment requires recognition of the importance of space 'as real territory, as "imagined" space, as social space, and as epistemological space and to the difference that space makes'.[3] Exploring some of these disjunctive intellectual, metaphorical and experienced spaces is the primary concern of this book. Inspired by recent work in cultural and historical geography, the history of science and urban history, particularly the spatial turn in social and cultural theory, and focussing particularly upon the English provinces, this study examines aspects of the geographies of Georgian scientific culture. The case studies explore various Georgian encounters with nature and the production and consumption of scientific ideas, exposing the English Enlightenment admixture of reverence and reinterpretation, tradition and change, deference and modernity.[4]

Although its interpretation remained contested, the Newtonian system provided the potential to manage understanding of these new vistas by maintaining order, limiting plurality and uncertainty using probability and calculus, reinforcing aspects of natural theology and inspiring visions in poetry, prose and painting. Metaphors of space and society further accommodated and encouraged measured change such as the popular concepts of the economy of nature and improvement applied to everything from turnips to roads and bridges – a spatial metaphor transferred from agricultural and horticultural spheres into urban space. The relationship between scientific culture and improvement in different contexts is one of the main themes of this book. With their emphasis upon chorography and relationship to antiquarianism, Georgian natural history, meteorology and related sciences such as astronomy and electricity were fundamentally situated and regionally differentiated endeavours. Enlightenment physiology and natural philosophy strove to bring mind, body and environment together with psychological systems that, visualised spatially, brought the potential for order and control in schools, hospitals and prisons. The associationism of John Locke and David Hartley provided an explanation for ethics and aesthetics whilst offering the potential to control mind and behaviour by ordering sensory perceptions and experiences through control of external spaces, attractive to revolutionaries and reformers alike by the 1790s.

English scientific culture developed as part of the creation of new public spheres of rational debate founded upon autonomous individuals. England's rise as a major global trading and imperial power and the degree of political stability experienced after the ruinous wars

2

of the seventeenth century fostered new kinds of public culture associated with urban renewal. Fire devastation created the opportunity to build a modern beautified and fashionable city whilst London's position as the greatest market and a global, imperial, political and trading centre enabled the growth of a prosperous and confident middling-sort merchant and professional class.[5] The multiplicity of scientific cultural spaces such as coffee houses and taverns allowed science to flourish in public spheres as part of polite knowledge and entertainment. Although English natural philosophy and natural history did not have the same role in shaping national identity as in Scotland, encouraged by the place of England within the British State at the heart of a growing empire, English natural philosophers and scholars (and their Scottish, Irish and Welsh counterparts residing in England) regarded themselves as participating in broader supra-national republic of letters. Whilst some celebrated the relative English constitutional freedoms and liberality, the international scholarly Enlightenment ideal offered powerful incentive for scientific pursuits. Although natural philosophy did play a role in fostering English identity and international rivalry, many Scots identified as British as much as Scottish and regarded themselves as participating in the kind of virtual European scholarly community investigated by Mayhew.[6] Scientific culture therefore contributed towards the development of multiple overlapping religio-political and spatial identities shaped by diverse local factors as well as national and international cultures. Divergence, contradiction, contingency, context and place were important, but only alongside the perception and experiences of international scholarly engagement.

Of course it was not all about serious rational learning and discovery. The pretensions of polite science on the part of everyone from parsons and preachers to monarchs, milords and merchants inevitably also attracted ridicule. One mid-Georgian coloured print shows aristocrats using astronomical instruments and models against a deliberately conventional background of classical ruins, heightening the ridiculousness of the scene. A lady peers down the wrong end of a telescope, one man looks through another and two others seriously contemplate an armillary sphere and celestial globe. One magnificent cartoon entitled 'The Quacks' pokes fun at public gullibility for science and novel electrical and magnetic medical treatments as well as James Graham and Gustavus Katterfelto, the objects of their seemingly misplaced trust. Similarly James Gillray's superb 'Scientific Researches! – New Discoveries in PNEUMATICKS!' (1802) depicts a varied assortment of socialites marvelling at Thomas Garnett and Humphry Davy's 'scientific' demonstration of some unintended consequences of nitrous oxide respiration at the fashionable Royal Institution lectures.[7]

Recent work on the spatial aspects and manifestations of social theory and cultural activity by historical geographers, historians of science and culture has helped to shift the emphasis from somewhat disembodied large-scale class models, grand unitary stories and theories of social causation. Post-modernist and post-structural scepticism towards traditional historical narratives has stimulated greater attention to the complexities, contingencies and divergences of phenomena as complex as identities and affiliations, which has had a major impact on – and indeed was partly driven by – recent work in the history of science. The older emphasis on the importance of conscious, rational and practical applications of scientific knowledge to technological and industrial innovation has been replaced by greater recognition of subconscious motivation, wider social and cultural influences, the breadth and subtlety of scientific 'audiences', irrational aspects of creativity and the multiplicity of formal and informal scientific sites and contexts. Jankovic, Livingstone, Withers, Ogborn and others have emphasised the cultural constructedness of spaces, places, institutions, identities and science itself as well as the interrelationships between natural philosophy and other aspects of intellectual endeavour and practice such as chorography, geography and meteorology.[8] One important example of this 'spatial turn' has been the mapping of the Habermasean public sphere and theories of modernity onto townscapes.[9] With respect to scientific culture the spatial turn was a further result of the sociological study of science, with its appreciation that the process of scientific change was mediated and negotiated through various fragmented, competing and localised spatial dimensions. In Shapin's words, 'science was not one thing – conceptually and methodologically unified … it was a variety of practices whose conceptual identities were the outcomes of local patterns of training and socialisation.' Though the ideal of an objective unitary science remains an essential ingredient of models of scientific

4

development even if the reality can never be attained, a conception of scientific progress requires both an ideal of objectivity and a recognition that it can never be fully realised – to paraphrase Alfred North Whitehead, the way has to be the reality.[10]

Close attention to the spaces of provincial scientific culture can transform our understanding of science and social marginality and difference. We can see this in various ways including analysis of spatial uses within scientific associations and institutions, those of collecting and experiencing science in the field and through the production of scientific knowledge and development of wider audiences for science. Most obviously, the spatial turn has encouraged the critical appreciation of a much wider range of sites for the production and dissemination of scientific knowledge and practices including museums, private collections, hospitals, botanical gardens, arboretums, parks, natural history societies and so forth. On the micro level, it has emphasised the importance of the objects in experimentally validated science, the design and decoration of buildings, layout of collections, and the meaning, use and display of objects and equipment both ornamentally and functionally such as globes, microscopes, telescopes, skeletons and all 'scientific' paraphernalia.[11] Medical institutions are, of course, of special interest here because of the intimate relationship between clinical theories, the design and usage of medical buildings, attitudes towards the physiology and psychology of the body and social, political and cultural imperatives towards control, freedom, individualism, identity, gender, sexuality and domination. Laboratories and museums are also, of course, equally susceptible to spatial analysis and have been used to demonstrate how much meanings ascribed to scientific knowledge transcend places of production, facilitating and transforming processes of definition, replication, recognition and the dissemination of scientific facts.[12] In rather different ways this was evident in Fleck's attempt to explore the travel and translation of laboratory facts beyond their immediate places of production during the 1930s and the operation of Latour's scientific networks from local to global. Shapin and Schaffer defined the consensus around the early Royal Society as a culture of experiment determined by 'technologies of fact creation' and the demarcation of facts, opinions and implied practices.[13] Another aspect of the spatial turn has emphasised the 'external' activities of scientific societies, such as the scope of fieldwork, including the collection of geological and botanical specimens from particular sites, which enabled the creation of county natural histories and geological maps, underlining the localised nature of negotiations conducted over the content, validity and status of scientific knowledge (Figure 1). Such work illustrates the move from broadly static and taxonomic early-modern natural sciences to the more active, experiential, negotiated, interdisciplinary and integrated nineteenth-century sciences identified by Foucault, whilst revealing the importance of the sciences in defining local identities. Naylor, for instance, has shown through a study of the Penzance Natural History Society with its museum, field sites and lecture hall how a set of key spaces were integral to the operation and output of such societies and to the creation and assertion of Cornish identity.[14]

1 View of Selborne and the countryside around after a drawing by Fahey from G. White, *The Natural History of Selborne* (1859)

Greater attention to the spaces of knowledge and practices has also encouraged the creation of a broader geography of scientific culture at the local, regional, national and international level which can distinguish particular characteristics. Insightful examples include the attempts to formulate geographies of Enlightenment in which scientific activity is related to national characteristics and identity, an early example of which are Henry Buckle's mid-nineteenth-century reflections on distinctive national characters of science.[15] Urban spatial typology has, of course, long been a feature of sociological study, as Charles Booth's late-Victorian coloured socio-criminological maps of London and the urban sociology of the Chicago School remind us. More recently this has been manifest in historical-geographical studies of varied nineteenth-century urban educational provision. This is reflected at two levels: the differentiation of intra-urban locations, including a consideration of the location of scientific societies and members within settlements,[16] and attempts to define settlement types according to religious, socio-political, politico-juridical characteristics and forge urban scientific associational typologies. As we shall see, analyses of county towns, for instance, have suggested that their juridico-political status facilitated richer and more diverse literary and scientific public cultures than other types of town.[17]

The advantages of geographical interpretations of scientific culture are manifest in two stimulating studies of British Georgian weather, meteorology and Enlightenment. Responding to his exhortation to face up to the challenge posed by chorography to analyses of Georgian provincial scientific culture and natural history, Jankovic has emphasised the importance of locality, continuity and tradition in eighteenth-century English meteorology in contrast to a 'change oriented' 'whiggish' perspective that interprets the endeavour in terms of progress towards nineteenth-century instrumentalism, institutionalisation and professionalisation. He links the 'meteoric tradition to society at large within the context of provincial naturalistic and humanistic scholarship'. Meteorological reporting was 'less an embodiment of the consensus about the cultural role of eighteenth-century naturalists' than 'a force which informed and defined such a role' located within a 'larger body of ideas concerning the nature of place and the place of nature'. For Jankovic, shaped by locality and rural and urban differences, eighteenth-century meteorology was overwhelmingly 'an aspect of geographical thinking' reflecting concerns with 'the creation and sustenance of national and parochial identities with respect to the moral topography of the land'.[18] Golinski defines the Enlightenment as 'the era when fundamental characteristics of modernity and its symptomatic attitudes to the natural world were forged'. He employs a study of British weather to argue that a new understanding of climate and the weather appeared in the period through the application of natural philosophy and natural history, stressing the 'unmistakably modern' aspects of changing attitudes to weather and climate. On the other hand he contends that although through scientific investigation Enlightenment weather was detached from the realm of divine providence, yet the climate also 'reflected the regular actions of physical laws' as 'manifestations of god's providential benevolence'. Meteorological studies also demonstrated to natural philosophers and others – and forced them to confront – the problems of the limitations of rationality and scientific study, so producing a delicate mixture of certainty and uncertainty, encapsulating the contradictions of modernity.

In his incisive study of the geographies of Georgian London, Ogborn argues that the concept of modernity provides the most important historiographical framework with which to judge changing spatial and cultural experiences of early-modern metropolitan life. In contrast to Ogborn and Golinski however, the concept of 'improvement' is favoured here because it conveys the complexities of changing Georgian scientific cultural experiences more effectively and avoids the dangers of anachronism and totalisation inherent in notions of modernity. With the possible exception of a few sceptics and the mad even the most implacable Enlightenment doubters and satirists did not fundamentally reject the possibility of meaningful knowledge whilst the distraught and despairing sought solace in the divine. It is comforting to embrace kindred spirits from centuries past, but to discern admixtures of doubt and certainty in eighteenth-century science and culture unjustifiably elevates disagreements concerning the scope of knowledge and forms of reasoning or satire into post-nineteenth-century nihilistic anguish. Whilst, as Ogborn and Golinski demonstrate so well, there was an important sense of progress in British Enlightenment culture and the terms 'modern' and 'progress' were in use to designate and celebrate change, contested notions of improvement more effectively carry the Georgian sense of advance built upon tradition rather than effacing the past evident in neo-classicism and the associations with rural agricultural change. Of course like concepts of the modern, progress and Enlightenment, designations of improvement were highly rhetorical, but this only underscores the importance of paying close attention to context, and in this case spatial context, the differences of area and culture that shaped variations in British scientific cultural experiences. As Jankovic demonstrates, concepts of modernity and progress are difficult to apply to the inchoate eighteenth-century sciences which were seldom regarded as unified endeavours except perhaps in terms of natural theology but divided into loose conglomerations of natural history, natural philosophy, mixed mathematics, meteorology and chorography.[19]

Whilst these approaches offer considerable insights into Georgian society and culture, they tend to emphasise aspects of Georgian scientific culture at the expense of others. It is important to recollect that the application of post-modern, post-structural, feminist and other bodies of ideas within the sociology of science has faced a rearguard action from scientific 'realists' who reassert the ultimate objectivity of science whilst emphasising the lack of scientific qualifications of many of the 'constructivists'. The importance of tradition, patronage and deference, the celebration and reinterpretation of the past apparent in natural theology and antiquarianism and the power of Enlightenment universalism, can be overlooked if too much emphasis is placed upon the geographies of nation, locality and modernity. Notwithstanding the attacks upon cultural histories of science by scientific objectivists and realists, geographical approaches to the history of science and scientific culture need not presuppose fragmentation of objectivist aspirations.[20]

Domestic spaces of natural philosophy have received relatively little attention compared to public arenas despite the central importance of intimate familial relations to the Georgians. Fostered by the desire for scientific education and the acquisition of polite knowledge, the first chapter shows how some aspects of natural philosophy gained particular associations with the domestic sphere such as geography, botany, astronomy, meteorology and electricity. Philosophers such as Joseph Priestley exploited the availability of everyday objects such as glass tubes, wool and the intimate use of the body as an instrument for experimental and demonstration purposes whilst meteorology was experienced as a domestic-centred activity. Ornamental and usable barometers and thermometers adorned walls and sometimes provided opportunities for constant empirical observation melded into the routines of domestic life and participation in Enlightenment scholarly progress. This is also evident in the importance placed upon scientific education for children, purchase and positioning of scientific instruments, popularity of tutors, publications of textbooks and other introductory works enjoyed in the private and semi-private domestic spaces. Scientific instruments were obtained for private education and as ornaments, toys and symbolic objects with special significance to particular social groups in the context of the domestic sphere, including Anglican clergy (many of whom took up electricity), women and Dissenters.

2 Mid-nineteenth century view of Soho House from S. Smiles, *Lives of the Engineers: Boulton and Watt* (1904)

For women, the wonders of science offered the opportunity, mentally, symbolically and sometimes physically to reach beyond the confines of the domestic sphere to the natural world and even the cosmos and experimental demonstrations specially adapted for females at home were devised. For many Dissenters, scientific education was inspired by the utilitarian Protestant ideals, the requirements of natural theology and desire for edification and enjoyment. Certain areas within homes were also meeting places for lectures, informal literary and scientific groups and more formal associations and provided a comfortable, cheap and more private alternative to renting or buying rooms in public houses and other locations. In turn, aspects of domestic culture shaped the way that such associations were conducted in other more public places. This confirms that distinctions between perceptions and experiences of public and private spheres should not be exaggerated. The middling-sort Georgian home was both a private and public place used by servants and traders as well as family members and visiting relatives, helping to explain why domestic culture mediated between the public and private spheres of natural philosophy. Finally, perceptions and experience of domestic life help to shape perceptions and interpretations of space in natural philosophy as analogously the constitution, monarchy and state were often celebrated as a family. Idealisations of the household economy provided a model for the economy of nature whilst the vast spaces of the cosmos were experienced in domestic life through journeys of the mind, telescopes, celestial globes and the clockwork mechanism of the orrery.

Often treated distinctly in the historiography, under the general guise of improvement there was a close relationship between urban scientific culture and agricultural and horticultural improvement. Far from being disembodied and placeless, natural philosophy and natural history were tied in very special ways to the land helping to explain the distinctive combination of modernity and tradition in British society. The large estates of the gentry and aristocracy, of course, strove to dominate cultural life and, through images of the country in art and culture and the role of aristocracy in urban life, had an important impact upon scientific cultures, although elite notions of improvement and enclosure competed with traditional, alternative and radical visions of rural life and land usage. There were two principal aspects: encouragement of agricultural and horticultural productivity regarded as patriotic, nationalistic utilitarian endeavour and the aristocratic emphasis upon refinement, display and pleasure. The importance of images of productive rural landscapes fostered the development of animal breeding, horticulture, chemistry, botany and animal and plant physiology whilst the emphasis upon pleasure, display and aesthetics facilitated the collection and breeding of novel trees and shrubs. Landowners were encouraged by bodies such as the Society of Arts and county agricultural societies which awarded

premiums for agricultural improvement. In turn agricultural experimentation and improvement such as that conducted by Robert Bakewell and Thomas William Coke in Leicestershire and Norfolk respectively was celebrated in the anatomical realism of agricultural
paintings such as George Stubbs' depictions of horses and livestock.

The second chapter examines the gardens and horticulture of
Erasmus Darwin a physician, poet and Lunar Society member at Lichfield and Derby. It explores the relationship between experience of
these places and his published works, especially the hugely successful
poem *The Botanic Garden* (1791), which was partly inspired by a botanic garden created near Lichfield and *Phytologia* (1800) a study of
agriculture and gardening. As we shall see Darwin's encounters with
his botanic garden and other gardens and orchards in Lichfield and
Derby challenged his understanding of Linnaean botany, encouraged
him towards more 'natural' systems of taxonomy and inspired a
series of experimental observations of vegetable anatomy. In effect
Darwin undertook three Linnaean translations: transforming the
Swedish system into an English picturesque botanic garden; the
translation of the works of Linnaeus into English prose; and finally,
the translation of the Linnaean system into epic, popular poetry.
Darwin's gardens were more than just places where plants grew
however, and provided him with insights into geology (such as artesian wells), rock formations, soil creation and composition.

The third chapter continues the interrogation of places associated
with Georgian scientific culture by focussing upon scientific and geographical education in various forms of Dissenting school. Aspects of
natural philosophy were taught in numerous private commercial,
mathematical, military and naval academies and some grammar
schools despite the traditional presentation of these as moribund and
devoid of educational innovation. Whilst Dissenting academies were
certainly important, private and commercial academies were also
major seventeenth-century sites for the teaching of mathematics and
natural philosophy. The chapter combines approaches from recent
scholarship of Georgian society and culture, the history of science
and the history of education to reassess the nature of scientific education in Dissenting schools and by Nonconformist teachers. It also
examines the relationship between Dissenting academies and other
educational institutions in the light of recent research on the social

position of Dissenters in Georgian society. It argues that the position of Dissenters was much more nuanced than has hitherto been portrayed and that they did play a major role in the development of Enlightenment teaching of natural history and natural philosophy and related subjects such as mathematical geography, especially during the first half of the eighteenth century.

The impetus that clubs, societies and other forms of sociability gave to the European Enlightenment has often been recognised. Freemasonry was the most organised and widespread and gave the most attention to space. Given that freemasonry was idealised, realised, framed, developed and experienced in spatial terms, arguably more than most other Georgian associations, the subject is also ripe for historical-geographical study. As Money has argued inspired by analyses of seventeenth-century natural philosophy, Masonic lodges 'not only acted in a direct sense as vehicles for the dissemination of the new natural philosophy and its applications' but were 'memory theatres'. They were the 'social equivalent of the Boylean laboratory', private spaces 'with a public purpose', in which 'models were ritually copied and a shared cultural knowledge was created'. Freemasonry was the most widespread form of secular association in eighteenth-century England, providing a model for other forms of urban sociability and a stimulus to music and the arts. Many members of the Royal Society and the Society of Antiquaries were Freemasons, whilst Jacob has argued that freemasonry was inspired by Whig Newtonianism and played an important role in European Enlightenment scientific education. The fourth chapter illustrates the importance of natural philosophy in Masonic rhetoric and utlises material from Masonic histories, lodge records and secondary works to examine lodge scientific lectures. It contends that there were other sources of inspiration for freemasonry besides Newtonianism, such as antiquarianism, and that many other factors as well as the prevalence of Masonic lodges determined the geographies of English scientific culture.[21]

The fifth chapter returns to the theme of horticultural, gardening and botanical spaces by examining the development of Enlightenment botanic gardens. Informed by maps and plans, it argues that whilst the Oxford and Cambridge University botanical gardens were primarily intended to promote plant medical benefits, the organisational and spatial character of the gardens was shaped by their public urban character as part of national Enlightenment botanising. Enjoying only a precarious existence at Cambridge signified by the difficulties encountered by titular professors of botany in receiving support for lecture courses and lack of university funds, university botanic gardens were nurtured as much by metropolitan and English provincial scientific culture as academic institutional leadership. It was the requirements and expectations of English botany and the demands of local urban public culture that shaped the spatial character and utilisation of the Cambridge garden. The adoption of Linnaean botany by Thomas Martyn the Professor of Botany had a major impact on the design, layout and usage of the garden. Partly inspired by British and European university gardens, semi-public urban botanical gardens were developed by William Curtis in London and subscription-based institutions in Liverpool and Hull. As we shall see, a variety of activities were conducted at the latter including botanical and medical research, special meetings and demonstrations. The chapter concludes by examining the changing relationship between the production and dissemination of botanical knowledge and practices, usage and patronage and the management, design and character of the Hull and Liverpool gardens.

The sixth and seventh chapters use typological case studies to explore the role of natural philosophy in the Georgian urban renaissance and the development of public spheres. Although metropolitan emulation and national Enlightenment cultural imperatives were important stimulants, provincial scientific cultures had distinctive identities often concentrated through the lens of urban sociability and associations. Furthermore, scientific culture fostered and was influenced by multiple identities including those founded upon religion, gender and social class and shaped by the complexities and contingencies of local circumstances such as distance from London and proximity to Scotland. Industrial and manufacturing, port, leisure and spa, county, market and regional centres tended to foster public scientific cultures with different characteristics. The juridico-political significance of county towns such as Nottingham meant that they tended to have significant residential concentrations of lawyers, medical men and clergy, often the lynchpins of provincial natural philosophy. They were also often foci for county chorography, natural history and antiquarian studies, whilst county and market towns were imbricated with agricultural hinterlands. Equally, although acquisition of polite knowledge was important, the apparent utilitarian applications of natural philosophy underpinned much scientific culture in manufacturing and industrial centres. The requirement to provide polite entertainment as well as the concentrations of medical men and interest in the medical efficacy of water provided spa towns with a particular public intellectual character. Similarly, trading and naval requirements in ports such as Bristol and Hull incentivised acquisition of mathematical, naval and geographical knowledges as reflected in private school curricula and public lecture courses.

The powerful disjunction between Enlightenment ideas and everyday realities and its impact upon society and culture has already been emphasised. Recent work on the Enlightenment has underscored the impact of progressive ideas modelled upon the laws of natural philosophy beyond courts and capitals. The development of public spaces idealised as improved, rationalised, illuminated, socially differentiated and governable places fostered the civic and social interaction of urbane middling and aristocratic citizens. The eighth chapter examines the role of modern Enlightenment townscapes in shaping urban intellectual life and scientific culture and the reciprocal impact of scientific ideas upon town development. In the context of neo-classicism and a vastly enlarging worldview, scientific ideas helped to inspire, mould and justify urban improvement and through aesthetics, changing ideas in the sciences were reflected in architecture, building and urban design. Renaissance and Newtonian natural philosophy was invoked in support of Palladianism whilst mid- and late-eighteenth-century utilitarian empirical, psychological and medical sciences impacted upon the location, structure and design of urban public buildings. British Enlightenment ideas were couched in the discourses of neo-classicism, natural theology and improvement, providing a grammar of creativity and intellectual endeavour which smoothed the jagged fissures of modernity with tradition. Justified by an underpinning utilitarian aesthetic, natural philosophy provided a means to managing probability and a resource of knowledge for urban spaces and institutions. This encouraged the development of townscapes supposed to embody the principles of order, balance, harmony, symmetry, and regularity at the microcosm of small spaces

15

and individual buildings and the macrocosm of the townscape. Progressive Enlightenment ideas inspired provincial burghers and *savants* to promote, shape and contest urban improvement and enclosure.

The final chapter employs a case study of Abraham Bennet to explore some of the spaces of electricity and meteorology and the relationship between the two pursuits. From being merely a drawing-room plaything, electricity came to be regarded as an elemental power in the economy of nature and exemplar of Enlightenment scientific achievement, partly induced by the belief that changing weather conditions were driven by electric fire or effluvia. Hundreds of English natural philosophers, amateur experimenters, teachers, medical practitioners and *savants* became obsessed with electricity which was encountered in various contexts from domestic plaything and cure to dramatic and dangerous public spectacle. Frequently subjected to 'internalistic' analyses in the history of science, the chapter contends that from a historical-geographical perspective Bennet's experiences of electricity were clearly shaped by residence in the Derbyshire Peak, stimulation from provincial scientific networks and various forms of patronage. The impetus to study changing Peak weather patterns for the benefit of agriculture, horticulture and social improvement induced Bennet, with the encouragement of philosophical friends, to design and utilise atmospheric electrical detection apparatus, compile a meteorological diary and develop a theory of contact electricity which stimulated the work of Alessandro Volta.

This book is not a comprehensive historical-geographical study of Georgian scientific culture, which would be a major undertaking. For example, there is no attempt to map electronically the distribution and use of scientific instruments and objects, subscription lists, scientific lectures or national and international memberships of scientific societies. It demonstrates how Enlightenment sciences and ideas permeated all aspects of culture and society and can be read with my *The Derby Philosophers: Science and Culture in British Urban Society, 1700–1850* (Manchester, 2009), which provides a more detailed case study of scientific culture in one region. By focussing upon home, garden, town, countryside, county and a range of associations, networks, processes, behaviours and practices, this book strives to illuminate the importance of place and space in scientific cultural experiences.

16

1 Scientific Culture and the Home in Georgian Society

INTRODUCTION

The psychological importance of the home and the local has long been acknowledged in anthropology, cultural geography and the history of science. Love of the home or 'domicentricity' has been regarded as a product of psychological, ethnic, cultural and mythological rootedness of individuals and communities to the soil, providing meaning and security through attachment to place. In a classic essay, Sopher contended that the 'domicentric view' particularly as manifest in the English-speaking associations of the concept of 'home' was 'of course the powerful, established one, able to exert enormous pressure on behaviour'. 'Untold psychic damage' could result from 'rootlessness' which implied uncertainty, displacement and instability, and had therefore become associated with gypsies, tinkers, vagabonds and others often regarded as social outcasts. This has also led to those who were – or believed themselves to be – displaced or marginalised idealising the myth of a homeland where they would one day be able to live in peace and harmony such as the immigrant vision of North America as the New World. As Jankovic has emphasised, perceptions and practices mediated through experience and sense of the home or 'topophilia' impacted upon Enlightenment aesthetics, philosophy, natural philosophy and natural history, but the 'implications of local-patriotism and domicentricity have been largely left outside the generalisations of history'. Porter has argued that topophilia 'lay not just in its traditionally agreeable aspects [pastoral, etc.] but in new modes – in the sublime, the picturesque and the new romantic.' Naturalists 'viewed the environment through aesthetic filters, such as order and disorder, regularity, symmetry, organic form' with 'value-laden perceptions' serving as 'yardsticks of scientific truth'.[1] Placing and experiencing 'scientific' objects such as telescopes and globes around the house usually meant more than advertising wealth, signifying the ascription of meaning as well as the desire for emulation. In this chapter we will follow the lead of geographers and historical geographers and consider the importance of Georgian scientific culture in the micro-region or intimate spaces of the home, explaining how experiences and idealisations of the home shaped eighteenth-century cultures of natural history and natural philosophy.

OBJECTS, ORNAMENTS AND TOYS

During the late-seventeenth and eighteenth centuries, there was, of course, a major expansion in the production and consumption of scientific instruments and books intended to satisfy the demand for scientific knowledge and an associated rise in the production and consumption of luxury goods.[2] The production and consumption of scientific books and objects evident from newspaper advertisements, trade cards and other sources denoted the creation of a public scientific culture. Thomas Tuttell of Charing Cross, for instance, mathematical instrument maker to the king, produced globes, quadrants, sextants, octants, compasses and solar system models; similarly, by the 1780s, John Bennett of the Globe, Crown Court, London, royal instrument maker to the dukes of Gloucester and Cumberland produced optical and philosophical instruments including telescopes, spectacles, sun dials, quandrants, globes and the ever-fashionable orrery. The instrument trade was not, of course, confined to the metropolis but spread throughout the British Isles responding to varied geographies of demand, such as the need for mathematical and astronomical instruments as surveying or navigational aids.

Some scientific instruments were explicitly designed and marketed for display and education. Using an elaborate and elegant system of gears and motions, the orrery demonstrated the diurnal motion of the earth upon its axis, demonstrating the causes of night and day and the annual solar rotation 'by which the encrease and decrease of those days and nights, and consequently of the seasons, are also illustrated'. It became one of the most famous instrumental exemplars of Georgian scientific culture, and elaborate examples were created such as those constructed or improved by Thomas Wright and George Adams the younger.[3] According to essayist Richard Steele the orrery allowed astronomical studies that might have previously consumed 'a year in study' to become familiar 'in an hour and administered 'the pleasure of science' to anyone from those of 'any ordinary business', profession or occupation to women who will 'easily conceive what are the uses of sun and stars' and be 'better pleased in being compared to them for the future'. Every person would 'come into the interests of knowledge, and taste the pleasure of it' and 'any numerous family of distinction' ought to own an orrery 'as necessarily as they would have a clock.' This simple 'engine' would encourage 'pleasing ... obvious ... useful' and 'elegant conversation' and 'open a new scene' to the imagination and a 'whole train of useful inferences concerning the weather and the seasons'. By the time Joseph Wright of Derby painted his group scene entitled 'A Philosopher giving that Lecture on the Orrery in which a lamp is put in place of the Sun', the subject was a familiar one, as this name suggests. Demonstrating how responses to science had become a topical theme for Georgian artistic sensibility, it is striking that Wright chose to depict a small domestic gathering of family and friends rather than a form of public lecture, reflecting the origins of the painting as a private commission.[4]

The evidence of wills, probate inventories and sale catalogues confirms the importance attached to scientific books and objects in some Georgian households and can sometimes be used to place these within libraries, studies, drawing rooms or workshops and in relation to other objects and furniture, providing evidence concerning usages, meanings and the importance ascribed to them as part of domestic life. The importance of science in Norwich Georgian public culture is well known. The domestic context of this is evident from the number of weather glasses, barometers, telescopes, globes, thermometers and other scientific instruments listed in local household inventories and correspondence.[5] Similar evidence from elsewhere confirms this picture such as the will of attorney George Wegg (d.1777) of Colchester, Essex which lists books, manuscripts, globes, telescopes, quadrants, magnets and other mathematical and scientific instruments at his East Hill house. Another inventory for the possessions of Abraham Greenwood of Clapton (d.1813) Middlesex dated 1820 lists various scientific instruments amongst the contents of each room in his house, although they appear to have been in storage so it is difficult to be sure about original locations. These include various globes and stands, four different sized telescopes, a microscope, electrical machine and two types of air pump and apparatus.[6] Other inventories and sale catalogues confirm the close relationship between private study and experiment and public scientific culture. A sale catalogue listing the domestic scientific library and instruments of Derby attorney Charles James Flack (1799–1837) for instance, which lists a microscope, electrical machine, airpump, furnace and barometers, reveals the extent to which support for public scientific associations such as the Derby Philosophical and Derby Literary and Philosophical societies was related to private experiment.[7] Using such inventories, Georgian homes can be subjected to the same kind of spatial analysis applied to modern domestic and institutional laboratories, which has underscored the importance of objects in experimentally-validated science, design and decoration of buildings, layout of collections, and the meaning, use and display of equipment both ornamentally and functionally.[8]

The production, consumption, use and varied meanings of scientific instruments and the encouragement provided by popular authors such as Steele were stimulated by demands for the sciences as polite knowledge and signifier of middling-sort social status and aspirations as well as practical utilitarian concerns. According to Langford politeness was a 'logical consequence of commerce' encouraged in a society where 'the most vigorous and growing element was a commercial middle class involved in both production and consumption' which requiring 'a more sophisticated means of regulating manners', conveying gentility, Enlightenment and sociability to a wider moneyed elite. Although it was supposed to begin with allegedly superior aristocratic morality, it was through 'material acquisitions and urbane manners' rather than moral emulation that 'competition for power, influence, jobs, wives and markets' was permitted and controlled. Walters emphasises that polite science should be situated not just in the formal public sphere of lectures, but also in the 'comparatively informal domestic space of the home'. Georgian encouragement for scientific pursuits 'resonates with the language of politeness' which she defines as a contemporary notion embracing a set of social agendas and activities in the process of being articulated by commentators and literati such as Richard Steele and Joseph Addison.[9] The emphasis upon natural philosophy and natural history as polite knowledge for elite and aspiring middling sort families articulated by Steel and the deist Anthony Ashley Cooper, third Earl of Shaftesbury, can be interpreted as a Tory ideology and some social groups such as Dissenters tended to put more emphasis upon the practical, moral and natural theological importance of the sciences than others. Idealisations of the household economy provided models for the economy of nature whilst the vast spaces of the cosmos were experienced in domestic life through journeys of the mind, telescopes, celestial globes and the clockwork mechanism of the orrery. Similarly, perceptions and experience of domestic life help to shape knowledge and interpretations of space just as the constitution, monarchy and state were often idealised as a family. Natural philosophy and natural history came to be inextricably intertwined with the celebration and mythologizing of home life, as evident in the importance placed upon domestic scientific education.

However, whilst the desire for genteel status and politeness were important motivations for acquiring scientific objects and doing science other factors were also important. It was persistently believed that scientific education fostered industry and commerce, mechanics and agriculture. The natural theological importance of the sciences is apparent in all aspects of pedagogy including textbooks, schooling and lecture advertisements. Thermometers such as those manufactured by instrument makers and savants like the lecturer Stephen Demainbray (1710–82) were intended as practical instruments and in some cases were conveniently half the size of traditional examples so they could be easily carried around to conduct readings. Other instruments such as terrestrial and celestial globes crafted by George Adams (1750–95), instrument maker to the king, featured important new information and discoveries such as Admiral George Anson's voyage around the world between 1740 and 1744 and recently observed and

21

named constellations in the southern hemisphere.[10] Whilst the acquisition of such instruments served as symbols of polite sensibility and status, as one anonymous portrait of an unidentified gentleman from the mid-eighteenth century suggests, they were also intended to symbolise practical and patriotic aspirations towards useful science, mathematics and geography as the orrery, mariner's astrolabe, microscope, mathematical instruments and part of a celestial globe in the painting demonstrate.[11]

The importance of the Georgian house as a nexus of science, innovation, collecting and enlightened sociability and the interface of these activities with changing internal and external spatial organisations is evident when we consider the transmutation of Soho House by the industrialist Matthew Boulton and the Boulton family (Figure 2). As his business ventures grew, the Boultons moved their main residence from central Birmingham to Soho House which they strove to remodel in colonnaded neo-classical splendor, applying industrial mechanics to domestic economy and dramatically improving the gardens. Boulton accumulated natural historical and philosophical collections and a large library within and without, placing a laboratory and fossil and mineral collections in outbuildings. Plans were drawn up by the architect John Rawsthorne in 1788 to provide special rooms in the wings for 'wet chymistry', 'dry chymistry', natural history, botany and astronomy and an observatory. Scientific, geographical and meteorological apparatus were accommodated within the library. These underscore the extent to which as Jones has contended, the transformation of Soho House facilitated and maintained Boulton's 'self-invention as Enlightenment patron, entrepreneur and savant.'[12]

Domestic natural philosophy and natural history were stimulated by other sociable activities which are harder to subsume into progressive Enlightenment rationality like astrology, gaming, clubbing, drinking, feasting and gambling. Gambling and gaming encouraged the application of mathematical and probability theories in textbooks whilst, despite being condemned, astrology was often indistinguishable from meteorology and astronomy, also inspiring detailed empirical observations, instrument production and mathematical calculation.

DOMESTIC SCIENTIFIC EDUCATION AND FAMILY LIFE

The importance of domestic spaces in scientific education is evident from the positioning and usage of philosophical instruments within the home, the prevalence of tutors offering private tuition in scientific subjects and the publication of scientific textbooks. Secord has emphasised how by the mid-eighteenth century natural philosophy, usually defined in terms of Newtonian science, had 'entered the family circle' and was used to teach 'moral codes and good manners' being 'valued above all as an education in good manners and virtuous citizenship', the most important lessons concentrating upon issues of 'deference and social position.' Newtonian cosmology had become 'fit food for babies, an innocuous pabulum...served up to the most impressionable child' by enterprising commercial publishers who responded to – and encouraged – the changes in taste.[13] Recognising the popularity and potential of the scientific market, publishers, instrument makers and other craftsmen, traders and manufacturers directed many goods and services towards this burgeoning market including instruments, books and games such as juvenile natural histories, encyclopaedias, miniature herbals, jigsaws, charts, board games and dialogues, originally based upon French models, intended to provide education and amusement. Educational wall charts such as those designed by Joseph Priestley were hung on the walls of houses or schools where they were used to teach astronomy, geography and other subjects.[14]

Many introductory scientific children's texts, often read by all ages, were based upon older models and provided 'invaluable indicators of the changing social, religious and moral values carried by scientific knowledge in different circumstances'. Probably the most popular example was Tom Telescope's *Newtonian System of Philosophy* (1761) produced by London publisher John Newbery, which sold over 30,000 copies by 1800. Encouraged by Tom and his friends, household objects and toys of the nursery such as candles, cricket balls and whipping tops, even a captured rat, became philosophical apparatus and 'imbued with the marvellous'. They were used to illustrate eclipses, the double rotation of the earth on its axis and around the sun was likened to the motion of the carriage wheel on the drive. Similarly, many geographical, educational and scientific board games were based upon 'the Game of the Goose', a popular board game across Europe since the Renaissance and adapted to their particular subject. One surviving early nineteenth-century example was entitled 'Science in Sport' or 'the Pleasures of Astronomy' and featured a middle engraving of Flamstead House, at Greenwich surrounded by the heads of famous astronomers including Tycho Brahe (1546–1601) and Ptolemy of Alexandria (c. 90–168 AD). Instructions were provided and the game was played around the edge of the board with numbered squares depicting scenes of astronomical interest including eclipses, astronomers with astronomical instruments, celestial globes and a map of the solar system.[15]

As these books and games demonstrate, the market for domestic scientific educational products had a broad political and religious appeal. It has been maintained that women in late-Georgian Dissenting families, especially Quakers and Unitarians, received special encouragement to learn natural philosophy. Despite the fact that Dissenting scientific culture has traditionally been perceived in public institutional terms with a focus upon academies, most of these schools were primarily for the education of boys intended for the ministry whilst most, like other private schools, claimed to be modelled upon the best aspects of the family and domestic learning. Most Dissenting affiliations placed special emphasis upon the importance of domestic family education for both sexes and frequently justified this by citing the moral dangers of mismanaged public schools. For Dissenters, scientific education was inspired by utilitarian Protestant ideals, the requirements of natural theology and desire for edification and enjoyment. The importance placed upon domestic learning, self-improvement and the acquisition of knowledge, including natural history and natural theology is evident in many popular tracts such as Joshua Oldfield's *Essay Towards the Improvement of Reason* (1707). Oldfield strongly emphasised the importance of learning historical, geographical and scientific subjects at home including mathematics, astronomy, experimental natural philosophy and navigation alongside theology and languages such as Hebrew, Latin and Greek. Through experience of natural philosophy and other forms of 'useful knowledge' gentlemen would improve their minds, gain commercial advantages and increase their estates for themselves and their neighbours and 'be of common service to their country, or to the world'.[16]

24

Nurtured within Dissenting communities, the emphasis upon domestic scientific education spread well beyond until at least the more turbulent politics of the 1780s and 1790s. Books by Oldfield, Philip Doddridge, Isaac Watts and Joseph Priestley circulated beyond Dissenting communities, the works of Watts being strongly advocated by Tories such as Samuel Johnson. Like Doddridge and Oldfield, Watts and Priestley argued that mathematics, natural philosophy and other modern subjects were essential for education, emphasising that individual study improved morality whilst providing economic benefits and facilitating religious toleration.[17] Natural philosophy enhanced scriptural understanding revealing the myriad worlds and wonders of God's creation, temporal, microscopic, global and cosmological.[18] Science enlarged mental capacity allowing greater understanding of the numerous 'ranks of beings in the invisible world, in a constant gradation superior to us' opening up actual and metaphorical spaces by moving beyond the 'scenes of worldly business' on 'this little planet' to 'take a distant view of other remote worlds'.[19] Indeed, it was held to be sinful for citizens of all social ranks to remain ignorant of intellectual endeavours for without this they would be vulnerable to the dogmatism of the Roman church.[20] Everyone should be 'astonished at the almost incredible advances' made in science, and the 'hope of new discoveries' should encourage 'daily industry' as the exciting new worlds of 'modern astronomy and natural philosophy' revealed, and there would be future 'Sir Isaac Newtons in every science'.[21] Women too, as Watts emphasised, should 'pursue science with success' many being 'desirous of improving their reason even in the common affairs of life, as well as the men' so should be encouraged to 'apply and assume' his programme of education and 'accept the instruction, the admonition, or the applause which is designed in it'.[22]

Interest in domestic scientific education coincided with an increasing emphasis upon the ideology of the domestic sphere which, according to Davidoff, Hall and others circumscribed female behaviour. Although impacting upon many middling sort, gentry and aristocratic families, the significance of the home and private spaces varied according to social group as changing perceptions of motherhood and the domestic arena in scientific education for women and Nonconformists demonstrates.[23] According to the original model, the seventeenth and early eighteenth centuries constituted a period when there was, broadly, a balance between the public and private spheres. In the late eighteenth and early nineteenth centuries, the growing public sphere exemplified by print culture helped to confine women and children to the home which was celebrated ideologically as a special kind of space insulated from the public and male-dominated worlds of business, politics and associational activity. This was reflected in the changing nature of the physical spaces of the middling-sort Georgian home evident in the construction of terraced houses with subdivided rooms designed for specific social functions and other spaces such as attic rooms and concealed corridors intended for servants. Where once servants and members of the family had freely intermixed together within rooms that served various functions, the lives of servants and members of the immediate family became functionally, spatially and mentally segregated. These developments reached their apogee in the nineteenth-century detached villa, intended to be a miniature urban version of the aristocratic seat complete with walled and railed space, thoroughly private with a screen of planting and removed from other dwellings, allowing a retreat from the public world to private family spaces for families increasingly drawn to suburban fringes. The Georgian home had been split into a multiplicity of rooms for different functions. Of course as George emphasised long ago, such houses remained the fashionable aspiration of the middling sort and urban gentry, for many in large cities, especially London, the reality was very different with crowded courts, collapsing dwellings and myriad alleys beyond the purview of improvement commissions.[24]

The tendency towards separation of the spheres is evident in ideologies of domestic scientific education, although as Vickery has emphasised, public and private spheres centring upon the home remained interlocked. There was a significant expansion in the provision of private schools and tutors and the teaching of scientific subjects for both sexes and corresponding increase in the production of textbooks, educational games and other equipment. Although female education was dominated by the requirement for polite accomplishments for agreeable wives and companions, some parents and tutors encouraged girls to learn modern subjects including modern history, geography, modern languages, natural history and other sciences.[25]

This was encouraged by a Lockean and Rousseauian emphasis upon the importance of everyday practical activities and educational experiences and upon the dangers of premature book learning. The Welsh Dissenting minister and educationist David Williams urged that familiar objects and local walks were essential prerequisites for early education and superior to rote book learning. Domestic ideology was also reinforced by the Enlightenment sciences, which, as Schiebinger has argued using predominantly European examples, promoted domestic imperatives. Physiology, psychology, comparative anatomy and medicine tended to promote and naturalise gender differences although this should not be exaggerated. Other philosophical theories and practices, including associational psychophysiology, potentially minimised natural sexual distinctions whilst notions of complementarity suggested that aspects of science were especially appropriate for women.[26]

Through reading, attending lectures and engaging in scientific pursuits, women could travel beyond the domestic environment into the realms of wider creation. Notable examples include the 'astronomical' households of William and Caroline Herschel and Margaret Bryant. A stipple engraving of Margaret Bryant and her daughters used as the frontispiece to her *Compendious System of Astronomy* (1797) hints at the many wonders of scientific discovery, celebrating and underscoring to potential readers and clients the efficacy of such an education afforded and the domestic character of the Blackheath school.[27] However, there were limitations and the tensions between women's domestic status and pursuit of scientific activity are evident in the career of Caroline Herschel who progressed from an ill-educated dressmaker to an astronomer feted by the Royal Society and leading European natural philosophers, the first woman to publish her scientific results in the *Philosophical Transactions* who never quite escaped her domestic position. Her life is usually perceived, encouraged by her own emphasis, as being predominantly domestic assistant to her brother William at Bath and subsequently Windsor. Whilst Caroline is credited with the discovery of numerous comets and nebulae, this was perceived as an extension of her household duties. When William's greatest discoveries were made, Caroline managed the household, cooked meals and even placed pieces of food into her brother's mouth during his astronomical observations. The termination of her successful musical career to assist her brother's work seems to have been taken for granted, and it is perhaps significant that her first telescope was designated the 'sweeper', whilst her greatest work of later life involved re-cataloguing her brother's discoveries for the benefit of her nephew, a kind of domestic management of his astronomical legacy.[28]

3 Georgiana Duchess of Devonshire, keen experimenter, naturalist
and honorary member of the Derby Philosophical Society,
courtesy of Derby Local Studies Library

Aristocratic women such as Georgiana, Duchess of Devonshire on the other hand, in contrast, normally had the wealth and opportunities to pursue the sciences unavailable to the middling sort and lower orders, including unrivalled purchasing power, great houses and estates with resources and numerous servants to undertake menial operations (Figure 3). Georgiana challenged the boundaries of the domestic sphere as her political patronage and campaigning for the Whigs during the 1780s illustrates. However, her passion for the sciences still has a domestic quality to it and they seem to have gained in importance as she grew older and made fewer public appearances. Georgiana took a keen interest in natural philosophy and electricity, forming her own mineral collection using samples collected abroad and in the Peak, including the extensive family estates and famous Ecton copper mines. With her husband William Cavendish, fifth Duke of Devonshire, she took lessons from Bakewell geologist White Watson who catalogued the collections and sold them mineral cabinets and topographic inlaid strata using local samples, designing a tufa grotto for the Chatsworth gardens lined with minerals and fossils. The Duke and Duchess also patronised various works on natural philosophy and natural history such as Robert Townson's *Philosophy of Mineralogy* (1798) and Frances Arabella Rowden's *Poetical Introduction to the Study of Botany* (1801) was dedicated to Georgiana. After Georgiana visited the celebrated physician and natural philosopher Thomas Beddoes (1760–1808) in 1793, he told his friend the physician Erasmus Darwin that her knowledge of modern chemistry was superior to what he had expected and more than 'any duchess or any lady in England was possessed of'. As we shall see, the Duke and Duchess also employed Rev. Abraham Bennet of Wirksworth, one of the leading British electrical philosophers as their private chaplain, patronising his experiments and subscribing to his *New Experiments on Electricity* (1789). Georgiana's scientific interests are also evident in her correspondence with Erasmus Darwin. She was elected an honorary member of the Derby Philosophical Society and supported White Watson's application for membership. Motivated by her declining health she sought information in 1800 from Darwin concerning the newly invented Voltaic pile; however, Darwin's response far exceeds that typical of normal correspondence between physician and patient or between most male natural philosophers and women, demonstrating his respect for her status and knowledge of the subject. Darwin's description and sketch of the pile for Georgiana provide one of the fullest expositions of his understanding of the subject and includes details of construction and medical applications. Darwin offered to demonstrate the pile at Chatsworth or to send one of his philosophical or medical friends to do so and Georgiana asked whether one could be constructed for her own use or purchased.[29]

The importance attached to the domestic sphere as place of scientific learning is apparent in late-Georgian educational treatises such as Stephen Jones's *Rudiments of Reason* (1793) and Maria and Richard Lovell Edgeworth's highly influential *Practical Education* (1798). Jones's work claimed to be a 'series of family conferences in which the causes and effects of the various phenomena that nature daily exhibits, are rationally and familiarly explained'. There was no study 'more useful, nor more entertaining' for young persons of both sexes than natural philosophy, which would facilitate 'the acquisition of useful knowledge' and reveal 'the secret springs and movements by which nature operates, whose laws are unerring and invariable'. The reassuring familiar imagery of the clock and mechanics of the Newtonian cosmos illuminated one major theme, that 'such apparatus' and experience was 'either always at hand, or is very readily to be obtained from around the house' or immediate locality. Jones maintained, in the tradition of Tom Telescope that the principles of hydrostatics and hydraulics might be taught by observation and discussion of the passage of water 'even into our very apartments for the purpose of domestic convenience'. Likewise, the servant bell demonstrated the behaviour of sounds and matter, the ductility of metals, microscopic animalcules, and other aspects of physics were revealed by experiments with sulphur, oil, lard, butter, wax, copper, spirits of wine, vinegar, quicklime, inks and other everyday household objects. Similarly, motion was illustrated using clocks, musket balls, pendulums and air and sounds, musical instruments, barometers, clocks, bells, the buzzing of flies and thong of the whip, whilst the conference on optics utilised sight defects, spectacles, glow-worms, chandeliers, mirrors and magnifying glasses.[30]

The rhetorical role of the home in experiencing science is asserted equally as emphatically in Maria and Richard Lovell Edgeworths' *Practical Education*. Commonplace experiences should be utilised and simple everyday understanding as it was 'not necessary to make everything marvellous and magical to fix the attention of young people' for 'if they are properly educated' they will find 'more amusement in discovery or in searching for the cause of the effects which they see than in a blind admiration of the juggler's tricks'. For the Edgeworths, contemporary natural philosophy enunciated by the works of Priestley, Benjamin Franklin, the Manchester physician Thomas Percival and the transactions of the Manchester Literary and Philosophical Society abounded with 'a variety of simple experiments which require no great apparatus' yet which simultaneously amuse and instruct, such as papers on repulsion and the attraction of oil and water which simply required oil, water, a cork, a needle, a plate and a glass tumbler. If habits of observation were acquired then everyday events and experiences became 'a source of amusement and natural history' of great interest to children when young despite discouragement from tutors and parents who complained that it dirtied the house. Although unreasonable that the experiments of a 'young philosopher' should 'interfere with the necessary regularity of a well-ordered family', a room could be allotted for children to 'learn chemistry, mineralogy, botany or mechanics', taking sufficient exercise 'without tormenting the whole family with noise'. Geology, chemistry and mechanics were ideal subjects for domestic pedagogy. To provide education in fossils and mineralogy, the Edgeworths advocated that the children should be provided with specimens 'of ores ... properly labelled and arranged in drawers' with empty shelves in the cabinet for them to fill with their own acquisitions so that they learnt the principles of classification, although there was a danger that little naturalists' collections might fall 'sacrifice in an instant to the housemaids undistinguishing broom'.[31]

The boundary between the everyday, commonplace and domestic sphere and the progress of natural philosophy was removed in the rhetoric of Priestley, Jones and the Edgeworths, enhancing and mythologising the status of the domestic sphere. Like Priestley, the Edgeworths emphasised the democratic and egalitarian nature of natural philosophy, claiming that many great scientific discoveries had 'often been made by attention to slight circumstances' such as 'the blowing of soap bubbles, as it was first performed as a scientific experiment by the celebrated Dr [Robert] Hooke before the Royal Society' as portrayed by Priestley. This 'trifling amusement' had therefore 'occupied the understanding, and excited the admiration of some of the greatest philosophers', and just as 'every child' could observe colours

31

seen in panes of glass windows, so they could appreciate the observations of Priestley, Robert Hooke and George Louis Leclerc, Comte de Buffon. Similarly household objects were used to construct machines such as 'card, pasteboard, substantial, but not sharp pointed scissors, wine and gum', and might 'supply the want of carpenter's tools' enabling the construction of simple model engines. Working from objects and machines such as wheelbarrows, carts, cranes, scales and pumps they claimed it would be 'easy to proceed gradually to models of more complicated machinery'.[32]

THE HOUSE OF EXPERIMENT

Georgian domestic natural philosophy not only served as a vehicle for expressing cultural values and social and political ideologies, but was also instrumental in the development of some sciences. Before the nineteenth century, when the division between educated lay person and specialised scientific practitioner or 'scientist' became more established, reflecting broader changes in the nature and institutions of science, it remained possible and normal to advance national and international science from houses of experiment. Of course, it was also an important rhetorical strategy of scientific promoters such as Priestley for their own ideological purposes to emphasise the importance of private democratising scientific activity in the public sciences. The attention to everyday, domestic and family spaces and associations of the sciences also helped to allay suspicions concerning religious and political challenges. However, this rhetoric would have been superfluous if not reflecting genuine intersections between private and public scientific activity.

The importance of domestic life is evident from researches in natural history, meteorology, chemistry and electricity. As the number of introductory botanical texts and dialogues for women and children demonstrates, botany was generally regarded as a fitting female pursuit. The taxonomic and visual aspects of the subject (illustrations of plants were sewn, drawn and coloured) meant that to commentators such as Thomas Martyn, Professor of Botany at Cambridge and translator of Jean-Jacques Rousseau's letters on botany, it was 'of use

to those of my fair countrywomen' who wished 'to amuse themselves' for instilling wonder of god's creation whilst being 'simple'
and 'pleasurable'. For James Edward Smith, President of the Linnean
Society botany was 'all elegance and delight' with nothing 'painful,
disgusting' or 'unhealthy' about it, and 'none but the most foolish or
depraced' could derive anything from it but 'what is beautiful' or
'pollute its lovely scenery' from 'corrupt taste or malicious design'.
Hence, as Shteir has emphasised, 'with notable exceptions' female
botanical activity 'contributed towards the diffusion of knowledge
rather than to its creation' and was represented primarily by works
for juveniles, women and elementary introductions rather than contributions to learned bodies such as the Royal and Linnean societies.
Through its associations with the garden and domestic healing, the
acquisition of dried collections or herbariums, often stored in books,
conveyed a strong domestic ideology. According to botanist and tutor
Frances Rowden, who adapted Erasmus Darwin's *Loves of the Plants*
(1789) for her girls' school, as the 'situation of the female sex devotes
[women] to a retired and domestic life, it is necessary they should
acquire the great art of depending on themselves for amusement' and
learn happiness by concentrating 'their pleasures and pursuits within
a narrow circle'. Not too 'immoderate or injudicious' or sedentary,
botany was the perfect activity improving the mind and constitution
by encouraging women from the drawing room yet firmly within the
domestic umbra.[33]

Meteorology was another important, and often domestically fo-
cused popular scientific endeavour in which local and domestic
knowledge remained crucial.[34] The the popularity of thermometers
and barometers designed for indoor use underscores the centrality of
domestic meteorological observations although these presented problems of interpretation, such as how to correlate internal and external
pressures and temperatures with changing weather conditions. Ornamental and useful barometers and thermometers adorned walls and
melded into domestic routines, providing opportunities for empirical
observation and participation in Enlightenment scholarly progress.
Even in small provincial towns there was usually at least one instrument maker able to sell such instruments. At his parent's house at
Stamford Hill, London during the 1780s, the young Quaker meteorologist Luke Howard created a meteorological station in the back

garden featuring a rain gauge, thermometer and recording barometers for recording wind direction, air pressure, temperature, rainfall and evaporation. Meteorological diaries became popular during the eighteenth century, providing information for county natural historical and chorographical works such as those of the Cornish clergyman, Rev. John Borlase. Similarly, Major Hayman Rooke, the Nottinghamshire antiquarian and retired military man, conducted observations through his study window for a detailed weather diary. Geologist, instrument maker and Lunar Society member John Whitehurst designed weather vanes that could be read in the study using dials mechanically connected to the roof for friends such as Erasmus Darwin and Josiah Wedgwood so that it was scarcely necessary to leave the house to undertake observations.[35]

Although Golinski has seen late-Georgian chemistry as instrumental in the development of a public culture of science, whilst the subject developed as a distinctive discipline within natural philosophy, its associations with domestic life were repeatedly emphasised. For Priestley, chemistry was an endeavour peculiarly suited to domestic life. It was entertaining to the young and, it was claimed, had special applications in domestic economy including medicine, cookery and household management. As Maria Edgeworth commented in 1795, chemistry was 'particularly suited' to the 'talents and situation' of women and was 'not a science of parade' but afforded 'occupation and infinite variety' that 'could be pursued in retirement' demanding 'no bodily strength'. It 'applied most immediately to useful and domestic purposes' and whilst 'the ingenuity of the most inventive mind' might be exercised, and the judgement improved there was 'no danger of inflaming the imagination' as the mind remained 'intent upon realities', the knowledge acquired was 'exact' and the 'pleasure of the pursuit is a sufficient reward for labour'.[36] In their *Practical Education*, Maria and Richard Edgeworth noted that 'in some families girls are taught the confectionary art' which might be 'advantageously connected with some knowledge of chemistry' as every 'culinary operation may be performed as an art, probably as well by a cook as by a chemist', although they warned that the children should not keep too much company with servants. Children were 'very fond of attempting all experiments' using 'small stills, and small tea kettles and lamps' which could be used in their room with parents as a family activity. Simple chemical operations such as expansion, crystallisation, calcination, detonation, effervescence and saturation could be demonstrated using 'water and fire, salt and sugar, lime and vinegar', which were 'not very difficult to be procured and ... to be found in every house'.[37] They acknowledged that domestic chemical experiments could be dangerous, but suggested that 'with proper direction' materials could be sold in a 'rational toy-shop'.[38]

The role of electricity, and sometimes magnetism, in forging mid-Georgian public cultures for science has been acknowledged, and the former became the most popular philosophical pursuit after astronomy. Accounts of electricity and electrical experiments were circulated in print and within scientific communities, whilst instruments and electrical machines were promoted and sold by provincial and metropolitan instrument makers. However, electricity also had very important associations with the domestic sphere reflected in the encouragement given towards the pursuit of experiments at home which helps to explain the popularity of electricity and magnetism in mid-Georgian society. This also helps to explain why electricity was regarded as a pursuit especially appropriate for women who were courted in textbooks and lecture advertisements. Variously described and regarded as a form of fire or an aetherial fluid, electricity was celebrated as the wondrous science, a power that exhibited the splendour and potency of the divine economy closely associated with the most powerful phenomena in the natural world such as earthquakes, violent storms and to some, plant growth, yet whose marvellous effects could be captured and observed by the most intimate domestic gatherings. In comparison with staring through a telescope, observing a plant or dried specimen, collecting minerals and rocks or looking at a globe or map, electricity was incredibly colourful, participatory, dynamic, enticingly dangerous, dramatic and engaging.

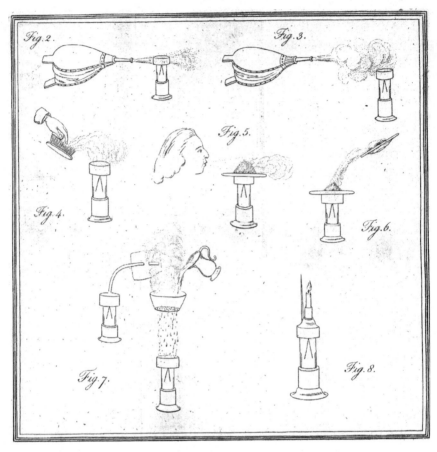

4 Domestic science: electrical testing of household powders by A. Bennet,
Philosophical Transactions, 77 (1787)

Electrical experiments were – and were promoted as – relatively
cheap and simple to replicate utilising everyday objects easy to procure from around the home such as wool, resin, cork, pith balls,
straw, glass jars, apothecary's bottles, spectacle lenses, cooking apparatus, pans and even stockings and joints of beef (Figure 4).[39] Like
magnetic and chemical experimenters, electricians were able to exploit the easy availability of quantities of good quality, and increasingly standardised and cheaper smooth glass containers, balls, jars,
plates, tumblers and tubes resulting from improved production and
greater consumer demand. Unlike older glass objects, these were
transparent and highly malleable, easy to observe through, yet strong

when cooled, easy to clean, seal and resistant to chemical change, therefore able to remain apparently relatively inert during experiments.[40] Globes, for example, were utilised in many electrical machines, such as the belt-driven devices where globes or cylinders were rubbed against wool or other material to produce a charge, whilst electricity was contained in Leyden jars. Invented in 1746, the Leyden jar which 'stored' and delivered greater quantities of artificial electricity than any other device, consisted of a bottle partly filled with water that contained a metal rod projected through the neck and was ideal for domestic use. Foil was placed inside and outside the bottle to prevent damage to the leaves. If the rod was connected to the prime conductor of a static generating machine and then the jar taken away, it was found that the charge could be kept and transported. Electrometers were also constructed from household objects. One of John Canton's electrometers, for example, utilised pith balls hung on fine linen threads, Volta used straw for an electrometer and developed another using cakes of resin whilst another device used gold leaves placed into adapted glass lanterns.

Other discoveries seemed to arise in equally mundane domestic circumstances. Excluded from membership of the Royal Society until 1730 and lacking expensive apparatus, the Canterbury dyer Stephen Gray made use of everyday household objects in order to investigate the electric effluvia using a series of simple experiments. He demonstrated to the august members of the Royal Society that hair, feathers, silk, paper and gilded ox guts were 'electrics', demonstrated electrification by communication over distance using various metal wires, a shilling, tea kettle, silver pint pot, a fishing rod, stones, bricks, tiles and vegetables. Robert Symmer's influential arguments for dualistic electrical theory during the 1750s were stimulated by observing that upon pulling off his silk stockings in an evening at home they 'frequently made a crackling or snapping noise' and emitted sparks of fire. Subsequently he tried expanding and moving the stockings electrically, chucking them against walls and various objects, repeating the experiments with white, black and coloured new and newly cleansed silk and worsted stockings, also testing the strength of the 'electrical cohesion'.[41] Similarly, as we shall see, Abraham Bennet's apparatus for magnetic and electrical experiments included a magnetometer constructed with spider's thread from his back garden and a sowing needle, an umbrella and pans for investigating electricity induced by household and culinary substances such as flour and chocolate. Bennet contrived a machine using a deal rod, electroscope and a 'doubler' to collect, augment and record atmospheric charge in the study to supplement external observations, data that was incorporated into an electro-meteorological diary.

Although mathematics was of course required for a greater understanding of Newtonian natural philosophy, little if any mathematical knowledge was needed for most meteorological and electrical observations and experiments. Few contrived electrical experiments with the mathematical rigour of Lord Henry Cavendish. For Priestley, electrical experiments were 'of all others the easiest, and the most elegant that the compass of philosophy exhibits'. They could be 'performed with the least trouble', there was an 'amazing variety in them', and they furnished 'the most pleasing and surprising apparatus for the entertainment of one's friends'.[42] Elementary introductions to the subject appeared and model electrical machines such as electric orreries were produced for domestic education and play. As Priestley observed, budding electricians could imitate 'in miniature' in their homes, 'all the known effects of that tremendous power, nay, disarming the thunder of its power of doing mischief' and 'without any apprehension of danger to themselves', draw 'lightning from the clouds into an private room' whilst 'amusing themselves at leisure' performing experiments with electrical machines. The domestic dimension of electricity was a key component in the promotion of the science by Priestley and was facilitated by experiments and demonstrations utilising the most intimate and apparently trustworthy of all scientific instruments, the body, just as Isaac Newton had tortured his eyeballs with fingers and bodkins for the sake of optical progress.[43]

Parts of the body and simple observations seemed all that was necessary for many electrical experiments, and striking and theatrical demonstrations by Stephen Gray and others featured the electrification of individuals suspended or in human chains. Bodies were used to pass charge between other objects, which became a favourite parlour game, and also provided a means of estimating the strength of the effluvia before electrometers, for instance, through the tongue. The direct use of the body and bodily senses provided great immediacy and contrasted with observations of distant heavenly bodies or relatively impersonal and methodical taxonomic sciences such as botany or mineralogy. This fostered the development and increasing popularity of medical electricity from the 1740s, which was made possible through the application of the Leyden jar and superior electrical machines, which provided the opportunity for portable and measurable charge to be administered to different parts of the body. Medical electricity and magnetism, in turn, encouraged domestic electrical activity and the acquisition of domestic electrical apparatus for self-medication. In his *Desideratum* (1760), John Wesley argued that the application of electrical treatment in the domestic environment offered the opportunity for ordinary people to avoid the expensive fees of medical professionals, placing treatment within the reach of those previously unable to afford it. The availability of electricity as a new egalitarian form of treatment was thus one of the most wonderful and beneficial applications of Enlightenment science.[44]

MUSEUMS

During the late seventeenth and eighteenth centuries public and semi-public museums grew from private cabinets of curiosity and wonder, and collections were assembled in domestic spaces by aristocrats, gentry, philosophers and middling sort collectors. These often comprised artistic and antiquarian objects as well as natural historical items such as plants, shells and fossils and scientific instruments which were accumulated, ordered, classified and displayed as microcosms of local or global wonders. Cleaned and tended by servants and sometimes family members and theatrically opened for friends and the curious, they became part of the domestic routine spilling from single rooms into halls, galleries and stairways. Mediating between private and public realms, as Livingstone has argued, collecting betokened achievement, education, taste, and discernment. However, whilst private and public accumulating, concentrating, showing and experiencing reinforced prevailing worldviews it also 'refashioned reality', challenged received classifications and taxonomies and informed new systems.[45]

Probably the most famous domestic museum was that of the Irish physician and naturalist Sir Hans Sloane (1660–1753) who accumulated a cornucopia of international natural historical and antiquarian objects in his house at 3 Bloomsbury Place, London. These occupied a large proportion of the building necessitating the employment of a curator and purchase of the house next door for extra space before the collection was removed to Sloane's Chelsea manorhouse. This became a semi-public museum visited by numerous British and international savants and the curious. President of the Royal Society from 1727 after the death of Newton, Sloane bequeathed the collection to Britain for £20,000. After the sum was raised by benefactions and a lottery it formed the basis of the British Museum created by Act of Parliament which opened to the public at Montagu House in 1759. Important provincial museums were also formed within private houses. The Lichfield apothecary Richard Greene (1716–93) accumulated a museum of curiosities from the early 1740s at his house on 12 Market Street consisting of natural history, antiquarian, artistic and other specimens some being provided by local friends such as Brooke Boothby, Erasmus Darwin, Matthew Boulton, Samuel Johnson and Anna Seward. The objects are listed in successive editions of Green's catalogue and the collection has been designated the museum of the Lunar Society by Torrens. Greene corresponded with other naturalists and collectors, exchanging information and specimens including Rev. John White brother of Gilbert White author of the *Natural History and Antiquities of Selbourne* (1788), Daines Barrington (1727–1800), Thomas Pennant (1726–98) and Sir Aston Lever (1729–88), creator of another large domestic semi-public museum at Alkrington Hall near Manchester and Leicester Square, London. After visiting Lichfield in March 1776 with Samuel Johnson, James Boswell noted that it was a 'truly wonderful collection' shown by Greene to visitors with 'obliging alacrity' with 'all the articles accurately arranged, with their names upon labels, printed at his own little press'. On the stairs leading to the collection 'was a board, with the names of contributors marked in gold letters' printed catalogues being available from local booksellers.[46]

A full catalogue of the domestic museum of antiquary George Allan (1736–1800) of Darlington, Durham also survives demonstrating why it became a source of great local fascination. Allan purchased the entire museum of his friend Marmaduke Tunstall FRS (1743–1790) of Wycliffe Hall, Yorkshire which included many 'curiosities brought by Captain Cook from Otaheite', many local specimens and an ornithological collection worth £5,000 alone. Allan moved the museum to his mansion at Grange where it occupied two large rooms on the north side, naming his objects 'scientifically in a very masterly manner' with labels 'in the neatest and most beautiful writing' with common and scientific names and references to authors. The larger room was 'principally filled with birds', the cases of which were 'so placed in partitions, back to back, as to form the room into three smaller apartments, through which you passed by two arches in the centre.' These had been utilised by Thomas Bewick amongst others as subjects for some of his ornithological studies. The collection grew within the house filling 'every panel' and 'gradually insinuating itself ... along the passages' and clothing 'the walls of the great staircase' as Allen added other departments. The museum was opened 'gratis' for 'public inspection in June 1792 and from that time to January 1796 (three years and a half) it had been viewed by 7,327 persons' which is a significant number given that the population of Darlington in 1801 was only 4,760 inhabitants. In November 1793 the *Gentleman's Magazine* noted that a society consisting of ordinary and corresponding members and including Allan's Grange museum had 'lately been instituted at Darlington, for the promotion of the knowledge of natural history, antiquities, etc.', which they believed 'from the public characters of several of its members' would 'flourish'. After the sale of Allan's museum under the terms of his will and purchase by his son, the museum remained in the Grange until 1822 when it was bought by the Newcastle Literary and Philosophical Society.[47]

SCIENTIFIC LECTURES AND ASSOCIATIONS

Although, as we have seen, a distinction between private and public spheres was widely recognised in Georgian society there was considerable interplay between the two spatially and culturally. The Habermasean notion of the public sphere as an arena of autonomous citizens depends, of course, upon a conception of private individuals although the home took on many characteristics of semi-public and public spaces. Domestic rooms and spaces served as museums, meeting places for lectures, informal literary and scientific groups and more formal associations, providing a comfortable, cheap and more private alternative to renting or buying rooms in public houses and other places. The domestic context of many scientific lectures and associations has received little attention, and they have often been regarded primarily as signifying the birth of modern urban institutional public culture and sociability. This is particularly marked in analyses of European Enlightenment culture, such as McClellan's study of eighteenth-century scientific societies, which treats continental scientific societies, academies and museums as evidence for a growing public, institutionalised and professionalised science.[48] Such a model applies more to Europe than to Britain where there were few, if any, academies on the continental model, where there was much less government interference, where many scientific associations remained informal and where national scientific associations such as the Royal Society did not dominate to the same degree as European equivalents. The domestic context of scientific learning and experiences helped to reinforce their status as polite pursuit. As Erasmus Darwin emphasised in a letter of 1791 to 16-year-old John Horseman junior, who had requested advice concerning how to gain a scientific education, domestic company was the best, preferably engaging in tea drinking rather than settling in taverns and public houses, which were associated with drunken and frivolous activity.[49]

Walters's analysis of scientific lectures follows common practice in emphasising the degree to which these were public events when many were, like museums, as much private family activities in domestic settings. Similarly Morton and Wess specifically equate 'science as public culture' with 'early scientific lectures'. Secord has emphasised how the discourses of Tom Telescope were 'obviously intended to recreate one of the most typical means for introducing science to children, attendance with their parents at the spectacular demonstrations by itinerant lecturers in natural philosophy'. Walters reminds us that the 'well-documented point of intersection between politeness and science is the public space of the peripatetic philosophy lecture'.[50] Yet Steele's celebration of the orrery, as we have seen, evokes private spaces as much, if not more than a public sphere of politeness suggesting a series of intermediate or interlocking spheres in between. Politeness requires a public display of taste and behaviour – no one can be polite on their own. This equally applies to workhorse scientific instruments such as thermometers, telescopes, microscopes and even orreries. Most of these instruments were designed, of course, for observation and experiment by individuals and were difficult to demonstrate to large public gatherings because of the problems of interpreting and projecting the readings and information they produced. The orrery was inherently a more public instrument, of little use for original scientific experiment, designed for demonstrating the plan and operation of the Newtonian solar system to individuals or small groups of people, which was why Steele recommended that families 'of distinction' should have one 'as necessarily as they would have a clock'. This is also reflected in Wright's painting of 'The Orrery' which was a private commission for a wealthy aristocrat depicting a small group of family and friends more widely circulated in the form of prints. Even the largest and most expensive early and mid-Georgian orreries were of limited use for public scientific lectures with large audiences as they required close observation, which was why new types of giant orrery such as diostrodons were created for later audiences.[51]

Most itinerant scientific lectures probably spent more time lecturing to small groups including women and children in domestic contexts than to large audiences in public places which was reflected in their advertisements. The Scottish mechanic and philosopher James Ferguson, for instance, emphasised that gentlemen in or near to London could have his course of lectures on astronomy, mechanics and other subjects read 'at his house, for twelve guineas, or any single lecture for one guinea, over and above the charge of carrying the apparatus'. He also taught the use of the globes for an hour a day except on Sundays, geometry, the construction of maps, the projection of the sphere, dialling and other subjects for two guineas at his own house or for four guineas at others' houses, although students had to provide their own equipment such as globes.[52] Likewise, during the 1770s, Pitt emphasised that 'ladies as well as gentlemen, without any previous knowledge of natural philosophy' could 'readily understand the principal phenomena of nature' whilst 'any select company' could have private tuition so long as they provided 'timely notice a day before'.[53] During the 1730s and 1740s, William Griffis, toured the midlands and south-west providing lectures on natural philosophy in domestic contexts at Birmingham, Lichfield, county towns, villages and 'gentlemen's houses' for groups of ladies and gentlemen. He had an extensive apparatus including pumps, water wheels, orreries, telescopes and other astronomical instruments. Most of these lecturers specifically encouraged ladies to attend and claimed that more difficult and abstruse aspects of natural philosophy and mathematics would be described in approachable everyday language. During the 1740s, John Arden emphasised every subscriber was 'entitled to introduce a lady; for whose sake all uncommon terms will be as possible avoided ... and always explained'. As the sciences were 'very entertaining, and frequently the subject of conversation (of which the fair sex constitute so agreeable a part)' so the 'greatest care' would be taken to 'explain any seeming difficulty ... in the most easy and intelligible manner'.[54]

Informal, semi-private and domestically oriented literary and scientific associations such as the Spalding Gentleman's Society of Lincolnshire were more typical of the eighteenth century than formal public institutions, especially in the English provinces. In turn, aspects of domestic culture shaped the way that such societies were conducted in other more public places. As private and public places, middling-sort homes mediated between the public and private spheres of scientific culture. Although Uglow exaggerates when she claims that the members of the Lunar Society nudged 'their whole society and culture over the threshold of the modern, tilting it irrevocably away from old patterns of life towards the world we know today', she also emphasises that they were a 'small, informal bunch' who 'met at each other's houses on the Monday nearest the full moon'. Jones similarly emphasises the informal nature of Lunar domestic gatherings with no formal rules, committee or premises. The Lunar men also, of course, enjoyed highly successful public careers in medicine, engineering and business, all playing major roles in wider communities and fostering the development of public scientific culture. Yet the essential quality of the society, which helped to ensure its success, was the nature of meetings in which the pleasure of informal banter and games mixed with serious discussion, experimental apparatus with cutlery, crockery and the chink of wine glasses. Meetings were usually conducted in members' houses, especially Darwin's residence in Lichfield, and included dinner at which wives and children were often present. This was noted by contemporary visitors such as the Scottish philosopher James Robison who emphasised the warmth, charm, 'unaffected ease and civility' of meetings which were far removed from formal institutional gatherings.[55]

Other Georgian philosophical societies behaved similarly. Sukey Wedgwood, who attended early meetings of the Derby Philosophical Society during the 1780s, described how, like the Lunar Society, these took place in member's houses in which women of the household were also present, despite the fact that none were apparently listed as members with the significant exception of Georgiana Cavendish, who was described as an honorary member. At one meeting at the Strutts' house in Derby, the Miss Strutts served tea despite the fact that they 'did not like this at all'. Seeing this, Darwin, 'with his usual politeness, made it very agreeable to them by shewing several entertaining experiments adapted to the capacity of young women; one was a roasting tube, which turned round itself'. At another meeting in August 1794 the Derby philosophers took tea on a Sunday afternoon at Darwin's house in Full Street where they were shown James Watt's new 'pneumatic apparatus' for making 'carbonic-hydrogen gas' in the recently refurbished 'elaboratory'.[56]

Even the Manchester Literary and Philosophical Society (1781) often held to exemplify the growth of public provincial scientific culture because of its formal organisation, publications, networks of communication and institutional associations, originated through informal meetings of a 'few gentlemen, inhabitants of the town, who were inspired with a taste for literature and philosophy' and who 'formed themselves into a kind of weekly club' for discussion. The first meetings were held at the house of Thomas Percival, the physician and first president as at Derby, only when the society became established did the members obtain institutional premises.[57] Reflecting the tight character of elite social groups in many Georgian towns and within *savant* circles, the provision of meeting rooms and libraries for regular ordinary meetings, regulations, behavioural expectations, social divisions and membership restrictions, tended to foster an intimate domestic – although increasingly masculine – rather than public institutional atmosphere. As Thackray pointed out, six Unitarian families provided almost 5 per cent of the membership of the Manchester Philosophical and Literary Society from the 1780s and 1790s, the Heywoods, Gregs, Henrys, McConnels, Robinsons and Philipses, who formed part of a 'closely intermarrying, almost dynastic elite' and held collective office for a total of 144 years.[58] Similarly, at Derby, individuals from the Darwin, Strutt, French, Haden, Spencer and other families tended to lead philosophical society activities for over seventy years. Books and apparatus were stored in homes and workshops whilst domestic servants sometimes assisted in performing experiments as when Richard Roe secretary of the Derby Literary and Philosophical Society, encouraged his maid to help with the organisation's public lectures. If anything, the domestic-centred nature of provincial philosophical scientific culture increased during the 1790s faced with Church and King hostility and the revolutionary wars. For most members, the Derby Philosophical Society settled down to being a private gentleman's library for the purchase and circulation of books.[59]

CONCLUSION

This chapter has discerned various patterns in eighteenth-century domestic-centred scientific culture. Encapsulated by Newtonian mechanics and astronomy, mathematics and geography, the Georgians inherited a profound belief in the utility of natural philosophy for national industrial, economic and military progress. This, along with the – repeatedly invoked – natural theological value of the sciences, provided important and continued inspiration for domestic scientific education and philosophical experiment. The Georgians believed that sciences such as mathematics, chemistry, astronomy, electricity and botany progressed through domestic education, discussion and experiment. The sciences were also valued as polite activities which promoted a domestic-centred ideology of aristocratic and middling-sort life. They remained inextricably interlinked with practical utility and natural theology, encouraging the production, marketing and consumption of scientific books, instruments, games and other objects. As we have seen private and public spheres were interwoven in the home despite an increasing ideological separation which helped to reinforce the relationship between aspects of scientific culture and the home, such as female botanising, a process reinforced by other developments within and beyond the sciences.[60]

2 The Garden: Darwin's Gardens: Place, Horticulture and Botany[1]

INTRODUCTION

INTRODUCTION

Erasmus Darwin was a physician, natural philosopher and member of the Lunar Society who later became famous for the publication of the *Botanic Garden*, a poem in two parts with philosophical notes. The *Loves of the Plants*, the first part of the *Botanic Garden*, which contained a poetic exposition of the Linnaean system, appeared in 1789, whilst the *Economy of Vegetation*, the second part, was published in 1791. Darwin was also largely responsible for an English translation of the *Systema Vegetabilium* and *Genera Plantarum* of Carl Linnaeus, published under the auspices of a botanical society at Lichfield in 1783. He also composed the *Zoonomia* (1796), a major study of human physiology and medicine, the *Temple of Nature* (1803), a grand epic poetical celebration of the wonders of life again with philosophical notes, and *Phytologia* (1800), a work on the philosophy of agriculture and gardening. In these works and encouraged by his medical practice, Darwin presented what could be regarded as a biological, geological and cosmological developmental theory in which, as we shall see, constant analogies were made between the animal and vegetable worlds. Although often described as an evolutionist, this was not a term that Darwin used and it carries the serious risk of 'Whiggish' anachronism by conflating him with nineteenth-century evolutionists including his grandson. Exploring the question of the relationship between the evolutionary ideas of Erasmus Darwin, his grandson Charles, and Jean Baptiste Lamarck, Gruber suggested that it would be more accurate to describe the elder Darwin's 'general world view' as one of 'pan-transformism'. This was because of his emphasis upon the changes of individual growth rather than species transmutations.[2] (Erasmus) Darwin does refer to 'progressive improvement' and the gradual acquisition of 'new powers' by animal life to 'preserve their existence', considering that 'innumerable successive reproductions for some thousands, or perhaps millions of ages, may at length have produced many of the vegetable and animal inhabitants which now people the earth'.[3] The places that Darwin lived, worked, socialised in and experienced in other ways helped to shape the content and character of this philosophical work.

After initial failure at Nottingham, once Darwin's Lichfield practice had taken off he gained a good income and became a relatively wealthy individual with an income and lifestyle commensurate with the landed gentry, a position that he cemented by marrying into the Poles, an old Derbyshire family of landed gentry. Between 1756 and 1781 Darwin resided in a townhouse and practised medicine in Lichfield before moving to the Pole's seat of Radburn Hall in Derbyshire in 1781 to live with his new wife Elizabeth and her family, and then to another townhouse in Derby from 1783. Finally, just before his death in 1802 the Darwins moved to Breadsall Priory near Derby, a house that they inherited from Darwin's eldest son Erasmus, who had died in 1799. An omnivorous reader and highly clubbable socialite, he was largely responsible for the foundation of three philosophical associations, the Lunar Society of Birmingham, the Derby Philosophical Society and the aforementioned Lichfield Botanical Society. Although his mind wandered around the world through reading and his communications with philosophical contacts, Darwin's experience of place was largely confined to an English Midland triangle between Elston in Nottinghamshire, the family home in the east, the Derbyshire Peak in the north-west, and Birmingham in the south. However, there was considerable variety within this domain which embraced everything from fertile well-watered agricultural lands and genteel county and market towns of the Midland plain, manufacturing and industrial centres such as Birmingham and the mills of the Derwent Valley to the craggy eminences of the Peak.

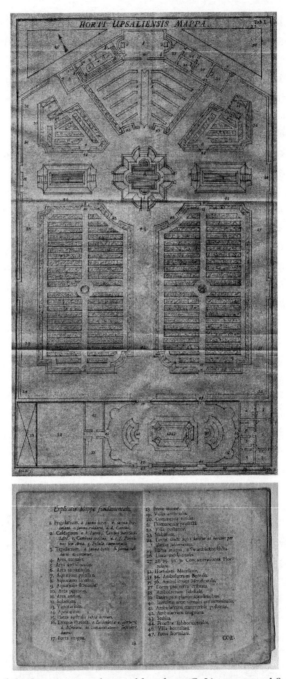

5 Plan of the Linnaean garden and key from C. Linnaeus and S. Nauclér,
Hortus Upsaliensis (1745), courtesy of the Department of Special Collections,
Memorial Library, University of Wisconsin–Madison, USA

Encouraged by the medical potential of botany in addition to aesthetic enjoyment, and particularly after 1778 by his second wife Elizabeth, a keen gardener, Darwin developed a series of gardens and advised his friends on gardening matters. By 1787 he admitted to his Lunar Society friend Richard Lovell Edgeworth that gardening had become his favourite 'hobby horse' as he laboured on 'a theory of gardening', 'wonderful, and useful and delightful' which grew into *Phytologia*. This chapter explores the relationships between Darwin's gardens at Lichfield and Derby and his work on natural history, primarily represented by the *Botanic Garden* and *Phytologia*. It is well known that the *Botanic Garden* was inspired by a botanic garden that Darwin created near Lichfield. However as we shall see, Darwin's encounters with this and other gardens and orchards in Lichfield and Derby challenged his understanding of Linnaean botany, encouraged him towards more 'natural' systems of taxonomy and inspired him to undertake a series of experimental observations of vegetable anatomy. In effect Darwin undertook three Linnaean translations and one translocation: an initial attempt to translate the Linnaean system into an English picturesque botanic garden; the translation of the works of Linnaeus into English prose; the translation of the Linnaean system into epic, popular poetry; and finally the translocation of Linnaean botany into English horticulture in *Phytologia* (Figure 5a and b). Darwin's gardens were more than just places where plants grew, however, and prompted insights in geology (such as artesian wells), rock formations and soil creation and composition.[4]

DARWIN'S GARDENS

The fullest account of the Lichfield botanic garden appears in Anna Seward's *Life of Darwin* (1804). According to Seward around 1777 Darwin purchased

> a little, wild, umbrageous valley, a mile from Lichfield, amongst the only rocks which neighbour that city so nearly. It was irriguous from various springs, and swampy from their plenitude. A mossy fountain, of the purest and coldest water imaginable, had,

51

near a century back, induced the inhabitants of Lichfield to build a cold bath in the bosom of the vale. *That,* till the doctor took it into his possession, was the only mark of human industry which could be found in the tangled and sequestered scene.[5]

Within this a 'dripping rock in the centre of the glen' which dropped 'perpetually', thrice each minute and aquatic plants bordered 'its top and branch from its fissures'. The rock was allegedly unaffected by winter frosts, summer drought and heavy rains. Darwin went to some effort to landscape the garden, the changes that he made are described most fully by Seward. According to Seward, Darwin made various changes widening the brook in some places

> into small lakes that mirrored the valley; in others, he taught it to wind between shrubby margins. Not only with trees of various growth did he adorn the borders of the fountain, the brook, and the lakes, but with various classes of plants, uniting the Linnean science with the charm of landscape.[6]

Charles Darwin noted in 1879 that, according to a guidebook some aspects of the garden still survived. It then formed part of an adjoining park and was described as 'a wild spot, but very picturesque; many of the old trees remaining, and occasionally a few Darwinian snow-drops and daffodils peeping through the turf, and bravely fighting the battle of life'.[7] The fact that the land took the form of a valley with the only rocky outcrop around Lichfield must have attracted Darwin to the site because of the picturesque possibilities that it had. The poet William Shenstone (1714–63) seems to have provided an important inspiration, having designed a celebrated garden at the Leasowes, Hampshire which he depicted in his poetry and which became an important and much-visited model for later Georgian picturesque gardens. In his remarks on gardening, Shenstone argued that it was 'pleasing the imagination by scenes of grandeur, beauty or variety'. This suggests that the aesthetic aspects of the garden were always as important to Darwin as the intention was to present a systematic collection of botanically significant plants. Shenstone's garden was a model for Darwin's creation in a number of ways. Shenstone created his garden on a relatively small site consisting of a couple of wooded valleys with pools which he shaped to achieve maximum effect, introducing very varied planting with irregularly interspersed groups of trees and shrubs and green slopes. He deliberately tried to invoke historical associations through walks, gothic ruins, follies, alcoves, cascades, grottoes and inscriptions and all on a relatively modest income of only about £300 per annum.[8]

Strongly imbued with contemporary ideas of the picturesque, gardens and landscape beauty meant more to Darwin than the aesthetics and science of plant configuration. He also advanced a psychophysiological explanation for the impact of landscape beauty which reinforced his belief that lower flat situations were less aesthetically pleasing and more likely to produce ill-health or disease for animals and plants. It provided an explanation for the power of the picturesque in landscape gardening and even, by association and analogy, landscape painting, and, through the pleasure excited by utility, the delights of practical horticulture. It also perhaps helped to explain why Darwin found his botanic garden at Lichfield and last garden at Breadsall Priory to be more beautiful than his urban Derwent-side garden and orchard in Derby. According to Darwin one of the earliest and most fundamental infant delights was experiencing the joy of being pressed next to its mother's bosom with its sense of warmth and smell of milk, whilst its 'sense of touch' was also engaged, the breast being perceived as the 'source of such variety of happiness'. These 'various kinds of pleasure' subsequently became 'associated with the form of the mother's breast' with which it had so much intimate experience. Hence, when older, 'any object of vision' that 'by its waving or spiral lines bears any similitude to the form of the female bosom, whether it be found in a landscape with soft gradations of rising and descending surface, or in the forms of some antique vase' would generate a 'general glow of delight, which seems to influence all our senses ... as we did in our early infancy the bosom of our mother.'[9]

An important part of Darwin's gardens were devoted to growing herbs and other plants for culinary and medical use. They were also places for his family, servants and friends. Of course although often removed from landscape descriptions and paintings, much of the laborious work was carried out by labourers and gardeners who had their own networks, connections and degree of tacit knowledge and experience. Darwin was able to utilise his network of friends and local landowners in order to obtain suitable workers such as William Mathers, whom he employed during the 1780s and who had been employed on the Duke of Devonshire's estate at Chatsworth and Sir Robert Wilmot's estate at Chaddesden. He also interviewed and investigated gardeners for friends such as Josiah Wedgwood to whom he recommended Mathers, a Mr Talkington and the Scotsman James Downs in May and June 1788, partly on the grounds that the former had been employed by Rev. Thomas Gisborne of Yoxall Lodge, Staffordshire and the latter by Brooke Boothby for his elaborate gardens and hothouses at Ashbourne.[10]

Darwin's gardens were prominent in the lives of his wives Mary and Elizabeth and their children. Darwin's first garden was that of his townhouse in Lichfield. Subsequently, after marrying Elizabeth Pole and a trip to London in 1781, the couple moved first to Radburn Hall close to Derby and this immediately impacted upon his gardening and botanical experimenting, whilst he continued to work on the Linnaean translations for the Lichfield society. Henceforth, given that, as Darwin remarked to Thomas Day, they both 'loved the country and retirement', they became more domestically focussed which included a greater focus upon their gardens, as represented by the increase in volume of Darwin's correspondence. During the trip to London, Darwin met Joseph Banks, who had been President of the Royal Society for three years, and on his return to live at Radburn he began a series of important investigations into vegetable anatomy and physiology, subsequently detailed in the *Economy of Vegetation* and *Phytologia*. Darwin corresponded with many botanists and publishers concerning the translation including Banks, Daniel Solander, Anna Blackburne of Orford near Warrington and William Curtis, whose *Flora Londinensis*, which he praised, Darwin was collecting in parts, and who was to create public botanical gardens in London as we shall see.[11]

Subsequently, Darwin acquired the house at 3 Full Street Derby in 1782 and, after carrying out some building work on it, conducted his medical practice there during the winter whilst still residing at Radburn before moving there fully with the family in the summer of 1783 (Figures 6 and 7). Part of Darwin's garden lay between the house and the River Derwent, and he also owned an orchard on the bank opposite adjoining gardens that belonged to nearby Exeter House.[12]

The route to the gardens via St Mary's Bridge was circuitous and Darwin's solution was the construction of a ferry operated by pulling on a rope strung across the river, a sketch of which was later provided by his son Sir Francis Darwin. Darwin's orchard on the opposite bank of the Derwent is clearly visible on Peter Perez Burdett's map of Derbyshire as reissued in 1791 and later maps, and also appears in the sketch, which shows quite tall fruit trees. A surviving brown leather notebook lists around a thousand plants, trees and shrubs in the Derby garden and includes a separate list of the trees in the orchard on the other side of the Derwent. This was the period when gardening became, as he told Edgeworth, his favourite 'hobby horse' and much of *Phytologia* was researched and composed. Changes in the notebook reveal alterations in the gardens during the 1780s and 1790s, and the thousand or so trees and plants were not all in existence at once. The list of hardy plants shows that Darwin combined highly unusual plants, which could only be identified by experienced botanists with herbs typical of physic gardens such as St John's wort and saxifrage and more common English garden and wild varieties such as gladioli, foxgloves, sunflowers, forget-me-nots, lily of the valley, daffodils, narcissi and roses. The garden must have been a riot of colour in the summer months. The many changes in the notebook, references to plants dying and at least one major relisting, reveal alterations made in the garden and the difficulties Darwin experienced in growing some kinds, partly due to the low-lying boggy nature of the riverside site.[13]

Despite the centrality of the Linnaean system to Darwin's conceptualisation and experience of the garden it is significant that he did not choose to arrange his Derby garden according to a taxonomic scheme. In his botanic gardens at Lichfield and in print Darwin had striven to unite botany and gardening, yet the Derby garden and other observations prompted a more critical appraisal of the Linnaean system culminating in the subtle Linnaean critique of *Phytologia*. The notebook lists hardy plants, trees and shrubs by spatial location defined according to other features within the garden during a typical perambulation rather than by Linnaean taxa. Types of tree in the orchard located using numbers painted on rails opposite to the espaliers were identified with assistance from Dr William Jackson (1735–98) who had been a member of the small Lichfield Botanical

Society and who cared for the Lichfield garden after Darwin's departure for Derbyshire.

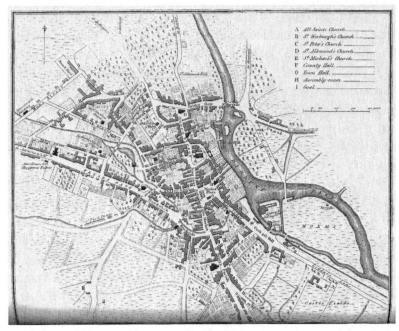

6 Map of Derby from D. and S. Lysons, *Magna Britannia, Derbyshire* (1817)

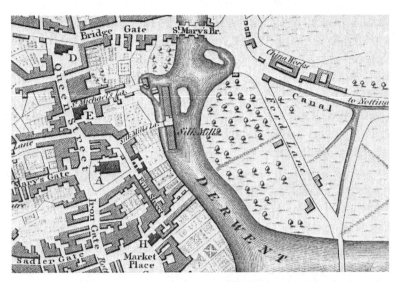

7 Detail from map of Derby, D. and S. Lysons, *Magna Britannia,*
Derbyshire (1817)

8 'Flora attired by the elements' from E. Darwin, *The Botanic Garden* (1791),
courtesy of Derby Local Studies Library

Whilst identification and naming of Darwin's garden plants still depended, of course, upon Linnaean and post-Linnaean terminology, they were not listed according to taxonomic groups. Location and identity was defined by workaday descriptions such as 'beyond the shed', 'beyond the fountain', 'around the fishpond', 'return to the well', 'opposite the stone table', 'return from the Derwent' and 'turn up towards the summer house'. The well was an artesian well that Darwin had sunk just after acquiring the house to provide cool clean water for domestic and garden use, which was commemorated with a plaque, and an account of this was published in the *Royal Society Transactions*. He was thus able to employ his geological knowledge and exemplify the utilitarian applications of Enlightenment science to create a water supply for his plants akin to the spring and perpetually dripping rock in the Lichfield botanic garden. Information concerning the qualities of plants and trees is also provided in the notebook with notes concerning colour, scent or beauty and fruit trees being described by their cooking, eating and preservation qualities. The notebook also reveals the sources for some plants including those sent from the family seat in Elston by Darwin's brother Robert Waring, author of the *Principia Botanica* (1787), another introduction to the Linnaean system. Another source of plants was Maria Jacson, author of the *Botanical Dialogues between Hortensia and her four Children, by a Lady* (1797) and various other botanical works, who corresponded with Darwin.[14]

Darwin's final garden was that of Breadsall Priory which he moved to with Elizabeth and family in 1802, having inherited the house from his son Erasmus after his tragic death in December 1799. In November 1799, Darwin told his son Robert that Erasmus had bought 'a place 5 miles from Derby call'd the Priory. Here was 'a large old house, and a farm of 80 acres, where he intends to live, but not to occupy the farm; and to sleep away the remainder of his life!' Erasmus had paid £3,500 for the Priory' which was 'thought a very cheap purchase' because it had 'long been exposed to sale' and it was 'a fine situation with 3 fish-ponds descending down a valley' with views of the River Derwent and the tower of All Saints Church in Derby'.[15] Whilst the Darwins were grief-stricken, the land surrounding the Priory presented a major opportunity for reconstruction, landscape gardening and gardening, and they sold the Full Street house, orchard and farm adjoining the new residence. Darwin described to Richard Lovell Edgeworth how they had all

> removed from Derby for about a fortnight, to the Priory, and all of us like our change of situation. We have a pleasant house, a good garden, ponds full of fish, and a pleasing valley somewhat like Shenstone's – deep, umbrageous, and with a talkative stream running down it. Our house is near the top of the valley, well screened by hills from the east and north, and open to the south, where, at four miles distance, we see Derby tower. Four or more strong springs rise near the house, and we have formed the valley, which, like that of Petrarch, may be call'd Val Chiusa, as it begins, or is shut, at the situation of the house.[16]

The way in which Darwin describes it as a small 'umbrageous valley' is reminiscent of the manner in which he portrayed the site of the Lichfield botanic garden and he even alludes to the fact that Maria Edgeworth might find equal literary inspiration by visiting. Although Darwin and his family lived at the house for only a short space of time, it appears that landscape improvements were begun before this. They were certainly continued by Elizabeth after her husband's death for the remaining 30 years of her life, and visitors to the Priory emphasised how much attention she devoted to the garden. With a letter to Barbara Strutt in the summer of 1801, Darwin sent a branch of a sugar maple, which he contrasted with an ash-leaved maple growing at the Priory and referred to the purchase of other plants for the gardens including a 'double ragged Robin' (*Lychnis floscuculi*) and a 'narrow-leaved Kalmia', the latter being a genus of evergreen American shrub.[17]

AGRICULTURE, HORTICULTURE AND IMPROVEMENT

As well as using his gardens for medicinal preparations, striving for beauty and studying taxonomy, inspired by the established British tradition of agricultural improvement, Darwin used them to conduct horticultural and agricultural observations and experiments. Agricultural experiment and improvement, especially from the seventeenth century included practical experiments with land drainage, enclosure, planting, soil improvement, tree planting, animal and plant breeding, including in the later case fruit cultivation. The desire for agricultural improvement was partly satisfied and stimulated by the application of mechanics and engineering and the accumulation of natural knowledge, particularly in county and regional histories from the second half of the seventeenth century. The English economy and society remained dominated by agriculture until the 1830s at least, as the fluctuations in food prices in towns led to crime, riots and public order problems. Most towns owed much of their status and economic character to local markets whilst livestock was still commonly kept in urban areas and horses and other animals were the principal forms of transportation. Paternalistic forms of urban government redolent of the countryside often remained in place and represented by the old economy with its magistrates fixing 'just' food prices, manorial government and landed estates, although enclosure patterns, commerce, manufacture and industrialisation were altering these relationships.[18]

There were numerous associational manifestations of the desire for agricultural improvement throughout Britain, drawing members from the gentry, nobility, professions and the universities, particularly in Scotland, where a perception of backwardness compared to England and other European countries motivated landowners and philosophers to support such measures. These included the Edinburgh-based Society of the Improvers of the Knowledge of Agriculture in Scotland (1723), the Aberdeen-based Gordon's Mill Farming Club (1758) and the English county agricultural societies.[19] The patriotic dimension to the scientific improvement of agriculture is evident in the prize-giving activities of the English Society of Arts and the Board of Agriculture, which were backed by many Tory and Whig landowners, whilst numerous philosophical groups and societies frequently discussed and sometimes published papers on agricultural subjects.[20] This was supported by philosophical and later agricultural societies, including the Royal Society, the Society of Arts, which offered prizes for agricultural and horticultural improvement, and especially from the second half of the eighteenth century, the county agricultural societies. During the 1790s county surveys were undertaken under the auspices of the governmental Board of Agriculture, although the quality of these varied considerably as they were undertaken by different individuals. Much of this experimental work was motivated by the potential utility of agricultural and horticultural improvements and these stimulated a large variety of sciences, including botany, anatomy, meteorology, chemistry and geology. In Leicestershire for instance, although a well-known Dissenter and wealthy benefactor of the Loughborough Unitarian congregation, Robert Bakewell, the horticulturist and animal breeder, mixed on equal and familiar terms with the gentry and nobility, dispensing advice on land management and breeding techniques. He also helped to found the Leicestershire Agricultural Society and kept his own museum of animal specimens and skeletons, which was made freely available to enquirers.[21]

The British celebration of agricultural improvement inspired Darwin who read many practical philosophical texts on the subject. He used the library of the Derby Philosophical Society to obtain many volumes produced by foreign scientific societies which helped him to write the books during the last ten or fifteen years of his life. Darwin was aware of important French taxonomic and practical botanical studies and imperial botanising, which rivalled and frequently overshadowed more amateur British efforts such as those encouraged by Banks and the agricultural societies. Darwin's efforts to translate the works of Linnaeus into English provided him with a profound and detailed knowledge of the current state of botany and he gained much inspiration from Philip Miller's *Gardener's Dictionary* (1724) and other works concerning the improvement of horticulture and agriculture such as Jethro Tull's *Horse-Hoing Husbandry* (1733), works on botanical taxonomy and classification, notably by Linnaeus and studies in vegetable physiology and anatomy such as Nehemiah Grew's *Anatomy of Plants* (1682) and Stephen Hales' *Vegetable Staticks* (1727).[22]

INDUSTRY AND GARDENING: DARWIN'S HOTHOUSE

One of the most important manifestations of Enlightenment horticultural improvement were hot houses which provided winter shelter for delicate varieties, demonstrating the impact of industrialisation,

changes in gardening taste, collecting and display and the importation of plants from warmer climates. Glass houses had a major
impact upon gardening, facilitated by industrialisation and transport
improvements, although they remained largely the preserve of middling sort and gentry because of the window tax, substantially increased by William Pitt's administration from the mid 1780s.[23]

As is clear from contemporary maps, the hothouse dominated Darwin's Derby garden, and fortunately he provided a very full description and sketch in a letter to Richard Lovell Edgeworth in 1788. Responding to enquiries that Edgeworth had made concerning hothouse construction, Darwin explained that it was 82 feet long and
about 9 feet wide with a glass roof of 8 inch square panes and brick
walls. He emphasised that the hothouse was divided into two so that
'one half may be a month forwarder than the other'. It produced
'abundance of kidney beans, cucumbers, melons and grapes' but not
pineapples and there were pots of flowers between the vines growing
up the 12-foot tall rear wall. At the front stretching from one end to
the other was a three and a half-foot wide bark-bed for melons
fronted by a two and a half-foot tall interior brick wall with flues
behind which was a wooden walk constructed from 'old barrel staves'
to prevent vine roots from injury. The hothouse was heated by two
fire places on the back wall and slow-burning 'athanor' stoves for four
months only, which consumed about six tonnes of coal. Two chimneys were on the much taller rear wall which backed onto his neigh-
bour's property. The heat was circulated around the building from
the stoves using a series of flues entering two feet below the ground
and passing through flues in the interior wall to the east and west
extremities, passing under the door step at one end before returning
from each side to the central chimneys on the rear wall.[24]

Darwin's Derby hothouse demonstrates the impact of mechanics
and industrialisation upon Georgian horticulture in various ways, including the construction of the glass roof and design of the heating
system which provided a highly controlled environment. The attention that he gave to the subject induced by the move to Radburn and
Derby is evident from various gardening inventions described in his
commonplace book including a 'melonometer or brazen gardener'
designed to open and close windows in hotbeds or hothouses according to atmospheric conditions, which King-Hele likens to a see-saw.

The device incorporated two four-inch copper globes joined under-neath by a lengthy pivoted horizontal tube, one filled with hydrogen gas and the other mercury, a vacuum existing in the top half of the globe. The tube was pivoted near the mercury sphere, so that the hydrogen sphere won the see-saw in cold weather and kept the win-dow closed whilst in sunny weather, the gas in the hydrogen globe expanded pushing the mercury up the tube into its globe overbalan-cing the see-saw and opening the window, clouding conditions in-ducing closure again.[25] Darwin did not attempt to introduce such a device in his Derby hothouse but did introduce oblique sashes rather than perpendicular ones.

The development of eighteenth-century hot-air heating systems paralleled – and was reciprocally stimulated by – the evolution of hot-air heating systems in textile mills including the Derby and Maccles-field silk manufactories and the Arkwright and Strutt cotton mills of the Derwent Valley. Inspired by these Darwin's friends such as the mechanic John Whitehurst and William Strutt produced versions for domestic and public buildings such as hospitals. Whitehurst designed chimneys and hothouses and oversaw the installation of a hot-air heating system in St Thomas's Hospital, London whilst Strutt de-signed domestic stoves and a hot-air heating system for the Derby-shire General Infirmary. Although Darwin's hothouse cost a hundred pounds to construct because of the expense incurred by the window tax and he had to order the 8-inch square glass panes from Stour-bridge (which he found easier to fit with each other), he declared to Edgeworth and his son Robert that he did not regret the outlay. The cost and effort expended by Darwin on obtaining the best glass from Stourbridge, the regional centre of the industry, is significant. The town had become integrated into the Midlands canal network during the 1770s, which facilitated the importing of raw materials for the production process such as coals, potash and lead and the export of glass products across England. Darwin's Lunar friend James Keir had moved to Stourbridge in 1770 and become a partner in a glass works, turning an old glasshouse to the rear into a laboratory, and striving to improve the industry experimentally and utilise its processes and materials for his chemical work which included improving the pro-duction of lithage (red lead) and translating Pierre Macquer's *Diction-arie de Chimie*. It seems likely that Keir would have offered advice – or possibly negotiated the Stourbridge sale for Darwin, who refers to him in the same letter. Much of the glass trade was devoted to ornamental ware and jewellery but great attention was also being given to producing larger, stronger, clearer and cheaper glass panes with immediate application to domestic, industrial and horticultural buildings where the need to maximise light and heat was paramount.[26]

TAXONOMY

In this section we will explore the relationship between Darwin's gardening and his presentation and reformulation of Linnaean taxonomy. During the second half of the eighteenth century botany became one of the most fashionable scientific pursuits as the number of popular works, herbariums or dried plants collections, flower painting and later formation of botanical societies demonstrates.[27] This process was aided by the contested dissemination, translation and adoption of the Linnaean botany in Britain after the 1750s (Figure 5a and b). The Linnaean system strove to rationalise the natural world by classifying living organisms using a binomial system of nomenclature according to a hierarchy of seven principal groups: kingdom, class, order, genus, species and variety, facilitating a relatively straightforward identification of plants in flower. Regarded as the most important or essential parts, reproductive organs of plants provided the basis for Linnaean 'artificial' taxonomy.[28] Nonetheless, tensions remained between practical botany and abstract taxonomy. Experiments on living bushes and trees by Stephen Hales, Darwin and Thomas Andrew Knight for example, which twisted, contorted and bled them in a highly 'unnatural' manner, demonstrated the control, manipulation and exploitation of nature for human benefit. Although Linnaean botany predominated in Britain until well into the nineteenth century led by Sir Joseph Banks as President of the Royal Society and James Edward Smith, President of the Linnean Society, rival systems, especially that of Antoine-Laurent de Jussieu in his *Genera Plantarum* (1789), claimed to approximate more closely to the natural world's divine plan. These disagreements partly stemmed from differing conceptions of nature which was variously understood and encountered as a fundamental order or harmony beneath the apparent diversity and disorder of the world and a manifestation of the divine plan.[29]

Natural history had long been an important partner of natural philosophy in British intellectual culture particularly because of its importance for natural theology and chorography. With its use of archetypal specimens, the Linnaean system provided standardisation and a sense of order and a universally recognisable and quite easily learned botanical framework, like a botanical alphabet, relatively accessible to amateurs, particularly after comprehensive translations and popular introductions to Linnaean botany appeared from the 1780s. As well as virtually creating a novel genre of natural historical literature Rev. Gilbert White's *Natural History of Selborne* (1789) encouraged practical botanising in the field as part of the celebration of locality (Figure 1). On the other hand the ubiquity of botany, accessibility of the Linnaean system and botanising of women and children helped to encourage the notion that it was not to be accorded the status of other more masculine sciences such as astronomy or Newtonian mechanics.[30]

Darwin's *Loves of the Plants* which was actually the first part of the *Botanic Garden* although published second, explored the sexual system of Linnaeus 'with the remarkable properties of many particular plants' (Figure 8). In this, Darwin strove to present a 'camera obscura' in 'which are lights and shades dancing on a whited canvass, and magnified into apparent life' and he urged his readers if 'perfectly at leisure for such trivial amusement, walk in and view the wonders of my enchanted garden'. As discussed with Seward, the premise of the poem was that just as Ovid had by poetry transmuted 'men, women, and even gods and goddesses, into trees and flowers' so Darwin undertook 'by similar art to restore some of them to their original animality, after having remained prisoners' in their 'vegetable mansions'. He now exhibited them in a manner likened to 'divers little pictures suspended over the chimney of a lady's dressing room, *connected only by the slight festoon of ribbons*'. Even if unacquainted with the originals, these would amuse 'by the beauty of their persons, their graceful attitudes, or the brilliance of their dress'. The general design of the *Economy of Vegetation* was to 'enlist imagination under the banner of science and to lead her votaries from the looser analogies, which dress out the imagery of poetry, to the stricter ones, which form the ratiocination of philosophy'. Darwin particularly hoped to 'induce the ingenious to cultivate the knowledge of botany by introducing them to the vestibule of that delightful science, and recommending to their attention the 'immortal works of the celebrated' Linnaeus. The *Economy of Vegetation* examined the physiology of plants and the 'operation of the elements as far as they may be supposed to effect the growth of vegetables'.[31]

The symbiotic relationship between place, experience, social networks and reading is evident in the formation of Darwin's Lichfield botanic garden and the important effects that it had upon his work. Knowledge of botany was required for medical education and the prescription of many medicines, and Darwin was taught the subject at medical school in Edinburgh, using and developing this knowledge for patient prescriptions as he travelled around the Midlands. The additional impetus for this botanising seems to have come from the physician William Withering's greater involvement with the members of the Lunar circle during the 1770s. Withering also obtained his medical degree from Edinburgh, visited Lichfield often and established a medical practice in Stafford, serving as physician to the Stafford Infirmary from 1766. He was taught botany by John Hope, Professor of Medical Botany at Edinburgh and one of the first academics to lecture on the Linnaean system of classification in a British university. Withering collected botanical specimens, encouraging one of his patients to undertake plant drawings, whom he subsequently married, and by 1772 had begun a treatise on British botany arranged according to the Linnaean system. This appeared in 1775 with the encouragement and help of Darwin and other Lunar friends as *A Botanical Arrangement of all the Vegetables Naturally Growing in Great Britain* and featured 'descriptions of the genera and species' presented 'according to the system of the celebrated Linnaeaus'. Darwin offered advice to Withering concerning terminology and symbolism although his suggestion to use a short and 'easily remember'd and distinguishable title was rejected in favour of a 24-line designation. Although initial reviews of the book were enthusiastic describing it as 'the most elaborate and complete' national flora 'that any country can boast of' and it remained in popular use in improved editions into the nineteenth century, the first edition consisted largely of translations of relevant sections of Linnaeus concerning indigenous British genera and species. Plates depicted plants and plant sections and botanical instruments and there were instructions concerning the preservation of specimens, whilst some changes to terminology were recommended, primarily on the grounds that sexual distinctions between classes and orders would be rendered in delicate language so as not to offend polite sensibilities.[32] Withering and Darwin fell out seriously concerning various issues, especially the fact that the Lichfield Botanical Society's translation of Linnaeus' *Systema Vegetabilium* (1783) appeared to the former to have pre-empted a translation of the same that Withering had been attempting.[33]

In translating the works of Linnaeus into English assisted by Brooke Boothby and William Jackson, Darwin and the members of the Lichfield Botanical Society discovered that whilst some of the Linnaean terms were easily translatable into English, other Latin terms and phrases required more original English coinage, for instance through compounding terms, and they consulted the 'great master of the English tongue Dr Samuel Johnson' on the 'formation of the botanic language'. One of the principal objectives was to bring, or rather return, the Linnaean system democratically from the confines of the study into the world of practical gardening and horticulture and so to improve the botanical education of landowners, gardeners, apothecaries and others with professional interests in the subject. They noted that 'like the Bible in Catholic countries', Linnaean botany had been 'locked up in a foreign language, accessible only to the learned few, the priests of Flora'. This meant that gardeners, 'the herb-gatherer, the druggist', farmers, and all 'concerned in cultivating the various tribes of vegetation, in detecting their native habitations, or in vending or consuming their products', were never able to obtain the knowledge to improve. Social interaction and the sharing of ideas and practices in gardens, shops and workplaces was stymied by such elitism, when these were equally 'capable of enlarging' the science of botany.[34]

Darwin was well aware of the hostility with which some botanists who had tampered with the Linnaean system were faced by Linnaeus and his followers. As we have seen, he was largely responsible for the Lichfield Linnaean translations which included the master's own summary of his principles of classification and distinguished between 'true' and 'false' botanists. As rendered by Darwin and his friends, Linnaeus emphasised how God had proceeded from the 'simple to the compound, the few to the many', forming as many different plants 'as there are natural orders' at the creation. He had then 'so intermixed the plants of these orders by their marriages with each other', that as many had been produced as were 'now distinct genera'. Although the time frame involved was not specified, 'Nature' had then 'intermixed these generic plants by reciprocal marriages (which did not change the structure of the flower) and multiplied them into all possible existing species'. 'Mule plants', were, however, excluded 'from the number of species ... as being barren'. Each genus was therefore 'natural', nature assenting to it, if not making it.' The character was never to constitute the genus but was 'itself diligently to be constructed according to the genus of nature'. The 'diagnosis' of the plant consisted of 'the affinity of the genus, and in the difference of the species', which was reflected in plain nomenclature which pro- vided 'generic family name' and specific trivial name', under which were 'the vague synonymies of authors'. 'True' botanists moved from genus determined by 'characters of the displayed plant or flower' to the 'appellation of the species by the differences of the larva or herb', and to 'its synonymies' and then finally to 'authors' who imparted the botanical wisdom of the ages.[35]

The fruits of Darwin's careful plant observations of plants and experience of translating the Linnaeaus and coining appropriate ter- minology are evident in the reformulation of the system that he attempted. The object of these changes primarily induced by the dis- parity between Linnaean order and the experience of planting and observing garden plants, was to make the system more 'natural', the meaning of which needs carefull considation. The extensive notes and additional notes in the *Botanic Garden*, really a set of philosophic- al essays in their own right, represented subjects that Darwin consid- ered necessary to supplement his presentation of the Linnaean sys- tem as well as more general scientific observations. Despite his fulsome praise for Linnaeus, the notes concerning natural history illustrate areas that Darwin considered were inadequately explored by the Linnaeans including vegetable physiology and anatomy. The experience of being able to stroll around his metaphorical garden and undertake philosophical detours helps to explain the appeal of the work for British readers. The Linnaean system as defined in the *Philosophia Botanica* (1751) and *Species Plantarum* (1753) aimed to chart the animals and vegetables of the natural world, and Linnaeus used the metaphor of a map, describing it as his *Geographia Naturae*. The globe was divided into five climate zones: Australian, oriental, Mediterranean, boreal and later the Alpine zone. The works of Linnaeus and his followers were received differently across Europe, but in Britain there was generally an enthusiastic reception represented by the number of partial translations and introductions that appeared.[36]

In a lengthy appendix to *Phytologia*, Darwin presented what he described as a 'plan for disposing part of the vegetable system of Linnaeus into more natural classes and orders' which exploited and paralleled his experience in designing and planting the botanic garden, doing the translations of the genera, orders and families and composing the *Botanic Garden*. He was able to observe how the system related to growing plants rather than dried herbariums and highlight discrepancies, anomalies and inconsistencies. Whilst admitting that he had 'often admired' the Linnaean classifications deduced from organs of reproduction, Darwin contended that 'some of the classes' appeared 'more excellent than others, as they seemed to approach nearer to natural ones'. Reading through *Phytologia* decades later, Charles Darwin thought it extraordinary that his grandfather had 'apparently ... never heard of Jussieu', whose natural system provided considerable help in the composition of the *Origin of Species* (1859), which he took to be evidence for how 'English botanists had been blinded by the splendour of the fame of Linnaeus'. However, the fact that, like most British botanists prior to 1810, Darwin did not draw upon Antoine-Laurent de Jussieu in the formulation of a more 'natural' taxonomy, demonstrates the importance of the inspiration he received from his gardens. His gardening bore fruit and resulted in three different but interrelated translations of the Linnaean system into an English picturesque botanic garden, English prose and English poetry.[37]

The Lichfield garden was defined by sections divided according to Linnaean taxa with additional flowers, trees and shrubs and landscaping for picturesque effect. For the list of plants in the Derby gardens, on the other hand, as we have seen, Darwin eschewed clearly defined Linnaean sections, treating them more according to experi-

mental and observational objectives and aesthetic effect than taxonomic representation. Whilst the variety of plants was large for such
small gardens, some of Darwin's favourite flowers reappear in various
locations determined by practical experience on the low-lying riverside site and because colour, beauty and other features marked them
out on his perambulations.[38] Darwin's gardening led him to conclude
that classes 'deduced from the proportions or situations of the stamina' or which included the number of the stamina along with
'proportions and situations' were 'more natural' than those merely
distinguished by stamina number. For example, the classes 'Dydy-
namia and Tetradynamia' derived from 'the proportions and situations of the stamina as well as their number', were 'wonderfully
natural' as were 'the classes Icosandria, and Polyandria', their diag-
nostic characters consisting of the 'situation of the stamina on the
calyx or petals in the former class, and on the receptacle in the latter',
although the reference to number was unfortunate given that these
were 'very variable'. Other of the Linnaean classes 'distinguished by
the situation of the filaments' such, as 'Monadelphia, Diadelphia,
Polyadelphia, and Gynandria' were more natural whilst Syngenesia,
'distinguished by the adhesion of the anthers', was a 'beautifully natural' class, 'except the last order'. Similarly, Darwin considered that
many of the orders in the Linnaean system were also 'natural classifications' such as 'the grasses in the class Triandria, the umbellated
plants in the class Pentandria, and perhaps the cruciform plants in
the class Tetrandria'. Most of those described as natural orders at the
end of the *Genera Plantarum*, 'might probably be discriminated by
some situation, or proportion, or form, of their respective stamina'.[39]

For Darwin, a natural classification enabled the greater exploitation of plants for practical purposes. More natural classification
meant 'classes founded upon the proportions, situations and sometimes additionally number of the stamina' which appeared to 'produce more natural distributions of vegetables than those derived
simply from their number'. On this basis he considered that it would
have been 'more fortunate for the science of botany' if Linnaeus had
'classed all of them from the proportions, situations, and forms of the
stamina ... or from these conjointly with their number', and to have
distinguished orders 'according to the proportions, situations, or
forms of the pistilli alone, or conjointly with their numbers'. It was

important to place plants within more 'natural classes' for identification to facilitate exploitation for culinary, artistic, medical or manufacturing purposes such as 'dying, tanning, architecture, ship-building'. This had 'already been happily experienced in attending to the genera or families of plants', which were 'all natural distributions', so that the 'same virtues or qualities generally exist among all the species of the same genus, though perhaps in different degrees.'[40] Another 'great advantage' of 'deducing the characters of the classes of vegetables from the situations, proportions, or forms of the sexual organs rather than from their number' was that such 'criterions of the classes and orders would be much less subject to variation'. From nurturing and observing wild and cultivated plants and recording features such as multiple flowering, petal number and situation and instances of leaf variegation, Darwin suggested various reasons why variation in the number of stamina appeared to have occurred which were liable to confuse novice Linnaean botanists. One factor was 'too luxuriant growth of many cultivated flowers' another was 'the duplicature or multiplication of their petals or nectaries' which meant that 'several ... species of plants' had 'but half the number of stamina which other species of the same genus possess'. This occurred so frequently that the 'defect of number' was often 'expressed as an essential character of the species'.[41] It is important to note that Darwin refers to classes and not orders and the importance of these parts as the most unchangeable organs in the plants and therefore the most reliable markers or sets of characters for identification. These were the most natural, invariable, immutable, essential and unchangeable, whereas, as he observed in the garden, other vegetable parts were much more mutable in response to environmental factors.

PLANT PHYSIOLOGY AND ANATOMY

Darwin also used his gardens and observations of plants elsewhere to try and determine optimum conditions for growth, primarily for the economic benefits that this would confer. Floods from the Derwent, for instance, provided information upon how plants behaved in low-lying conditions and recovered from inundations and mud burial. His gardens stimulated investigations into vegetable circulation, placentation, respiration and glandulation. Experiments and observations of plants within combined with observations of farming practices challenged or confirmed theories encountered in the agricultural and horticultural literature. With his friend Rev. Abraham Bennet for example, Darwin investigated the role of electricity in the germination and growth of plants. In *Phytologia*, Darwin suggested that water was decomposed into oxygen and hydrogen by the action of electricity in plants, 'by the decomposition of water in the vegetable system when the hydrogen unites with carbon and produces oil, the oxygen becomes superfluous, and is in part exhaled'.[42] To test this hypothesis, a Leyden jar or revolving doubler to augment the charge were of no use because the supply needed to be continuous, so Bennet invented the pendulum doubler portrayed in *Phytologia*, which kept flower pots 'perpetually subject to more abundant electricity'. The device incorporated three plates, two of which were fixed and a third that swung between them. A system of springy wires completed connections during different stages of the doubling process. The loose plate swung from the pendulum of a Dutch wooden clock and electricity was passed to the plant pot via another connection.[43] Another friend Dewhurst Bilsborrow, later a physician at the Derbyshire General Infirmary, subjected mustard seeds to both positive and negative electricity finding that they germinated 'much before' others that received none. Bennet's observations of meteorological electricity suggested that, because of the apparent role of electricity in promoting plant growth, artificial production of atmospheric electricity might benefit agriculture. Darwin therefore advocated the use of numerous metallic points on the ground to promote 'quicker vegetation' by supplying the electric ether and encouraging rain showers. Bennet's apparatus, including the pendulum doubler, would help compile journals recording the impact of the varied state of electricity upon agriculture and its 'influence' upon the human system.[44]

Darwin became notorious for eliding animal and plant psychophysiology mediated by his concept of the 'spirit of animation' which had close analogies to the electrical fluid. For Darwin, the spirit of animation was the power which allowed all kinds of life to react to changing environmental situations on the basis of individual ideas and experience. Plants were 'an inferior order of animals' and the motions of the Venus fly trap and Mimosa or sensitive plant demonstrated that there were 'not only muscles about the moving foot-stalks' of leaf claws and petals but that these 'must be endued with nerves of sense as well as of motion'. Furthermore, the sensitivity of the mimosa showed that 'there must be a common sensorium, or brain, where the nerves communicate' and that vegetable buds possessed irritability, sensation, volition and 'association of motion' though in 'a much inferior degree even than the cold blooded animals'. [45] For Darwin the 'sensorial motions' constituting the sensations of pleasure or pain, which 'constitute volition' and 'cause the fibrous contractions in consequence of irritation or of association', were not merely 'fluctuation or re-fluctuations of the spirit of animation', nor just 'vibrations or re-vibrations' or 'conductions or equilibrations of it', but were 'changes or motions of it peculiar to life'. [46] He therefore drew close parallels between vegetable anatomy the 'animal economy' in terms of physiology, some organs, arterial systems, organs of propagation or reproduction, secretions and muscles. Darwin conducted further experiments on different parts of plants and trees moistened to demonstrate the existence of absorbent vessels in all sections besides the roots. These parallels between plants and animals are also signified by the fact that the plan of the *Zoonomia* was modelled upon the Linnaean system.

The mutual interdependency of the animal and vegetable worlds in creation was fully evident to Darwin who claimed to have caught a 'glimpse of the general economy of nature' and could exclaim 'God dwells here'. On the other hand, he considered that animals and vegetables were perpetually warring with each other as well as being mutually interdependent. As vegetables were 'an inferior order of animals fixed to the soil; and as the locomotive animals prey upon them, or upon each other; the world may indeed be said to be one great slaughter-house'. As he maintained in the *Economy of Vegetation*, the mutual interdependency was founded upon the fact that 'digested food of vegetables consists principally of sugar', from which was produced their 'mucilage, starch, and oil', and since animals were sustained by these 'vegetable productions', he maintained that the 'sugar-making process carried on in vegetable vessels was the great source of life for all organised beings'.[47]

CONCLUSION

Principally through the success of the *Botanic Garden*, Darwin's work had a major impact upon Georgian culture and society. As his grandson remarked drawing upon the recollections of his father, 'its success was great and immediate', Darwin made much money from the publication with various British and foreign editions. Contemporaries such as Horace Walpole hailed Darwin's 'most beautifully and enchantingly imagined' creation whilst Richard Lovell Edgeworth claimed that some sections 'seized hold of his imagination' to such a degree that his 'blood thrilled back through his veins'. The young poets Samuel Taylor Coleridge, William Wordsworth and Percy Bysche Shelley were initially excited and inspired even if they later turned against Darwin's style, whilst the older generation of poets including William Cowper and William Hayley were equally enthused.[48]

However, changes in poetical taste and the association between Darwin's notion of progressive Enlightenment science, political beliefs and reforming friends caused greater hostility during the revolutionary period and put the late-Georgian celebration of horticultural improvement under strain. This is exemplified by the ridicule of Darwin along with his friend Priestley in contemporary cartoons and satirical poems such as the *Loves of the Triangles*, which parodied his style.[49] A poem entitled *The Golden Age*, ostensibly penned by Darwin to his friend Thomas Beddoes and published in 1796 satirised the close analogies between animals and vegetables and the notion that plants differed 'from animals alone in name'.[50] Whilst given the multiple classical and neo-classical precedents Darwin's plant personifications were acceptable in poetry, drawing analogies between animals and plants too precisely caused much more trouble, especially after the French Revolution and Reign of Terror. In his incredible panoramic cartoon 'New Morality; -or-the Promis'd Installment of the High-Priest of the Theophilanthropes with the Homage of Leviathan and his Suite' (1798), James Gillray depicted Darwin advancing amongst a grotesque procession of reformers led by the Duke of Bedford carrying a basket of revolutionary red-bonneted flowers labelled 'Zoonomia or Jacobin Plants'. The writer of *The Millenium, a Poem in Three Cantos* also went on to ridicule Darwin's revolutionary flowers in 1800, suggesting that plants deserved full legal and state protection. He claimed that it was the 'serious creed of Dr Darwin as a natural philosopher, as well as his system as a creative and vivacious poet' and had been enunciated in 'a late bulky quarto' (*Phytologia*) which was 'replete with entertainment and combines about an equality of fanciful and unfounded opinions' with 'novel and highly valuable facts'. The writer claimed to have heard that numerous individuals, 'particularly among the ladies, have of late become converts' and in 'more than one instance', a 'fair proselyte has extended it to so extreme an attitude as to give her geraniums and other plants an airing in her carriage wherever they appeared sickly and seemed to require exercise'. Darwin's suggestion that in the face of death and destruction in nature consolation could be found in the increase of the sum total of human happiness was likened to a 'hard-hearted wretch who exults upon having cleared his head' of lice being reminded that he had destroyed the lives of millions of microscopic animals – the sum total of whose happiness was 'perhaps greater than his could ever be, by whose destruction they have gained their existence!' Should this not be 'a consoling idea to a mind of universal sympathy' with the itch?[51]

Beyond the *Loves of the Plants*, the impact of Darwin's philosophical conceptions of botany, horticulture and gardening are less well known. This partly reflects the fact that two of his most important works, *Zoonomia* and *Phytologia* have attracted far less scholarly attention than the *Botanic Garden*, except for the sections that have been Whiggishly taken to 'anticipate' evolutionary theory. With regard to Darwin's work on vegetable physiology and anatomy, even sympathetic contemporary philosophers were bewildered and sometimes infuriated by the frequency and ease that he elided distinctions between animal and vegetable worlds. Joseph Banks knew Darwin's work very well having encouraged the Linnaean translations and experiments on plant physiology and anatomy, however, his candid view of Darwin as expressed to others was mixed. He observed to

Thomas Andrew Knight that Darwin mixed 'truth and falsehood, ingenuity and perversity of opinion, exactly in the manner we mix the ingredients of punch'. However, the regard in which Darwin's work in this field was held during the early nineteenth century is evident from the degree to which authorities such as James Edward Smith, President of the Linnean Society and most senior figure in British botany after Banks, cited him with approval in their works. In his *Introduction to Physiological and Systematical Botany* (1802), Smith described *Phytologia* as 'a store of ingenious philosophy'. Sir John Sinclair, President of the Board of Agriculture, who helped to inspire the composition of *Phytologia*, stated that it was 'highly gratifying to me, to have prevailed on so able a writer as the celebrated Dr Darwin ... to draw up a work on practical agriculture' and judged the book to be 'a most valuable performance', though 'of too philosophical a description' to qualify as practical agriculture.[52] Charles Darwin also seems to have maintained quite a high opinion of *Phytologia* and sought to remedy the fact that Ernst Krause made little mention of it in his *Scientific Works of Erasmus Darwin* (1879) by inserting a discussion of it.[53]

Darwin's gardens certainly were, as Anna Seward maintained, places that united 'the Linnean science with the charm of landscape', but they were also microcosmic worlds for observations and experiments in plant taxonomy, physiology and anatomy and much more.

3   The School: Dissenting Academies and Scientific Culture

INTRODUCTION

Inspired by perceptions of utility in addition to natural theology, aspects of natural philosophy and natural history were taught to boys, and some girls, in numerous private commercial, mathematical, military and naval academies and some grammar schools, despite the traditional presentation of the latter as moribund and devoid of educational innovation. Although often accorded a major role in the development of scientific education, the part played by Dissenting academies has recently been downgraded and the importance of private and commercial academies in the teaching of mathematics and natural philosophy emphasised instead. However, the attempt to counter exaggerated notions of cultural and educational progressivism amongst Nonconformists has gone too far and risks obscuring the distinctiveness of the Dissenting educational experience as part of the European Enlightenment. In many respects in these schools, the critical, rational and empirical Enlightenment was embodied and organised into a system through which 'practice and theory' were consolidated into a 'new encyclopaedic curriculum'. A 'new map of learning was drawn' and explored, encouraging the use of apparatus and pedagogical experimental demonstration, which impacted upon – and was shaped by – history and geography. The distinctive social and cultural position of English and Welsh Dissenters encouraged commercial, technological and scientific innovation and female participation in education and public life.

Combining approaches from recent scholarship on Georgian society and culture, the history of science, history of geography and history of education, this chapter explores how natural philosophy and related subjects such as mathematical geography were taught by Dissenting tutors and how this impacted beyond their communities and shaped female education. Nonconformist social and cultural position determined the character of their scientific teaching encouraging an emphasis upon utility, internationalism and the value of demonstration and self-discovery. The study of natural philosophy was pursued as part of a progressive Enlightenment pedagogy intended to foster moral, rational and civil education and religious freedom.[1]

Of course an emphasis upon modern subjects and teaching methods was evident in many other contexts besides Dissenting academies including commercial and mathematical and grammar schools.[2] Furthermore the apparent marginality of Nonconformists and the relationship between forms of Protestantism, capitalism and scientific and technological innovation has also been reassessed in the light of evidence for Nonconformist charitable and local governmental participation. Different forms of Dissent had distinctive changing geographies with varied concentrations of Nonconformists in the north, midlands and London and changing relationships and outlooks between metropolitan and provincial Dissenters. Numerically small groups such as the Unitarians had a disproportionate impact upon scientific education in towns such as Manchester, Nottingham and Norwich. Shifting and porous boundaries within Nonconformity and between Dissenters and Anglicans reflected subtle changes in affiliations and positive and negative identities, with occasional conformity proving popular and many regarded division from the Anglican Church as temporary.

Most tutors in early Dissenting academies were ejected Anglican clergy, and many pupils who attended 'open' academies were Anglicans from gentlemanly, aristocratic and wealthy merchant families. Some orthodox Dissenters during the 1780s were closer to liberal wings of the Anglican Church than to Socinians or Rational Dissenters. As Seed has pointed out the Anglican Church emerged from the 1730s in many ways as 'a powerful and hegemonic institution both locally and nationally'. Nonconformists reconcieved their identities through construction and reconstruction of historical narratives and as different practices in Dissenting academies demonstrate, there were important differences in receptivity to new ideas and methods. Whilst Unitarians, Rational Dissenters and some Presbyterians and Congregationalists were more favourable to modern subjects in education and are the primary focus of this chapter, many Baptists, strict Calvinists and at times Quakers were more conservative. The term Dissent as employed here is therefore used for convenience and should not be interpreted as ignoring differences between – and changes within – Nonconformist denominations.[3]

Recent scholarship has highlighted complex interconnections between the teaching of natural philosophy, natural history and other subjects perceived as 'modern' in the Georgian context, moving away from concentration upon text books, individual institutions and the notion of linear development. This has been replaced by an appreciation of the multiplicity of arenas in which different kinds of science teaching were occurring stimulated by the cultural demands of the Georgian middling sort, empire and trade. Equally it has become clear that supposed divisions between the sciences and mathematical, ancient and modern geography and other disciplines changed according to local circumstances. Building upon this work, this chapter examines how and why Dissenters taught scientific subjects, arguing that, particularly during the first half of the eighteenth century encouraged by tutors such as Isaac Watts and Joshua Oldfield, some innovative practices were adopted which impacted beyond the academies. Although distinctions were recognisable, practical mechanics, mathematics and natural philosophy were often taught together with classical, biblical and historical studies inspired by natural theology. Whilst serving and being rhetorically justified as part of biblical and classical studies, from the late seventeenth century, the sciences quickly became exemplified by Newtonian natural philosophy, astronomy and mathematics, navigation and other related subjects. They combined practical mathematical and theological elements with the intention to offer basic components of polite knowledge for the educated minister, professional or layman.[4]

DISSENT AND EDUCATION

With their emphasis on mathematics, geography and scientific subjects within a biblical and classical grammar curriculum, Dissenting schools were similar to some earlier and contemporary mathematical academies, however, they placed greater emphasis upon the importance of modern subjects for personal spiritual development and social improvement. Dissenting social, political, commercial and cultural networks were supported by the relative security and consistency of national subscription funds and organization which distinguished them from other schools and encouraged a relatively standardised curriculum, helping to explain why despite civil restrictions and short life spans, they became significant in metropolitan education. In response to legislation against them and the charges of infidelity, heresy and materialism levelled against them the kind of social hostility whipped up by firebrand High Tory preachers such as Henry Sacheverell, Dissenters made public protestations of loyalty. In response they made public demonstrations of loyalty, openly welcoming the Hanoverian monarchy, practised occasional conformity, and took care to keep their academies in a conspicuous, unobtrusive manner. Endowment and subscription funds with boards of trustees modelled on the Presbyterian Fund such as the King's Head Society (1730) and the Lady Hewly Fund (1707) offered some security although they encouraged a preoccupation with ministerial education.[5]

The influence of Dissenting academies extended far beyond congregations, particularly prior to 1770 and there were considerable cultural and intellectual interchanges between different tutors. Texts such as Joshua Oldfield's *Essay Towards the Improvement of Reason* and Isaac Watt's *Improvement of the Mind* were popular beyond Dissenting circles and Nonconformist tutors expressed the hope that their systems would help reform the English universities. There was much more cross-fertilisation between schools and educators with Dissenters serving as grammar school tutors and Anglicans being educated in Dissenting academies and reading Nonconformist educational texts. Anglicans, including clergymen, also supported Dissenting schools that sometimes became established as dominant local institutions such as the Welsh academy established at Carmarthen which 'gained recognition as a permanent public institution' supported by bequests for student annuities, book and equipment purchases.[6] Stocks of equipment and some quite extensive libraries were accumulated by some academies providing a degree of continuity between changing masters more redolent of endowed schools than commercial establishments. Although there was some educational experimentation in commercial schools, they were much more susceptible to competition from endowed schools and the whims of fashion and changes in the economic climate than their Nonconformist counterparts. Dissenting academies drew upon a collective stock of experience and knowledge passed between pupils and educators in succession over decades with successive schools modelling their curricula, practices, use of equipment and educational texts upon previous institutions. Priestley, for example, was born in Yorkshire and educated at Daventry, before becoming tutor at Warrington. He ministered in Leeds, Cheshire and Birmingham and associated with London academies before going to the United States of America, having also been a member and founder of other associations such as the Leeds Subscription Library and the Lunar Society.

Because of religious connections, civil restrictions and the need to seek additional education and professional qualifications elsewhere, Dissenting academies were often more responsive to developments in geography, the sciences and foreign scriptural studies than other English educational institutions. National and international commercial and cultural Dissenting networks facilitated the exchange of ideas and methods with other countries especially those with the closest religious connections such as Scotland, Holland and Switzerland. The curriculum at Tewksbury/Gloucester academy under Samuel Jones from the late seventeenth century, for instance, was modelled on that at Leyden where Jones had been educated and differed from that provided in most commercial academies. It included sacred and modern geography, Jewish antiquities, Hebrew and other biblical languages, the study of geography being justified by classical, theological and cosmological arguments. With their emphasis upon the importance of natural science, mathematics and other modern subjects in the context of a Calvinist classical and scriptural education, Continental pedagogical practices provided a model for the English and Welsh Nonconformist schools.[7]

The development of scientific teaching in Dissenting schools, especially open academies, was encouraged by the same forces of market competition, and the similar economic, technological and maritime demands of British imperialism that drove the development of private school curricula. However, metropolitan and provincial social and business networks and civil and theological imperatives provided Dissenters with a collective strength denied to most of their commercial rivals. The success of the Nonconformist academies in centres such as London, Glasgow and Manchester was clearly driven by the strength of Dissenting professional and mercantile communities, whilst smaller communities in the midland and north also supported schools. As Bryant has argued, 'Nonconformist academies were a national phenomenon' and provincial and metropolitan schools reciprocally adopted practices from each other. This also reflected 'shared assumptions' concerning the 'nature of knowledge, the authority of reason, the challenge to precedent and privilege' which allowed the Dissenters to transcend their immediate locality and the division between metropolis and provinces as part of a wider national grouping.[8] Thus although London Nonconformist schools were influential at various times such as the late seventeenth century in encouraging curriculum development, provincial academies in market, county and manufacturing centres such as Manchester were also centres of educational innovation. Innovative ideas and practices were communicated through printed texts, shared lecture notes and verbal communication, whilst ex-pupils and staff living around the country remained in communication with their old schools, recommending pupils and making bequests that included special funds, books and equipment. This meant that provincial Nonconformist schools did not simply assimilate metropolitan culture and values but were sources of educational experimentation and innovation in their own right.

TEACHING NATURAL PHILOSOPHY
IN DISSENTING SCHOOLS, C.1680–1760

Many Dissenting academies in existence between 1680 and 1750 taught natural philosophy and related subjects such as mathematical geography, a terraqueous counterpart to astronomy, the geography of the heavens. These subjects were regarded as having important utilitarian functions for chorography, navigation, surveying, cartography and mechanics. The emphasis upon experimental demonstration and scholarly validation in Newtonian mathematics and astronomy helped to make these subjects attractive to Nonconformist educators where they were regarded as a British Protestant corrective to Catholic Continental philosophies. The appeal that this mathematical education offered helps to explain why a disproportionate number of natural philosophers, merchant entrepreneurs and technological innovators were Dissenters. In comparison, endowed grammar schools were much slower to adopt the subject although their curricula were not as fixed as used to be contended. A national comparison between commercial and Dissenting schools is more difficult to make, partly because of the considerable overlap between the two. However, the evidence suggests that up to 1750 many of the former were stimulated by competition from the latter to offer modern subjects such as natural philosophy and geography and to adopt methods advocated by Nonconformist educators.[9]

It is not hard to discern why natural philosophy became so popular in Dissenting schools given the strong advocacy of the subject by Nonconformist educators Joshua Oldfield, Philip Doddridge, Isaac Watts, James Gregory, Stephen Addington and Walker. Educated at Cambridge and Edinburgh, Oldfield taught at an academy in Hoxton Square from 1699 to 1729 and emphasised the importance of historical, scientific and geographical subjects including mathematics, astronomy, experimental natural philosophy and navigation alongside theology and classical and biblical languages. Scientific and geographical knowledge would enable gentlemen to increase their estates and 'be of common service to their country, or to the world'. Natural philosophy, natural history and geography provided the knowledge necessary to understand contemporary political and civil affairs whilst globes and maps facilitated the study of religion, government,

83

astronomy, navigation, arithmetic and geometry.[10] Echoing Oldfield's *Essay*, in his textbook on astronomy and geography and the *Improvement of the Mind*, Watts argued that it ought to be common practice to introduce geography, mathematics and natural philosophy into teaching. Educated at Thomas Rowe's Academy in Newington Green which has been described as 'one of the most potent forces in shaping eighteenth-century Nonconformists thought', Watts became one of the most influential Georgian educationists. Because of his eminence as a classicist and popularity as a hymnodist, Watts's ideas reached far beyond Nonconformist circles, and he was venerated by Johnson as a greater teacher than Locke or Milton.[11] Like other Dissenting educationists such as Doddridge and Oldfield, Watts believed in the importance of individual study as a moral duty which would improve society by helping to forge good citizens through moral and intellectual improvement. This ecumenicalism would consequently bring social betterment for all and incidentally improve the social and legal position of Dissenters themselves as the rational discourse of international scholarship demonstrated. Oldfield, Watts and Doddridge all emphasised that education was necessary for citizens of all social ranks and status, and that it was sinful to remain in ignorance. According to Watts, individuals were 'accountable to god' the judge for 'every part' of their conduct and this included women several of whom were intent upon pursuing 'science with success' and 'improving their reason even in the common affairs of life, as well as the men'.[12]

This philosophy was manifest in a practical curriculum that placed great emphasis on the importance of scientific subjects. Watts believed that the original inspiration for studying astronomy and geography was personal study, amusement and theological education, as 'without commencing some acquaintance with these mathematical sciences, I could never arrive at some clear a conception of many things deduced in the Scriptures'. Like Oldfield and David Jennings at Hoxton, he celebrated the new worlds uncovered by modern science, temporal, microscopic, global, planetary, cosmological and universal. Jennings emphasised the importance of geography in the advancement of scriptural and historical studies and the possibilities for opening up actual and metaphorical spaces by moving beyond the 'scenes of worldly business' on 'this little planet' to 'take a distant view of other remote worlds'. Watts' course of study included topics such as the natural description of the earth and waters, natural and political divisions and geographical terms. However, he believed that his astronomical and geographical curriculum was distinguished from others because of its greater attention to 'manual operation' using diagrams and visual demonstrations. The Dissenters' emphasis upon the individual learning experience rather than upon dogmatic peda-gogies stemmed from their distrust of overbearing authority which was also evident in the attention they gave to practical classroom layout. The emphasis upon assiduous daily devotion to education was encouraged by theological imperatives which tended to distin-guish Dissenting academies from other schools.[13]

Watts recommended that marked globes, celestial globes and maps of the world be used, represented by brass and wooden work in order that the 'learner' appreciate the circles of the globe, equator, ecliptic, zodiac, lesser circles, points and poles of the ecliptic. These were demonstrated using globes and maps, sea charts and gazetteers and it was hard to escape geography as he recommended it was best memorised by 'having those schemes and figures in large sheets of paper, hanging always before the eye in closets, parlours, halls, cham-bers, entries, staircases'. It was 'incredible' how much geographical subjects could be learnt using 'two terrestrial hemispheres' and by maps and charts 'happily disposed around us'. He also criticised col-oured maps or projections 'daubed so thick with gay and glaring colours' which were hung up to be seen 'as though their only design were to make a gaudy show upon the wall' and 'merely to cover the naked plaster', which should have been coloured sparingly.[14] The use of equipment such as globes and maps to teach astronomy and geog-raphy was encouraged in other Dissenting academies including those that taught Anglicans. The Newington Green Academy for example, established by Charles Morton in 1675 and surviving until around 1706 took some Anglican pupils and deliberately tried to provide an alternative to university education. Pupils included Daniel Defoe and Samuel Wesley whilst the school had a library and a laboratory equipped with mathematical and scientific instruments such as an air pump. Morton had a strong reputation as a natural philosopher and the curriculum included in addition to the classics, Hebrew and oriental languages, mathematics, natural philosophy, logic, the principles of politics designed to encourage good intellectual and moral habits. Another influential academy at Moorfields was conducted by the natural philosopher John Eames (d.1744), an assistant to Newton who had been introduced to the Royal Society by his mentor and accumulated a large stock of mathematical and scientific apparatus. Assisted by James Densham, he taught geometry, geography, applied mathematics, mechanics, statics, hydrostatics, optics and surveyor's trigonometry.[15]

Oldfield and Watts regarded natural philosophy and geography as integral to natural theology and biblical study. As Watts argued without studying these he would never have arrived at 'so clear a conception of many things deduced in the Scriptures' whilst his educational works were suffused with philosophical metaphors to emphasise the growth of the knowledge in the modern age. Just as Bacon utilised his famous geographical metaphor of travel beyond the Pillars of Hercules to express movement by the moderns beyond the limitations of ancient thought, so Watts celebrated the new worlds uncovered by modern science, temporal, microscopic, global, planetary, cosmological and universal. Such physical and intellectual exploration of creation remained a religious act, which would encourage a more civil, tolerant and welcoming society for the Dissenters. Geography and astronomy also enlarged the capacity of the mind by demonstrating the chain of being and providing an understanding of the 'many various ranks of beings in the invisible world, in a constant gradation superior to us', those who confine themselves to localised knowledge having a 'narrowness of soul'. This could be cured by travelling which enlarged the mind or 'reading the accounts of different parts of the world, and the histories of past ages' especially 'the more polite parts of mankind'.[16] Watts recommended that the sciences and geography be pervasive in the classroom, dictionaries and other reference works to be 'consulted upon every occasion'. He repeated Heylyn's familiar image of geography as 'the eyes of history' and emphasised that with the natural sciences, these subjects were also essential for biblical study and had 'valuable and excellent uses' for improving the faculties of the mind and in later life. As Lockean association demonstrated, the least abstract aspects of astronomy, natural science, geometry and mechanics could 'easily be conveyed into the minds of acute young persons, nine or ten tears old'. They were also 'entertaining and useful to young ladies' and could be intermingled with 'the operations of the needle, and the knowledge of domestic life' with lasting benefit'. It was 'incredible' how much astronomy and geography could be learnt using two terrestrial hemispheres, maps showing other nations 'happily disposed around us' and coastal charts. These subjects were useful and practical and depended upon 'schemes and numbers, images, lines and figures and sensible things', so that the 'imagination or fancy' would greatly assist in understanding it. The sciences also made the 'dry labour' of classical grammar 'more tolerable to boys in a Latin school'. Astronomy and geography were best memorised by placing schemes and figures upon large sheets 'hanging always before the eye in closets, parlours, halls chambers, entries, staircases'. This would impress 'the learned images' upon the mind, keeping the information 'alive and fresh ... through the growing years of life'.[17] Watts claimed that his textbook was composed because he could find no other suitable for his school as existing works were 'much wanting' in diagrams and visual demonstrations. He argued that astronomy and geography could be taught through a series of problems on the sphere illustrated with figures, which would 'render the description of every thing more intelligible' and tempt 'younger minds' towards 'the higher speculations of the great Sir Isaac Newton and his followers'. He recommended that marked and celestial globes and maps of the world be used in order that equatorial, ecliptic and other divisions could be understood and provided precise instructions on moving and positioning of the globe for the 'learner'.[18]

Watts' *Heavens and Earth* explored topics including the natural description of the earth, natural and political divisions, geographical terms and the use of figures on the globe or map. Problems on the globe were designed to elucidate bearings, the basis of maps and sea charts and the projection of the sphere, whilst political divisions and natural and physical features of countries were treated individually following the pattern of a gazetteer. He hoped that the book would encourage students to pursue further geographical study using works such as Gordon's *Geographical Grammar*. The first part of the book concentrated upon 'the speculative part of this discourse' and examined the first principles of astronomy and the general part of geography with an initial survey of the divisions of the earth. The second or special part of geography treated of global divisions, states and governments and provided an account of the 'manners, temper, religion, traffick, manufactures [and] occupations'. It also described the various settlements and towns, physical features such as mountains, rivers and forests, products of national economies, 'rarities of art and nature' and various aspects of history. Astronomical and geographical problems were set for resolution upon the globe using household equipment, such as finding longitude and latitude, geographical features, the position of the sun at given times and the distances using a variety of measuring equipment.[19]

The importance of natural philosophy in Dissenting schools prior to 1750 is also evident from the teaching offered at another academy at Findern and Derby. This can be reconstructed from various editions of James Gregory's *Manual of Modern Geography*, which originated as a textbook for the school and was marketed as a cheap alternative to those already available (Figure 9). Under Ebenezer Latham and Gregory the academy gained a national reputation, the main subjects in the four-year course of study being medicine, divinity, classics, logic, mathematics, shorthand, natural philosophy, anatomy, geography, chronology and Hebrew antiquities. A Latin manuscript book by Latham survives containing notes and exercises in Latin on natural philosophy, geometry and mathematics utilising authorities such as Locke, Whiston, Le Clerc and 'sGravesande.[20] There are discussions of chemistry and the elements, the earth and first principles, the nature of the body, the physics of fluids and solids, minerals, metals, rocks, animal sensation and astronomy.[21] Like Watts and Oldfield, Gregory believed that geography, natural philosophy and chorography were pleasurable and valuable in teaching theology, government and polite knowledge. He wanted his scholars to 'take off their minds from those follies and vanities to which youth is generally addicted' and 'fit them for conversation' whilst contemplating and admiring 'the power and wisdom of the supreme being'. Knowledge of other nations and their features and productions demonstrated 'the peculiar happiness of Great Britain their native country', the value of her location secured by sea from enemy attack and the 'excellent constitution' that 'preserves all our rights and liberties'. There was a glossary of geographical terms, an appendix on hydrography, 'sketches of history and curiosities', and civil and political information concerning different states including their 'situation, extent, product, government, religion' and customs. Like Oldfield and Watts, Gregory cited Locke's *Thoughts Concerning Education*, in support of the argument that the sciences should be taught with Latin, given that as Locke said, it was 'an exercise of the eyes and memory' so a 'child with pleasure will learn and retain [it]'. Like Watts, Gregory emphasised the value of hanging maps around the schoolroom and recommended that geography be taught for at least half an hour each day.[22]

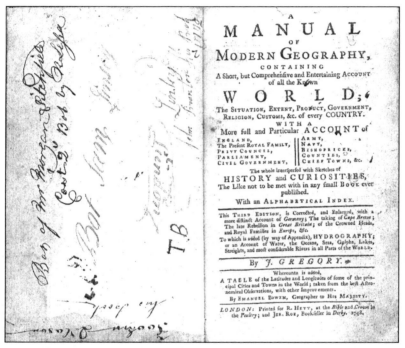

9 J. Gregory, title page with signatures of students from *Manual of Geography*, third edition (1748), courtesy of Derby Local Studies Library

Whilst not unique to them, the Dissenters' emphasis upon the use of apparatus and experimental demonstration in teaching natural philosophy reflected the importance of the senses in Lockean pedagogy, the value placed upon personal intellectual development and the distrust of dogmatism and authority. This helps to explain why such extensive apparatus was procured for academies such as those at Moorfields and Carmarthen which also symbolised the dedication to learning and the enthusiastic embrace of Baconian and Newtonian experimentalism. When he first arrived to teach at Northampton in 1729, Philip Doddridge was presented with a pair of globes, quickly acquired a microscope and was instructed by David Jennings to mount it above a large mirror so that experimental demonstrations could easily be viewed by pupils. At the Baptist academies founded by John Collett Ryland (1723–92) at Warwick, Northampton and Enfield from 1750, geography was taught using a range of apparatus along-side natural history, mechanics, astronomy, history and languages with the assistance of James Ferguson the itinerant lecturer in natural philosophy. Similarly, at Moorfields Academy, the natural philosopher John Eames (d.1744), an associate of Newton, assisted by James Densham taught geometry, applied mathematics, statics, hydrostatics, optics, surveyor's trigonometry and geography using an extensive mathematical and philosophical apparatus. At Carmarthen, despite a shortage of funds, supported by a bequest from an Anglican Vicar and the Presbyterian fund, a large apparatus was acquired. This included an air pump and glasses, a reflecting telescope, a microscope, astronomical quadrant, a planetarium, a Lunarium, a tellarium, theodolite, a barometer, a pair of Senex's 12-inch globes, a Hadley's quadrant, a hygrometer and a universal dial. Such a range of apparatus was comparable to that used by contemporary itinerant lecturers in natural philosophy such as John Warltire and Ferguson. This is also evident in the importance placed upon works of reference by later Dissenting educators. Abraham Rees (1743–1825) for example, taught elementary and applied mathematics, mechanics, statics, hydrostatics, optics and the use of applied mathematics in navigation, geography and astronomy. In 1778 he edited Chamber's *Cyclopaedia* and later the 39-volume *Cyclopaedia or Universal Dictionary*.[23]

NONCONFORMITY, THE SCIENCES AND GEOGRAPHY: DISSENTING EDUCATION, C.1760–1800

The growth of private academies and the popularity of private tutors meant that by the second half of the eighteenth century natural philosophy and geographical subjects were being taught by hundreds of English and Welsh educators. Dissenting academies therefore became in this respect, much less distinctive institutions largely the preserve of students intended for the ministry.[24] Furthermore, as Dissenters' perception of their own identities changed and they increased in social confidence so some felt able to pursue overt political campaigns against civil and political disadvantages which replaced the fulsome pronouncements of loyalty towards the Hanoverian crown that had characterised the previous decades. Dissenting groups such as the Presbyterians formed one of the strongest and most organised political lobbying groups in England and Wales and were able to utilise their extensive commercial and social national and international networks.[25]

Presbyterians, Unitarians and Rational Dissenters led by divines such as Price, Priestley and Walker campaigned for social and political reforms with the support of many liberal Whigs. Although as we have seen, Dissenting writers such as Watts and Doddridge had always regarded education as an important opportunity for the promotion of their own view of a tolerant civil society, this argument was expressed more vociferously by figures such as Priestley, Walker and Price. Encouraged by their Lockean association psychology, they believed that education would promote what they regarded as the rational principles of morality, civil tolerance, parliamentary reform which would help to end abuses of government by inculcating such beliefs and values. As Watts had emphasised, studies of the terrestrial and cosmological spheres, which many believed to be teaming with other beings, stretched minds, destroyed narrow prejudices and encouraged belief in the efficacy of progressive rationality over blind prejudice and dogma.[26]

Progressivist ideas were not of course, as Spadafora has emphasised, confined to some Dissenters but enunciated in a variety of forms and a large variety of texts constituting 'a virtual cross-section of the educated and literate strata of British society'. The reformist Enlightenment outlook of many Presbyterians, Unitarians and Rational Dissenters was popular in many political clubs and literary and philosophical societies. Although Dissenters played a key role in the activities of philosophical associations in Manchester, Newcastle, Nottingham, Derby and Birmingham, their Enlightenment philosophical outlook was shared by some other members, including the emphasis upon the role of education in facilitating social progress. Scientific and social progressivism was justified in Nonconformist theology as a realisation of the divine plan, in contrast to the more static hierarchical Platonism of older organised religions, which tended to find social and intellectual innovation more difficult to accommodate. As Fitzpatrick has made clear, for Rational Dissenters such as Priestley, natural philosophy was accorded a similar status to the fundamentals of theology, supported by the notorious religious heterodoxy of Newton. Encouraged by a perception of Newtonian scientific progress through philosophical dialogue and discourse, Unitarianism placed the greatest emphasis upon intellectual leadership as the motor of social progress. Study of natural philosophy and geography enabled one to appreciate the importance of striving for objective knowledge, which would provide a glimpse of divine cosmology.[27]

Although most Dissenters favoured a modern education that included geography, natural history and modern languages, many hesitated to send children to universities or academies for moral reasons, choosing local academies or private tutors instead. Joseph Strutt a wealthy Unitarian cotton manufacturer who subsequently served as president of Manchester College, York, worried that although it was 'a duty' for every parent to give to their child 'the best which his circumstances will allow', severe moral problems were presented by university residence away from home and family. Education could lay a foundation for human intellect and character, but an individual 'must learn to restrain and to instruct *himself*' where not under the control of others to 'complete his own education' which was difficult surrounded by 'all the blandishments of dissipation and debauchery'.[28] One important result was the creation of Hackney New College in 1786, intended to help fill the gap created by continued exclusion from English universities and concentration upon clerical education in the Dissenting academies. This offered an omnivorous plan of liberal education that included natural philosophy, modern languages, theology, modern geography and the use of the globes, natural philosophy, modern geography and the globes being taught by Rev. Abraham Rees the encyclopaedist and Rev. George Cadogan Morgan.[29] The Hackney Academy was able to exploit national Dissenting networks to obtain patronage and bequests despite the fact that many leading Dissenters favoured a midlands location on the basis that 'folly, extravagance, dissipation and vice' were 'not so triumphant' as in London.[30] The importance of the classics was undeniable; indeed the quality of classical education at academies such as the Hackney College under Rev. Gilbert Wakefield was second to none. As Strutt emphasised they did not despise, or undervalue knowledge of ancient languages 'connected as they often are with refinement and elegance in literature and morals'. However, they questioned whether sacrificing 'the best part of a boy's life, in the *minute* acquisition of what we daily see is of little use in the general affairs of the world, or which even contributes so little ... to the individual happiness of the professor' was worth the effort. Too much time was 'sacrificed in the acquisition of the learned languages', too little given to 'the general improvement of the understanding', to the sciences and to the 'formation of ... moral character'. Edinburgh and the Scottish universities were therefore generally favoured because they taught natural philosophy and other modern subjects whilst 'the methods of teaching were superior' to their English counterparts.[31]

Whilst mathematical geography, of course, continued to be taught in co-ordination with natural philosophy, after 1760 Nonconformist educationists took a special interest in the teaching of human and scriptural geography. It was believed that the subjects would foster civil and constitutional improvements, modelled and largely inspired by Enlightenment natural philosophy. Furthermore, as Seed has emphasised, Nonconformists were fascinated by narratives of their own development, which served as a means of forming their identities and reinventing themselves through historical study. Comparative international economic, cultural and constitutional studies served as a means of encouraging rational study of British society and the promotion of social and political improvements, whilst justifying the Nonconformist position. This also helps to explain why English and Welsh Dissenters so enthusiastically adopted continental educational theories and teaching methods such as those of Jean-Jacques Rousseau and Johann Heinrich Pestalozzi, which placed considerable emphasis upon geographical education using practical experience as much as books. Similarly, whilst the economic benefits of geography had often been noted, the importance of the subject for promoting business and international trade received more emphasis. Both forms of study were also, of course, intended to promote commerce and consumerism through which many Dissenting families had acquired their wealth. It is important to note, however, that overt political campaigning by Rational Dissenters and Unitarians such as Priestley also managed to alienate some moderate Tories and conservative Whigs who had been happy to recommend Watts and Oldfield for children and tutors. Hackney New College was forced to close in the straightened political climate of the 1790s because of the activities of Price and Priestley which caused parents to withdraw their children and patrons to cancel subscriptions.[32]

Some Dissenters, notably Presbyterians and Unitarians, emphasised the role of education in moulding citizenship for a modern rational society and history and geography had a special place in this pedagogy. For Priestley, geography and history ought to be taught together especially commercial geography, which exhibited 'the state of the world with respect to commerce' and taught what 'articles in the natural history of countries' were the 'proper subjects' of trade. Though he contended that this was hitherto seldom taught, gazetteers, maps and accounts of voyages provided the opportunity to engage the attention and encourage future merchants and captains to 'bring home specimens'. Men were 'losers' for want of learning natural philosophy and commercial geography for it was 'more than probable' that trade could be increased with foreign nations 'capable of furnishing us with commodities of much greater value'.[33] Priestley's work encouraged the teaching of natural philosophy in Dissenting schools, especially those with which he had close connections such as Warrington Academy, where he was employed as a lecturer between 1761 and 1767. Whilst at Warrington Priestley applied his teaching and scientific experience to compose his *History and Present State of Electricity* (1767) which aimed to introduce readers to the subject through practical demonstrational experiments, although science teaching was conducted by his colleague the surgeon and chemist Matthew Turner. In existence from 1756 to 1787, the Academy was patronised by Rational Dissenters and some liberal Anglicans and it was closely associated with Manchester Unitarian networks and subsequently the scientific circle around the physician Thomas Percival (1740–1804) and the Manchester Literary and Philosophical Society, established in 1781. Warrington was succeeded by the Manchester Academy, which moved to York in 1803, and New College, Hackney, which were patronised by Unitarian families mostly located in the north and midlands, and south of England. A high proportion of students went into the medical profession reflecting the close links with Scottish universities and the interests of some staff members such as the physician John Aikin (1747–1822). Scientific subjects were at the core of the curriculum in both these schools where tutors included the Quaker John Dalton, and they relied upon the patronage of wealthy Unitarian commercial and manufacturing families from the north and midlands including the Heywoods, Cromptons, Shores and Strutts.[34]

Dissenters such as Priestley were stimulated by their religious convictions and social position to advocate political reform whilst skilfully exploiting patriotism for state and constitution. Some questioned the inequity of wealth ownership and land distribution on moral, religious and economic grounds, and there were therefore important political as well as economic and utilitarian reasons for studying scientific and geographical subjects. James Pilkington a Presbyterian minister, tutor and author of a *History of Derbyshire* (1789), argued that excessive inequality was pernicious and could not be justified on moral or theological terms. Similarly, although theologically more conservative than Priestley, the Presbyterian minister Rev. Stephen Addington used his knowledge of local enclosures to attack inequality of land ownership as unpatriotic. Reflecting two major aspects of Enlightenment philosophy, he condemned indiscriminate enclosure both 'as a man, and a Briton' as irrational and an encroachment upon individual rights. Enclosure for the public good was sometimes appropriate but ought not to be blindly encouraged, especially where the soil was 'generally rich and fruitful'. Although it might 'oblige some few' it was 'highly injurious to nine-tenths of the inhabitants' if not 'to an equal proportion of the whole kingdom' reducing poor tenants to penury, destroying families and swelling the poor rates. With his knowledge of the rapid enclosure of Leicestershire and Northamptonshire, Addington supported estate and agricultural improvement as beneficial to the wider economy but argued that landowners did not set aside enough land for planting preferring to devote far more for grazing. As a result 'many a sturdy oak, and stately elm and ash' were 'falling yearly sacrifices to the new-inclosures' being 'sawn asunder' for 'mere temporary fence parts and rails' with very few being constructed in replacement. Addington condemned the unpatriotic aristocracy and clergy who benefited from enclosures and those who acted as 'mere tools' by signing supporting petitions that impoverished many to enrich a few and drove the poor into the market towns. As the ruins of farms revealed, it was wrong for the 'true interest of the nation, the authority of government, and the liberties and propertys of the subject' to be destroyed by forcing families into 'mean and miserable vassalage' akin to the Highlands of Scotland.[35]

Civil and political campaigns were still balanced by carefully coded and no doubt usually sincere protestations of loyalty and patriotism and, as we have seen, geography and the sciences provided an important means by which this could be accomplished. Thus alongside his political campaigning Addington used natural philosophy, natural history and geography at his academy in Market Harborough to emphasise the benefits of the British constitution. For Addington England was 'on the whole' a 'most beautiful and fruitful country' its inhabitants wanting nothing but 'a more powerful sense of religion but to render them the happiest people under heaven'. In contrast even other civilised nations were perceived as inferior. Of the Chinese he remarked that 'they are, in general, superstitious, fond of astrology, and very deceitful.' Comparative international studies of religion and government would underscore the justness of the protestant Dissenting position. Addington glossed over divisions between Nonconformists claiming that all agreed 'in acknowledging no head of the church but Christ' in obvious distinction to the Anglican Church 'established by law'. Likewise they concurred that 'his church is not national, and that the civil magistrate is not authorised by him to impose articles of faith or rules of worship'. Furthermore, it was made clear that Dissenters disapproved of the discipline of the Church of England, asserting 'the people's right to choose their own ministers, and that every one may and ought to think for himself in religious matters'.[36]

Natural philosophy, natural history and geography were also regarded as central to progressive Enlightenment philosophy by David Williams the Welsh deist, Unitarian, follower of Rousseau and friend of Franklin. Minister of High-Gate meeting house between 1769 and 1773, Williams opened an academy at Chelsea in 1774 where he re-duced the pedagogical authority of the schoolmaster by encouraging older boys to teach the younger children below them. Williams' interpretation of Rousseauian pedagogy was fostered by his extensive metropolitan and provincial philosophical networks, which included membership of the club of 13 with Richard Lovell Edgeworth, Thomas Day, Josiah Wedgwood, Thomas Bentley and the botanist Daniel Solander. They helped Williams to draft his *Liturgy on the Universal Principles of Religion and Morality* (1776) which tried to create a universal creed, synthesising world religious beliefs. He also main-tained contact with German educationists such as the deist and theo-logian Karl Friedrich Bahrdt and Johann Bernhard Basedow, founder of the Dessau Philanthropinum, whilst his *Treatise on Education* was translated into German in 1781.[37] Encouraged by his interpretation of Rousseau, Williams tried to teach natural history and geography dif-ferently from received Georgian practice. His pedagogy was flexible and included study of the classical languages, French, Italian, moral and political philosophy and poetry. Mathematical subjects and the sciences were very prominent including astronomy, natural history and chemistry, especially for older boys aged between 12 and 15, who were encouraged to conduct experiments in natural philosophy as early as possible. Children were not introduced to geography through astronomy as had been the case in many earlier English Dissenting academies and recommended by Rousseau, nor through history, but through elementary observations of the local topography and natural history. He recommended that pupils begin with a survey of their own house before proceeding to excursions in the neighbourhood and district. Once the boys had begun to study global features such as oceans and mountain ranges, they proceeded from a study of climatic variations to examine cosmology, which differed from the older method of teaching the terrestrial and celestial globes through con-ventional sets of problems. Pupils constructed their own maps and globes and studied the construction of timepieces and related mech-anical devices. Once this stage had been reached then geography be-came closely interrelated with history as Priestley had advocated, em-phasising the individual qualities and relationships between global divisions. This served as an introduction to history beginning with an examination of the 'fabulous origin or settlement of men' and civilisation from ancient settlements on the Ganges and Euphrates to modern European states and global exploration.[38]

The political repression and onset of European war during the 1790s curtailed most open Nonconformist political campaigning, but the impact of Priestleyan ideas continued to inspire the teaching of natural philosophy in late-Georgian schools such as those of William George Spencer in Derby and Robert Goodacre in Nottingham. Whilst they could be described as successors to the Dissenting academies of the eighteenth century and were run by Nonconformist teachers, in practice many were patronised by Anglican as well as Dissenting families, although with a preponderance of Whigs. They catered for the demand for useful and practical education for boys, and to a lesser extent, girls, amongst the middle classes, skilled craftsmen and wealthier traders of industrialising towns such as Derby and Nottingham.[39] Goodacre's school provides an excellent example of the application of Priestleyan methods to teach scientific subjects in the classroom and through his book *An Essay on the Education of Youth* the justification offered for this. A prominent schoolmaster in late-Georgian and Regency Nottingham, like Spencer in Derby, Goodacre was also a leading figure in local scientific culture and applied his schoolroom knowledge and techniques in public lecturers on scientific subjects in Nottingham and other towns in the Midlands and North. Although he began teaching in Nottingham in 1797 and was employed by the grammar school, it was an academy established on Standard Hill in the town in 1808 that made Goodacre's reputation, especially amongst Dissenters and reformers. He handed it over to his eldest son Robert when lecturing in the United States before it was continued by William another son with his cousin Thomas Cokayne until 1859.[40] The Standard Hill Academy continued under Goodacre junior to teach a large range of mathematical and scientific subjects during the 1820s and 1830s, whilst regular courses of public lectures were also offered, especially during the winter months.

In his *Essay on the Education of Youth* which served as an educational treatise and an advertisement for his school, Robert Goodacre emphasised the importance of Latin, English grammar and modern languages such as French in his curriculum, but devoted much space to the natural sciences and geography. The 'delightful study' of geography and the globes was 'man's peculiar regard', and neither 'passages of a common newspaper' nor the political state of Europe could be 'accurately comprehended' without such learning. Study of natural philosophy and the 'investigation of nature' allowed the 'student to the attainment of new heights' and expanded the mind'.[41] Following Priestley, Goodacre placed much emphasis upon the importance of apparatus and experimental demonstrations as 'knowledge obtained by these means is of infinite importance ... in the disposition which it gives the mind to admit things which at first glance appear incredible'. Electricity, hydrostatics and pneumatics were all 'equally deserving the attention of the tutor', but astronomy was his favourite science being 'admirably calculated to elevate the minds of youth, by a contemplation of the sublime and majestic'.[42] The school had a library and many maps and charts including Priestley's charts of history and biography. By 1810 the equipment included an orrery, a pair of twelve globes, a compound microscope, an electrical machine, an air pump, a prism, meteorological and land-surveying instruments, and an observatory equipped with telescopes.[43] Goodacre emphasised the conventional intellectual, theological and utilitarian reasons for studying natural philosophy, but the Priestleyan ethos of self-discovery and rational enquiry as the basis for religious belief was also attractive to the Dissenting and reforming families who patronised the school. Although Goodacre used 'familiar lectures on the properties' of the natural world incorporating 'experiments on instruments', students were 'expected to make every enquiry they think proper, and to give some account at the next lecture of the subjects previously discussed'. Abstract theological discussion in religious education 'designed to induce a belief of tenets deemed essential' was regarded by Goodacre as 'useless' as 'truths' needed to be 'proved to their understandings'. Only then would the scriptures become 'a subject of consideration' and the 'evidences of their authenticity' be 'calmly investigated' so that 'belief of their divine origin' could be 'placed on firm and rational grounds'.[44]

CONCLUSION

The emphasis upon the importance of natural philosophy, natural history and related subjects such as geography in Dissenting teaching was closely related to the mathematical and commercial academies that flourished in the seventeenth and eighteenth centuries. However, the degree to which education was intended for personal spiritual and social improvement and fulfilment tended to distinguish Dissenting schools from their commercial rivals. Nonconformist academies had strengths that commercial academies did not have supported by social, political, commercial and cultural networks and the relative security of national subscription funds and organisation. Dissenting schools drew upon a collective stock of experience and knowledge developed by pupils and educators over decades, successive academies modelling their curricula, practices, equipment usage and selection of texts upon those of other institutions. Ancient, modern and mathematical geography were pursued assiduously in many Dissenting academies, providing programmes that were adopted in other educational institutions, spreading beyond the Dissenting community through works such as Watts' textbooks. The sciences were perceived as important components of progressive Nonconformist teaching intended to foster social and political change, including the eradication of civil restrictions. Furthermore, because of international religious networks and civil restrictions Dissenters were generally responsive to developments in natural philosophy, geography, rational theology and scriptural studies. Dissenting educators remained inclined to experimentation as Williams' development of Rousseauian methods of teaching geography, the Dissenting response to Pestalozzianism and prominence given to female education suggest.[45]

4 The Institution: Freemasonry

INTRODUCTION

Despite being probably the most pervasive and influential form of secular voluntary organisation from the 1720s and inspiring a multitude of imitative, derivative and parodying fraternal organisations, aspects of freemasonry remain hardly studied. Jacob has noted that despite 'the importance of freemasonry for the Enlightenment' and its British origins freemasonry has received 'scant attention from British academic historians', a 'particularly unfortunate gap' in the intellectual and political historiography of the eighteenth century.[1] Similarly, Clark has stated that 'the social history of Hanoverian free-masonry' still 'cries out for detailed attention'.[2] Given that free-masonry was idealised, realised, framed, developed and experienced in spatial terms, arguably more than most other Georgian associations, the subject is also ripe for historical-geographical study. Inspired by recent analyses of seventeenth-century natural philosophy, John Money has argued that Masonic lodges 'not only acted in a direct sense as vehicles for the dissemination of the new natural philosophy and its applications' but were 'memory theatres'. They were the 'social equivalent of the Boylean laboratory', private spaces 'with a public purpose', in which 'models were ritually copied and a shared cultural knowledge was created'. Likewise, Clark has contended that there is 'considerable evidence, at least for the early eighteenth century, that masons were actively concerned with mental improvement' and that they provided lectures on natural philosophy and other subjects at lodges.[3] The association between intellectual endeavour and the rhetorical and experiential spaces of freemasonry was also celebrated in the rhetoric of Georgian Freemasons. Speaking to the members of a Masonic lodge in Newcastle in 1776, Rev. R. Green argued that 'there is no useful science, art, or mechanism with which masonry is not closely connected.'[4] Similarly, addressing the members of the new Derby Philosophical Society in 1784, Erasmus Darwin argued that their body would be, like the freemasons, a 'band of Wampum', or 'chain of concord' collecting together the 'scattered facts' of philosophy and converging them 'into one luminous point … to exhibit the distinct and beautiful images of science'.[5]

Money's contention that the part played by speculative, that is non-operative or craft freemasonry, in the diffusion of Enlightenment ideas has been acknowledged but little explored in the English context remains true although the importance of freemasonry in the careers of prominent natural philosophers such as John Theophilus Desaguliers has received greater attention.[6] Jacob and Weisberger have contended that freemasonry played a major role in promoting social and intellectual Enlightenment values in Europe, whilst Bullock has argued that Masonic lodges played a similar role in the American colonies and United States.[7] Hans emphasised the close relationship between new developments in British eighteenth-century education, scientific ideas and freemasonry and it has been recognised that many of the early members of the Royal Society were masons. The importance of the relationship between antiquarianism, history and natural philosophy is suggested by the careers of masons and savants such as Christopher Wren, Newton, William Stukeley, Francis Drake, William Hutchinson and John Wood, but the subject requires much greater elucidation.[8]

As the most popular and widespread associational model, freemasonry inspired new forms of civic life and urban improvement, stimulating British scientific culture and provided a model for some of the earliest provincial philosophical societies. As we shall see, an emphasis upon the geographies of English freemasonry helps to illuminate national patterns of Georgian scientific culture. The self-improvement imperative fostered personal study and intellectual endeavour manifest, as we shall see, in the lectures promoted at local lodges. At another level, encouraged by the semi-mythical origins of freemasonry in building, spatial metaphors were of profound importance from the celebration of the divine architect of the universe to the symbolism employed in Masonic rituals and carved into local lodges. Prosopographical research making detailed comparisons between lodge membership and that of other bodies, examining the activities of prominent Masonic natural philosophers and Masonic rituals and texts allows us to interrogate the claim that freemasonry was inspired by Newtonian natural philosophy and promoted forms of scientific education.[9] This chapter begins with a detailed assessment of Jacob's work on freemasonry and scientific culture and Clark's analysis of freemasonry in the context of other Georgian associations before demonstrating how a geographical approach informs analysis of freemasonry and scientific culture.

ROYAL SOCIETY, GRAND LODGE AND WHIG OLIGARCHY

In series of 'wide ranging and stimulating' studies, Jacob has made the most determined attempt to argue that speculative freemasonry was stimulated by Newtonian natural philosophy and that in turn freemasonry helped to shape Enlightenment scientific culture by promoting secular and progressive forms of civic culture.[10] She has contended that freemasonry as 'the most avowedly constitutional and aggressively civic' form of sociability with 'the philosophical society' and 'the scientific academy', underpinned the formation of modern society. Masonry transcended international boundaries and spread into Europe with continental lodges being 'replicas of British lodges' bringing 'over with them forms of governance and social behaviour developed within the distinctive political culture' of Britain.[11] According to this model, early eighteenth-century England was still beset by a 'thicket of religious tension' and in order to navigate through this 'tolerant-minded Newtonians' such as Desaguliers, William Stukeley (1687–1765) and others 'invented a new form of ritual to worship the Grand Architect of the Universe'.[12] In Jacob's view the formation of official English freemasonry involved a repudiation of the republican tradition and a transformation of the craft under Whig Newtonian leadership centred upon the Royal Society and facilitated by Newton's presidency. Whig grandees and their supporters such as Sir Hans

Sloane, Desaguliers, and Martin Folkes fought a 'political struggle' and 'emerged as victorious'. With over a quarter of the Royal Society being Freemasons, they were able to imprint their Whig Protestant scientific and political philosophy on the character of the grand lodge, which subsequently served as a model for provincial lodges.[13]

Freemasonry was most widespread federal association in British culture, with an increasingly metropolitan organisation of lodges, and it had spread into most large provincial towns by the 1740s. Although initially local lodges enjoyed a large degree of autonomy, by the 1730s most copied the constitutions, procedures and rhetoric of the metropolitan grand lodge established by 1717. The model was primarily disseminated through editions of James Anderson's *Constitutions*, originally published in 1723 by a group of London lodges who had gathered together to form the grand lodge. The *Constitutions* specified that freemasonry was a 'royal art' practised by free born individuals who enjoyed the benefits of peace and liberty whilst quarrels concerning religion, state policy or party politics were prohibited from meetings for the good of the lodge.[14] Although freemasonry became more secretive during the twentieth century, prior to this it took a major and open part in public civic affairs and urban government acting through elite patronage, and staging grand processions, religious services, concerts and charitable activities.

Events were often reported in local newspapers which also provided details of support for charitable ventures such as schools and hospitals with philanthropy being in Jacob's view where 'the movement had its greatest impact'. In some respects masonry took over the role of guilds in facilitating social and business contacts between initiates providing business and employment opportunities, and charitable support at times of economic difficulty.[15] The spaces of English freemasonry were therefore hardly limited to lodge buildings and meeting rooms but embraced numerous parts of civic society, culture and institutions, helping to form local urban and county identities, a fact that narrow attention to Masonic texts and the meaning of rituals has tended to ignore. The relatively democratic and egalitarian nature of freemasonry was embodied in the layout of lodges and meeting practices. Elements of Masonic organisation such as the official exclusion of political and religious arguments were adopted in other urban institutions, although some lodges had particular political affiliations. However, social hierarchy impacted upon relationships between initiates and the poorer sort were usually excluded by subscription and initiation. According to Jacob, freemasonry allowed individuals of different political and religious affiliation to give 'institutional expression' to the 'new scientific culture'. The lodges 'stood as a metaphor for their age' ruled by grand masters usually drawn from the aristocracy and 'strictly hierarchical in structure yet curiously egalitarian at their meetings and banquets', governed by a series of rules and 'mirrored a larger social and ideological consensus.'[16] British freemasonry after 1717 was a 'new form of social gathering complete with ritual and costume', very different from the older Masonic clubs from which it originated and was largely inspired by a 'cultural synthesis based on science, religion, and social ideology'. Masonic gatherings were the 'quintessential popularization of enlightened culture' in which 'literate gentlemen of substantial means' and a variety of religious affiliations 'worshipped the "great Architect", the god of the new science' which favoured advancement according to personal merit.[17]

Desaguliers is a pivotal figure through his articulation of the religious, political and Newtonian scientific concerns of Whig grand lodge freemasonry and his role in spreading this 'from London to the English provinces and the Low Countries'.[18] According to Jacob, Desaguliers and his associates used organised freemasonry 'as a social nexus' in order to promote 'specific cultural and ideological goals', especially stability under the constitutional monarchy, social mobility 'under aristocratic patronage', religious toleration and 'Baconian experimentalism' dedicated to 'the cult of the new science'. Desaguliers was employed as demonstrator and curator of the Royal Society and became probably the leading Newtonian propagandist of the period, giving courses of public lectures on natural philosophy in England and the Netherlands. Jacob and Weisberger have emphasised the importance of his importance in facilitating the transition between operative and speculative masonry and in helping to compose the Rosicrucian constitution published in 1721.

Desaguliers' career as a scientific propagandist extended to using Newtonian imagery to promote the Whig ministry in his poem *The Newtonian System of the World, the best Model of Government* (1728) and he served as chaplain to the Duke of Chandos and later to Frederick, Prince of Wales. Desaguliers served as grand master in 1719 and deputy grand master to the Duke of Wharton, the Earl of Dalkeith and Lord Paisley during the 1720s whilst giving scientific lectures to George I, George II and other members of royal family. His *Newtonian System* extolled the virtues of the British political system employing the analogy of a harmonious Newtonian universe to the delicate equipoise of the balanced British constitutional system. This New-tonian philosophy provided 'the foundation for the worship of limit-ed monarchy, for the overthrow of foreign philosophies like Des-cartes's with their atheistic implications' and for the 'maintenance of order and stability'. It was therefore 'little wonder that Whig oli-garchs and their placemen flocked to Masonic lodges whose spokes-men offered so much confirmation to their self-esteem and insured the comic significance of their activities'. [19] Desaguliers attempted to imitate directly 'the harmony and order of the universe as revealed, of course, by science' bringing the 'burden of Newtonian science to bear against the radicals of the Enlightenment' whilst his contribu-tion to freemasonry 'may' have been 'vitally important' in leading opposition to atheists and freethinkers and aligning freemasonry with 'court Whiggery and the Royal Society'.[20]

Stimulated by natural philosophers such as Desaguliers, other fel-lows of the Royal Society became Freemasons during the 1720s and 1730s, and many of these had strong provincial as well as metropolit-an interests. These included John, Second Duke of Montagu (1688–1749), grand master in 1721–2 the antiquarian and mathematician Martin Folkes FRS FSA, and Martin Clare FRS (d.1751), both of whom served as deputy grand master in 1724 and 1741 respectively with Folkes serving as president of the Royal Society, 1741–53.[21] Other masons held important offices in the Royal Society such as the suc-cessive presidents, George Parker, second Earl of Macclesfield (1697–1764) and James Douglas, Earl of Morton (1702–68), the chemist John Brown (d.1736) who served on the council between 1723–5 and Henry Hare, 3rd Baron Coleraine, FSA who was vice-president of the Royal Society. Other examples include the physician John Woodward (1665–1728), the natural philosopher John Ward (1679–1758) and William Rutty (1687–1730) and John Machin (d.1751), two secretaries of the Royal Society.[22] The close links between Scottish and English freemasonry and scientific culture are apparent in the career of

Morton, grand master in 1740, who was an intimate friend of Colin Maclaurin and the most prominent supporter of philosophical associations in Scotland including the Edinburgh Medical Society and the Scottish Society for Improving Arts and Sciences. He was also, like Macclesfield, an astronomer and mathematician who contributed several papers to the *Philosophical Transactions* and taking an active part in the preparations for the transit of Venus in 1769.[23] Equally important as a supporter of public scientific culture and as a natural philosopher was another mason, John Machin (d.1751), who served as secretary of the Royal Society for 29 years and as Professor of Astronomy at Gresham College, London from 1713 to 1751. Another mason, the mathematician and astronomer Brooke Taylor (1685–1731) served as first secretary of the Royal Society between 1714 and 1718, publishing many papers in the *Transactions* and it is likely that Taylor's work on linear perspective in architecture and the fine arts received stimulation from freemasonry.[24] Other Masonic natural philosophers and fellows of the Royal Society served as government officials such as Charles Delafaye (d.1762) who joined the Horn Tavern Lodge at Westminster and took an interest in horticulture, botany, mathematics and mechanics.[25]

In his *Course of Experimental Philosophy* published by John Senex a fellow freemason and natural philosopher, Desaguliers exclaimed that besides its theological importance, the primary purpose of natural philosophy was to 'discover causes from their effects'. This would allow 'art and nature' to become 'subservient to the necessities of life, by a skill in joining proper causes to produce the most useful effects is the business of a science'. Much of the work emphasised the importance of practical mechanics providing details of mechanical devices such as steam engines, pumps, pulleys, cranes and railways, whilst the entrepreneurial and industrial applications of natural philosophy are emphasised at every turn. According to Jacob, this utilitarian, mechanical emphasis appealed to Freemasons and 'yielded a variety of compelling analogies' given their 'urban based and heavily mercantile' lodges and 'lies at the heart of the Newtonian Enlightenment'. Desaguliers' emphasis upon the value of observations of skilled craftsman, despite his contention that they did not fully understand the principles that they were exploiting, meant that 'freemasonry with its roots in the mechanical and artisan crafts, would naturally appeal to mechanically minded Newtonians like Desaguliers'.[26] The 'providential order proclaimed by Newtonians such as Desaguliers could be imitated by art and science through the application of mechanical devices in industry, through the exploitation of human labour' or 'the imposition of order in the machinery of the state'. Hence servants of the 'public machine' were inspired by the apparent 'cosmic significance' of their actions. Their 'dedication to the principles and practice of Whig government' and devotion to the 'cult of Newtonian science' provided 'a new and powerful vision of their role as the masters and shapers of society and government – all in imitation of the Newtonian universe'.[27] This emphasis upon the utilitarian applications of mathematics and natural philosophy to engineering, surveying, navigation and manufacturing was commonplace in the eighteenth century and long predated organised English freemasonry as the activities of the founders of the Royal Society and Francis Bacon demonstrate. It probably attracted individuals to join lodges where such rhetoric was prominent and scientific lectures were given. The fact that some 45 per cent of the fellows of the Royal Society were masons during the 1720s demonstrates the proximity between the two organisations.[28] According to Jacob they had both become pillars of the Whig establishment, which retained its ability to tolerate and even support some religious and political opponents. There were, of course, other common intellectual interests shared by Masonic fellows of the Royal Society such as antiquarianism which shaped the character of – and audience for – scientific culture represented by the foundation of the Society of Antiquaries in 1717.[29]

Although it had unusual aspects, freemasonry was just one form of club and society, and a relative newcomer in England centred upon the taverns, clubs and coffee houses of the post-Restoration metropolis and somewhat later, provincial towns.[30] 'Clubbable' savants such as Hume argued that in contrast to the solitude of 'barbarous nations' forms of polite association allowed the 'middling rank of men' to become refined in the arts and sciences. Attracted to urban society they would 'receive and communicate knowledge' whilst demonstrating their superior wit, breeding and taste and forming associations where each sex could 'meet in an easy and sociable manner' so that 'the tempers of men, as well as their behaviour' would 'refine apace'.[31]

This middling-sort associational culture, where 'industry, knowledge, and humanity' were 'linked together by an indissoluble chain', was driven by economic progress. It countered the relative uncertainty and insecurity of middling-sort life by offering avenues for polite social interaction where trust could be built.[32] Eighteenth-century organisations, including freemasonry, tended to be modelled on older forms of urban association such as trade guilds, religious denominations, local corporations and vestries which was reflected in the similarity of their lodges and meeting houses. These contained features that were adopted by other associations such as special membership qualifications based upon property, wealth and place of residence. Shared occupational, religious, intellectual and civic spatial concerns were also used to forge collectivism, and common traditions and patterns of behaviour. Most such associations embraced aspects of democratic organisation such as subscription and membership payments, elected officers, governing committees, rules and regulations and egalitarian commitments, and they helped to shape and promote civic identities.[33]

This underscores the difficulty of distinguishing Masonic lodges from other organisations especially in a clubbable age when many associations shared some of its characteristics. Jacob has been criticised for adopting a 'sloppy use of the word Masonic to describe anything combining radical ideas and secrecy' and regarding 'a network of publishers and journalists, English radicals and French refugees' as Masonic with insufficient justification. In Stevenson's view, many 'other secret societies of the age ... had rituals and secrets as well as a convivial side to their activities' and 'obviously had something in common with Masonic lodges, but to call them Masonic lodges is illogical and confusing'.[34] Similarly, Thomas Munck has warned that freemasonry may not have been 'more influential than so many other eighteenth-century societies and networks in the promotion of sociability and new ideals.'[35] The social geographies of freemasonry, particularly the national and international networks and forms of organisation, even if lodges remained formally unaffiliated, set it apart from more localised clubs and societies, promoting shared ideals and rituals. Many of these bodies met in the same coffee houses and taverns as Masonic lodges and shared members in the rich nexus of Georgian conviviality, but had no official affiliation. One of the earliest was the Robin Hood Society which by the eighteenth century met in the Essex Head Tavern, London, and then the Robin Hood and Little John from 1747, which notoriously allowed the debating of religious and political questions. It fabricated a history tracing its origins back to 1613 modelled on that of freemasonry complete with accounts of debates, enjoyed a large membership and debated topics as wide as trade, navigation, politics and the sciences. The Kit-Kat Club whose members included writers such as Steele, Addison and Congreve and natural philosophers such as Samuel Garth and John Somers, also had strong deist, Rosicrucian and Masonic leanings.[36] Another example is the 'Club of Thirteen' which met at Old Slaughter's coffee house just before the outbreak of the American war and has generally been held to have been a Masonic lodge. The membership included John Whitehurst, Thomas Day, Josiah Wedgwood and Thomas Bentley, James (Athenian) Stuart, Daniel Solander the botanist, Franklin and the geologist and antiquarian Rudolphus Raspe, the former five, of course, having major commercial, industrial and mechanical interests.[37]

FREEMASONRY AND SCIENTIFIC EDUCATION

Under the guidance of English grand lodge freemasonry, lodges became 'places where gentlemen, whether lowly or titled, could receive a minimal instruction in mathematics' and listen to scientific lectures.[38] Freemasonry also promoted scientific education and research and provided an institutional model for philosophical associations. Some Georgian Masonic writings emphasised the value of scientific lectures as part of a broader Enlightenment culture in which natural philosophy was valued for its natural theological importance, as a means of polite cultural expression and as a form of entertainment.[39] Whilst addressing a Newcastle lodge in 1776, Rev. Green claimed that there was 'no useful science, art or mechanism with which masonry is not closely connected'.[40] Likewise, speaking to a lodge at Ramsgate in 1798, Rev. Jethro Indwood argued that the mason's lodge was the 'school of true, useful, and universal science' and a 'house of wisdom' and 'improvement' in which geography, astronomy, philosophy, with all the liberal arts, are, or ought to be, the objects of our lectures'. According to Indwood it was 'the peculiar province of Masonic science, to direct her improving sons to study nature' and he instanced the valuable 'practical effects' of the study of Newtonian astronomy and geography which revealed the divine principles underpinning 'ideas of the earth's construction'.[41] As we might expect given the semi-mythical foundation of freemasonry in ancient building, spatial metaphors, astronomy and the mathematical sciences were especially favoured, some lodges providing lectures on these subjects. It may also be that, as Jacob has argued, following the lead of Newtonians such as Desaguliers and stimulated by their interest in skilled craftsmanship, Masonic natural philosophers took a special interest in subjects of use in practical mechanics such as surveying, hydrostatics and pneumatics. Examples include lectures on subjects such as geometry, architecture, optics, anatomy and chemistry given in London lodges during the 1720s and 1730s and lectures on astronomy sponsored by Bristol freemasons in 1789 utilising a large transparent orrery.[42] Some prominent Masonic natural philosophers such as Desaguliers and his pupil the mathematician and astronomer Stephen Demainbray FRS (1710–82) offered courses of lectures on natural philosophy, the latter teaching science to Prince George and serving as astronomer at the Royal Observatory, Kew after his accession to the throne.[43]

Freemasons were also involved in the publication of learned reference works that enjoyed wide national circulation. Andrew Ramsay, a Jacobite grand master of the 1730s and 1740s, urged that the masons should create a dictionary of all the liberal arts and useful sciences which could be funded by subscription from all the brothers of Europe. John Senex, who served as grand warden whilst Desaguliers was deputy grand master in 1723–4,[44] published geographical and scientific works and constructed scientific instruments, whilst making improvements to the construction of celestial globes. He also published important and influential Masonic texts such as Anderson's *Constitutions* and moved in a metropolitan Masonic circle of publishers, instrument makers and natural philosophers. Assisted by Desaguliers as editor or translator, Senex printed the 1719 edition of s'Gravesande's *Mathematical Elements of Natural Philosophy* and a translation of Nieuwentyt's *The Religious Philosopher*.[45] Another member of this coterie was Ephraim Chambers (1680?–1740), an apprentice to Senex and a fellow freemason who appears to have belonged to the Richmond Lodge. The first edition of his popular *Cyclopaedia*, which contained one of the first published accounts of masonry, was partly published by Senex.[46]

Freemasons also projected learned institutions modelled partly upon masonry. The astronomer, natural philosopher and architect Thomas Wright (1711–86) of Durham planned a new national university or 'Pansophium' with a hierarchical system of government, committees, a museum, library, gymnasium and theatre. This posited radical geospatial alignments both within institutional confines and as a challenge to the ancient and prevailing pedagogical domination of the metropolis and the two ancient universities. Wright's new university would be located in different provincial centres around the country, diffusing the power of learning away from the established duopoly by removing some of their colleges to York, Exeter and Shrewsbury. Members were to be elected from the colleges, academies and schools of the nation and formed into fraternities and societies governed by a hierarchy of academic potentiaries under a grand patron or 'panarchon'. There was also a new royal academy of sciences with three colleges, nine schools and twenty seven professors to embrace 'all the respective branches of mathematics, philosophy and beaux-arts'. Wright planned a 'new Order of Wisdom or Knights of Minerva' suffused with Masonic symbolism and imagery and visualised by triangles such as the trinity in unity and the infinite and eternal unity.[47] Freemasonry may also have provided some inspiration for John Toland's projected learned societies, although the importance of masonry in his life has recently been questioned. In his *Pantheisticon*, Toland described a fraternity called the Socratica and the brothers Pantheistae. Apparently adopting aspects of freemasonry, they held rituals similar to lodge ceremonies, members being admitted to different degrees guided by reason as 'the sun that illuminates the whole' and being regarded as equal and free.[48]

Examination of the London Old King's Arms lodge records suggest that scientific lectures were regarded as quite an important part of lodge meetings. The lodge minutes reveal that philosophical lectures became part of the regular proceedings during the 1730s and 1740s when they appear to have faded out despite the introduction of penalties for those who failed to give their promised disquisition. The breadth of subjects was large including lectures on architecture, as might be expected, medical and physiological subjects, automata and clockwork, optics, microscopy, metal working, mathematics, painting, astronomy, magnetism and chemistry. The lectures on optics and microscopy in 1734 featured an examination on 'ye living creatures in ye lower parts than ye naked eye could not perceive; ye skin of man, ye down on a butterfly wing; ye proportions of a louse'. The company was, however, 'deprived from seeing ye circulation of ye blood, not being able to fetch a surgeon'. The minutes reveal that the lodge was more than merely a debating society as experimental and other types of demonstration were utilised during lectures and that these stimulated the work of natural philosophers. The degree of historical awareness of the masons is also apparent with most lectures, whatever the subject, beginning with a historical introduction often from antiquity. The lectures on automata and clockwork in December 1733 for instance, began with the automata of the Greeks and Arabs. According to Calvert the practice of reading papers was a 'custom much in vogue at the time among the higher class of lodges' but, again, much further work is required to substantiate this assertion.[49]

The most prominent natural philosophers associated with the lodge were the physicians William Graeme FRS and Edward Hody FRS, and John Clare FRS another associate of Desaguliers. All three men held offices in the lodge and gave lectures and demonstrations. Graeme's lectures on chemistry between December 1735 and February 1736 included examinations of distillation and a dissertation on the 'very curious subject that of fermentation wherein he showed that all vinous and intoxicating liquors were only to be found in the vegetable kingdom'. Clare was a schoolmaster and authority on the history and practice of masonry who gave frequent lectures to the King's Arms and other lodges. He founded an academy in Soho Square c.1719 where French, arithmetic, geography, surveying, navigation, astronomy and other subjects were taught. Students were entitled to attend the public lectures given at the school on moral, religious and philosophical subjects including experimental and natural philosophy utilising 'a large apparatus of machines and instruments'. Clare's address on the advantages of freemasonry was reprinted many times in the eighteenth and nineteenth centuries and emphasised how inspired by the structures of the universal grand architect they had 'made it their business and their aim to improve themselves, and to inform mankind'.[50] Clare's Masonic activities played an important practical and intellectual role in the composition of his popular book *On The Motion of Fluids* which was dedicated to Thomas Thynne, Lord Viscount Weymouth, grand master and 'the patron and encourager of useful arts'. Clare thanked Desaguliers and various fellows of the Royal Society and aristocratic patrons for supporting the venture which was designed to assist 'the young philosopher' and make them 'better prepared for receiving lectures in natural and experimental philosophy'. The book was based upon a series of lectures 'privately read to a set of gentlemen', who had encouraged its publication 'for the use and advantage of those' who wished to inquire into 'natural causes'. Although the minutes do not survive for the period prior to August 1733, these lectures were probably given to the King's Arms lodge where Clare gave dissertations on subjects including architecture, Masonic history and magnetism. In October 1735 he examined the manner in which magnetism was 'communicated to other bodies capable of receiving it' supported by experimental demonstrations.[51]

At lodge meetings, brothers engaged in scientific experiments and provided lectures which helped to give 'ritualistic expression to the fraternity of the meritorious', encouraging personal intellectual fulfilment. The 'value of science' was an important theme and the stoical serenity evident in Masonic writings was partly derived from an obsession with imitating universal order and harmony as revealed by natural philosophy. This perception of divinely ordained and ordered space was projected onto external nature just as it ostensibly guided the design of lodges and managed improvement embodied in ritual progression as fulfilment. Freemasons adopted the Newtonian emphasis upon 'order, stability ... the rule of law' and a spiritual hierarchy within nature, whilst the 'belief in order proclaimed by science' encouraged 'fantasies' about forging international fraternities from the microcosm of the lodge.[52]

FREEMASONRY, ANTIQUARIANISM AND NATURAL PHILOSOPHY

As we have seen there is evidence that promoters of natural philosophy and particularly Newtonian science during the 1720s significantly transformed English freemasonry and promoted the sciences with other intellectual endeavours. Many Masonic natural philosophers were also antiquarians and probably attracted to masonry by its vision of the past and belief in the rediscovery of ancient wisdom, and Anderson's *Constitutions* promoted an elaborate historical narrative that traced English freemasonry back to the Anglo-Saxon period and beyond.[53] Furthermore, the tradition and development of county natural history, topographical and antiquarian studies like freemasonry provided membership of scholarly networks and potentially after 1717 recognition from a royally approved metropolitan national association. Subjects prominent in county antiquarianism, chorography and meteorology such as geometry, astronomy, surveying and cartography held obvious attractions to provincial freemasons given their relationship to the imagery of the divine architect and Newtonianism as celebrated in Masonic texts.[54] As Money has said in the 'plausible syncretism of sacred and civil history by which it fabricated its own past', masonry 'ministered to the common desire for a believable universal meaning behind the diversity of the past'. This scholarly syncretism helped to inspire the foundation of some of the earliest provincial philosophical societies such as the gentlemen's societies at Spalding, Stamford, Peterborough and Lincoln.

Given that the Enlightenment promoted the application of rational study to the past as a prerequisite for modern progress, the philosophy of freemasonry held obvious attractions to antiquarians. Philosophers such as Newton, Locke and Toland searched for a new spatially realised rationality from the past and the Bible as well as in the natural world.[55] The general Masonic interest in 'mental improvement' and Enlightenment as part of an organised national, secular, though ritualised structure, promoted the value of both natural philosophy and antiquarian study. Speculative masonry's emphasis upon a mathematical deity accorded well with the Newtonian conception of the universe with its grand mathematician or architect at the centre. The semi-mythical history of freemasonry revered the grand architect in ceremony and celebrated the geometry, harmony and proportion of the greatest architecture, which they held to be embodied in the Newtonian universe as a whole and in special buildings such as Soloman's Temple which Newton himself believed. Hence many Masonic natural philosophers followed Newton in their studies of biblical chronology, physico-theology and antiquity which they combined with their researches in natural philosophy. These included Folkes, Hare and Ward whom as we have seen, held offices in the Royal Society and also the mathematician and geographer Charles Hayes (1678–1760), the botanist Richard Richardson (1663–1741) and the physicians Woodward and John Arbuthnot (1667–1735).[56]

Antiquarianism also provided the major inspiration for various intellectual coteries in the provinces with Masonic interests such as those at York and Spalding. At York, masons such as the physician Francis Drake FRS (1696–1771) ensured the elevation of the city into a regional capital and Masonic centre to rival the metropolis in the north partly by celebrating the history, antiquities and natural history of the city and county. Drake founded a society with John Burton (1719–71) 'for inquiring into the productions of nature, and improvements of art' which communicated its observations to philosophical networks around the country during the 1740s.[57] The Masonic formulation and rediscovery of the past helped to ensure that York's Georgian status was buttressed by its venerable antiquity and institutions as the design and splendour of Burlington's famous Egyptian, assembly rooms encapsulate.[58] At Spalding, most members of the Gentlemen's Society founded in 1714, many of whom moved in both metropolitan and provincial circles, engaged in antiquarian and chorographical studies, which may have provided the primary inspiration for their embrace of freemasonry rather than Newtonian natural philosophy. Maurice Johnson the founder of the Spalding Society for instance, was, of course one of the founders of the Society of Antiquaries and an inveterate promoter of intellectual societies.[59]

The relationship between antiquarianism, natural philosophy, natural history and collecting is also evident in the careers and ideas of Masonic natural philosophers such as William Stukeley, Thomas Wood and Thomas Wright. Through their practice of architecture and landscape gardening and inspired by their antiquarian studies, Wood and Wright strove to impose upon the world the Masonic ideal of ordered, managed spaces and to reinvent a harmony with nature once embodied in ancient civilisations that they maintained had been lost. As initiation into freemasonry and advancement through the hierarchy of degrees successively provided the keys to unlocking the secrets of Masonic texts and rituals, so patient study of antiquities and druidic temples would reveal the sophistication of ancient mathematics, geometry and astronomy. As Stukeley stated in his autobiography it was this 'curiosity' that led him to be 'initiated into the mysteries of masonry, suspecting them to be the remains of the mysterys of the ancients'.[60]

Surrounded by Masonic friends and initiated into the brotherhood at the Fountain Tavern, Strand in 1721, Stukeley has been credited with helping to shape the rituals and practices of organised masonry during the 1720s. He was certainly an enthusiastic mason and founded a lodge at Grantham in 1726 on his removal there which perhaps functioned somewhat like the nearby Gentlemen's Society at Spalding and lasted until 1730.[61] Wood believed that he was rediscovering the ancient laws of proportion and was inspired to construct Bath's Royal Crescent by the ancient 'druidical' temples of Stonehenge, Avebury and Stanton Drew. Equally fascinated by such stone circles such as Stonehenge and Stanton Drew (which he surveyed and drew for the Earl of Sandwich), Wright was perhaps the most enthusiastic mason of the three. His significance as probably Britain's leading landscape gardener and architect during the 1740s and 1750s has only recently become clear. After initially training as a clock and instrument maker he travelled around the country giving lectures and publishing on philosophical subjects producing an important theory of the universe and worked on various architectural and landscape projects which were strongly imbued with his philosophical and Masonic theories. Wright's surviving manuscripts and publications contain many Masonic references and drawings, notably the theory of the universe that was to so influence Kant, which projected a plenum of universes across creation represented by eyes of providence and bounded by the Hermetic Rosicrucian symbol of the ouroboros serpent eating its own tail, which symbolised the unity of matter (he also had it encircling his own portrait).[62]

MASONIC NETWORKS

As we have seen, Jacob has argued that whilst Tories loyal to the Hanoverian succession were welcomed, the Royal Society and Grand Lodge 'became social and intellectual forums that were imitated in provincial cities throughout the course of the century', with freemasonry spreading 'rapidly from London into the provinces' and thence to Europe.[63] Stevenson however, has argued for a less Londonocentric geography of non-operative Masonic cultural diffusion in which aspects of Scottish freemasonry were adopted in England by the early eighteenth century, suggesting that the passages relating to Scottish masonry in Jacob's *Radical Enlightenment* 'are sprinkled with misunderstandings'. He follows Knoop and Jones in emphasising the fact that the foundation of the English grand lodge, which merely represented the union of four lodges at the time, was 'almost an irrelevance in the long process of development of the movement'.[64] Stevenson instances the relative richness of source material for Scottish seventeenth-century freemasonry compared to the paucity of documents and references available for English freemasonry in the period. The Scottish system of permanent or stated lodges keeping records, performing rituals and grading members according to a two-tier system of apprentice and fellow craft master superseded more casual and occasional type of English freemasonry that met informally and irregularly. The case for such a model of Masonic geographical diffusion is supported by Anderson's *Constitutions* which distinguish between the occasional lodges typical of the south and more permanent bodies in the north. This is supported by the location of the earliest well-authenticated records of English lodges which generally relate to northern centres such as Alnwick, Warrington and Chester.[65]

Stevenson's model of cultural diffusion is paralleled by recent scholarship on British Enlightenment scientific culture. This recognises that whilst the metropolis and its institutions was, of course, a major cultural facilitator, there was a rich and varied provincial scientific culture as the popularity of scientific texts, many public scientific lectures and philosophical society numbers demonstrate. Philosophical culture in northern centres such as Newcastle or York looked as much to the Scottish Enlightenment science or to the Netherlands as to the distant metropolis. Many of the medical men and Dissenters who were the lynchpins of provincial scientific culture had received their education in Scotland or Holland, whilst English intellectual culture was regarded as part of a wider European intellectual Enlightenment tradition. Although the metropolitan grand lodge tried to assert its seniority over Scottish, Irish, Welsh and English provincial lodges from the 1720s, the geographies of Masonry mirrored and helped to shape the complex geographies of British scientific culture. According to this model, the geographies of freemasonry and scientific culture are determined by local factors and wider, social, religious and intellectual affiliations such as the relative importance and economic character of regional centres, and the prevalence of Dissenting groups.[66]

Divisions within freemasonry and the degree of autonomy enjoyed by many provincial lodges pose further difficulties to a Londonocentric cultural geography of British freemasonry often obscured by the metropolitan bias of surviving Masonic sources. Despite the prominence of freemasons such as Desaguliers in Hanoverian natural philosophy and the overlap between the Masonic brotherhood and institutions such as the Royal Society, the full implications of this remain unclear. In 1751 a schism occurred between 'modern' lodges who continued to adhere to the authority of the metropolitan grand lodge and 'ancient' lodges who tended to be more reformist, egalitarian and ritually experimental. This split reduced the authority of the grand lodge in Britain and especially the American colonies, and during the 1760s the moderns responded by trying to restore the primacy of the London lodge through defining a universal constitution and requiring that all provincial lodges register and list members.[67] Given the organisation and geography of English freemasonry by the 1740s, it probably provided a model for some philosophical societies. Literary and scientific associations, like Masonic lodges and debating clubs, often emphasised their democratic nature enshrined in constitutions or laws defining rates of subscription and fines, duties of elected offices, and permissible subjects for discourse. Of course, these were not always adhered to in practice, but the model of neutral fraternal comradeship evident in this gentlemanly ideal was also one shared by Masonic lodges.[68] Such rules were adopted by philosophical societies such as those at Newcastle and Bath, it being warned that otherwise they might 'too frequently degenerate to drinking clubs' and thereby become 'schools of sedition and infidelity'. Meetings of the Newcastle Philosophical Society were held in a private house with 'every sort of liquor' being 'absolutely excluded'. Subjects for debate were not allowed until they had been agreed by ballot providing the opportunity for the rejection of topics likely to lead to arguments such as those that 'too freely and incautiously' questioned 'the fundamental principles of religion and good government'.[69]

At Norwich a Masonic lodge founded by the 1720s may have served as a model for local literary and scientific societies. Members of the Society of Friars for example founded in 1785 underwent an initiation ceremony before being placed into 'orders'. In addition to acting as a debating society and supporting various improvement and charitable ventures like the masons, the Friars were keenly interested in agriculture and natural philosophy, building a library and collection of scientific apparatus, conducting experiments, and hosting courses of lectures such as those by Adam Walker.[70] Likewise, philosophical culture in the Derby area was stimulated by the patronage of Washington Shirley, fifth Earl Ferrers (1734–78), who served as grand master of the metropolitan grand lodge between 1762 and 1764. Shirley was most closely associated with the influential original no. 4 Horn Lodge and served as master in 1762. Brothers at the lodge by the 1760s included many members of Parliament, nobility and royalty, Prince William Henry, Duke of Gloucester and his two brothers the Dukes of York and Cumberland.[71] Shirley's social and scientific aspirations were supported by his election as a fellow of the Royal Society in December 1761 on the basis of his observations of the transit of Venus made upon his Leicestershire estates and the gift of 'an orrery or transitarium invented by his lordship, to facilitate the conception of future transits'.[72]

Members of the Derby philosophical circle who enjoyed Shirley's patronage during the 1760s and 1770s included the clockmaker, geologist and engineer, John Whitehurst (1713–88), the artist, engineer and cartographer Peter Perez Burdett (1734–93) and the artist Joseph Wright (1734–97).[73] Whitehurst resided in Derby between 1737 and 1778, probably joined the local Virgin's Inn Lodge and became a member of the Lodge of Emulation no. 21 in London by

1770. The Virgin's Inn Lodge was never affiliated to the grand lodge and no official lodge was founded in Derby until the Tyrian Lodge appeared in 1785.[74] Freemasonry helped to provide access to an important network of social and intellectual contacts enabling Whitehurst to move in both provincial and national intellectual circles. He was a correspondent and friend of Franklin, who was, of course, a leading fellow freemason and who visited Whitehurst at Derby.[75] Daniels has recently argued that the theory of central fires in Whitehurst's *Inquiry Into the Original State and Formation of the Earth* (1778) was partly inspired by the concept of central fire in the alchemical and Rosicrusian tradition, which also suggests that his freemasonry may have provided a stimulus and model for his philosophical study and associates the work with Wright's 'scientific' paintings of the period.[76] Perhaps the most evocative celebration of Enlightenment natural philosophy inspired by the activities of the Derby savants and Shirley's patronage was Wright's painting 'a philosopher giving that lecture on the Orrery in which a lamp is put in place of the sun' (exhibited 1766), which was purchased by Shirley. Often regarded as one of the most powerful depictions of the European Enlightenment, the painting celebrates the power of science and discovery and symbolised Shirley's philosophical and Masonic status and aspirations for intellectual improvement.[77]

The inspiration that freemasonry provided is also evident in the rhetoric of Derby a philosophical society formed around 1770. An anniversary address read to the society in 1779 is replete with Masonic language and distinguishes between 'wisdom's sons who nature's laws explore' and the 'vulgar', mere 'grovelling' slaves devoted to 'sloth and lust'. Wisdom's sons in contrast were devoted to 'the love of science' which filled them with rapture, ecstasy and 'inextinguishable joy'. These 'great souls', the 'learned', were engaged in astronomical observations 'in godlike thought ... mounted high, surveying regions in the sky' and observing new worlds opening before them. They exhorted the muses that each year would 'improve the flame' by obtaining new votaries to the society so that 'Derby's sons' would 'exalt the liberal arts, employ their genius and cultivate their parts'.[78]

The popularity of freemasonry throughout Enlightenment Europe and in the colonies, especially in America, encouraged the formation of a distinct republic of letters, with brothers sharing rhetoric, organisation and imagery.[79] Freemasonry facilitated the international spread of scientific ideas and furthered scholarly communication, integration, and debate, providing entry into intellectual circles for travelling philosophers and transcending cultural barriers. It is important to recognise, however, that freemasonry also reinforced the exclusion of large segments of society from public intellectual culture in what Jacob has called the 'paradox of Enlightenment'. Women were excluded from British but not some European lodges, and Jacob has suggested that the absence of women from aspects of the 'intellectual ... and social world' of the European Enlightenment may have been due to the example of freemasonry. Although Masonic literature emphasised the ideal of a virtuous, honourable and loving marriage, much literature emphasised the 'inferior sensibility of women', their essentially emotional rather than intellectual nature. Jacob discerns both elements of a modern egalitarian conception of marriage but also 'sentiments that advocate women's natural inferiority or that are downright misogynist'. Whilst emphasising equality for men within lodges, freemasonry also 'glorified domesticity and sexual fidelity, yet at the same time betrayed resentment and discomfort on the question of women and their rights and abilities'.[80] Given the extent of female participation in other aspects of British scientific culture such as botany, this argument is worthy of further investigation. However, the exclusion of women from freemasonry also, of course, reflected wider developments in society, and women's exclusion from major scientific societies such as the Royal Society pre-dated the rise of speculative masonry.

CONCLUSION

The importance of freemasonry as part of European Enlightenment culture is well attested. Despite some of the criticisms that have been made of Jacob's thesis that Whig Newtonian philosophy inspired the development of English freemasonry and that the lodges in turn promoted scientific education, her arguments have made a major contribution to our understanding of eighteenth-century intellectual culture. As the most widespread and influential form of secular association in eighteenth-century England and given the number of Freemasons who were natural philosophers and fellows of the Royal Society it is likely that there was a close relationship between the two in London during the 1720s and 1730s. Freemasons promoted scientific publications whilst their institutional organisation provided a model for philosophical societies. The distinctive federal and internationalist character of freemasonry differentiated it from other Georgian secular associations and appears to have facilitated international scholarly activity. However, considerable further work is required to determine the nature of the changing relationship between freemasonry and other cultural activities, especially in the English provinces. Here local geographical, political and socio-economic factors may well have given it a very different flavour from the Whig Newtonian character that is supposed to have predominated in the metropolis during the 1720s with northern English lodges for instance looking as much to Scotland as to London.

5 Public Science: Urban Botanical Gardens

Driven by the demands of private aristocratic and middling-sort gardeners and collectors, the importance of private and commercial nursery, market and botanical gardens and orchards to Enlightenment scientific culture has already been emphasised. The study of botany was also promoted by experiences of different kinds of public and semi-public gardens in flourishing urban renaissance culture from the burgeoning metropolis to provincial centres such as Norwich.[1] The experience of gardens and gardening was also enhanced by the creation of public walks and pleasure gardens by corporations, government bodies and private entrepreneurs, sometimes from the enclosure of common lands which became important features of the early-modern townscape as well as sources of local urban pride.

As centres of fashion and polite society, the metropolitan pleasure gardens of Vauxhall and Ranelagh, and to a lesser extent, the Royal parks inspired numerous provincial pleasure gardens, town walks and squares. Of the scores of London gardens offering varied forms of entertainment, Ranelagh and Vauxhall attracted the largest crowds and the greatest notoriety with their planted walks, lodges and assembly rooms, refreshments, music, illuminations and fireworks. Like the new forms of shops, piazzas and conspicuous commercialised consumption, as a new kind of public place, they signified and facilitated the transformation of the urban experience, fostering and facilitating novel ways of walking, viewing and interacting between different social ranks, extending public culture mentally, spatially and culturally into the evening and night and beyond the summer in heated lodges, refreshment and assembly rooms. Celebrated and parodied alike in novels, newspapers, prints and cartoons, some of the gardens also became notorious for high prices and the sometimes dubious quality of refreshments, drunkenness, quarrelling, crime and romantic and sexual liaisons. On the other hand, the ability to take afternoon walks also signified wealth and status whilst providing other benefits that were celebrated including exercise, fresh air amongst the tree-lined walks which offered direction, shade, shelter, sometimes springs and prospects to enjoy such as rivers, ruins, distant hills and fields or town walls.[2]

The creation of semi-public botanical gardens was partly inspired by the demand for other forms of public and semi-public town walks and pleasure gardens. Following the example of continental institutions such as the Uppsala Linnaean botanical garden, similar private and semi-public ventures were founded in the British Isles. These became cardinal and characteristic Enlightenment spaces, idealised microcosms where the concerns of improvement, medicine and systematic botany and botanical publishing intersected with commercial gardening, polite education and fashionable promenading. Although there was some significant resistance, from the 1760s English botanical gardens began to adopt the Linnaean system instead of older or rival taxonomies which began to shape their image, usage and design. Linnaean arrangements fostered an Edenic image and experience of rational order, natural harmony and purpose, helping to promote botany as a serious scientific endeavour, but, as we shall see, the process of planting and managing such collections could also expose and exacerbate anomalies and highlight the incommensurability or commensurability of rival international Enlightenment taxonomies.[3]

The older purposes of botanical gardens were not entirely forgotten and aided by patronage from medical men, the potential to exploit plants for physic remained rhetorically and practically important. Furthermore, there was resistance to the notion that botanical knowledge and experience was required for gardener's education or garden development. Inspired by the Oxford and Cambridge physic or botanical gardens (terms used interchangeably during the eighteenth century) and other private botanical gardens and commercial nurseries, semi-public Enlightenment botanical gardens were created by subscription-based botanical societies, dedicated to botanical education and horticultural improvement. Encouraged by greater acceptance upon botany as a polite science, the proliferation of botanical literature and the order presented by the Linnaean system, new forms of English botanical garden such as those at Cambridge (1761), London, Liverpool (1803), Hull (1812) and Whitby (1814) helped to develop and satisfy a new audience for botanical education, probably more so than Sir Joseph Banks's royal and imperial domain at Kew. As private and semi-private institutions dependent upon subscriptions, they had to satisfy the demands of patrons and their families and interested parties, including occasional visitors. In order to understand the spatial usage and character of these places it is necessary to consider both their functions – utilitarian medical and botanical roles and their status as urban pleasure gardens.

Georgian semi-public botanical gardens illustrate the complex relationship between urban public walks and gardens, idealised botanical systems, practical horticulture and gardening, the demands of medicine and scientific education and the production of botanical publications. They also demonstrate the varied interplay between microcosmic and macrocosmic spaces from small closely bounded worlds to regional flora and seeds and specimens conveyed around the terraqueous globe, from the changing growth and anatomy of individual plants to competing European botanical systems.

CHELSEA PHYSIC GARDEN

Established by a professional group for practical, medical and professional purposes, the Company of Apothecaries of London, the Chelsea Physic Garden had a major impact upon botany and horticulture, providing inspiration and a model for subsequent physic and botanical gardens. Apprentice apothecaries were required to identify and use 'simples' or medicinally significant plants as part of their apprenticeship from gardens and surrounding countryside. The Chelsea Garden was formally constituted in 1673 when the Company selected a site adjoining the river which it leased from Charles Cheyne,

Lord of the Manor at five pounds per annum that allowed easy access for boats carrying plants and visitors. The garden was managed by the Court of Assistants of the Company who appointed a 'Botanical Demonstrator' whose office was to 'explain to his pupils (Company apprentices) the names, classes, and medicinal uses of the plants' and make monthly summer 'herborizing excursions in the neighbour-hood of London' annually between April and September. In July a 'general herborization' led by the Demonstrator was held with mem-bers of the Court and interested patrons in attendance, whilst later in the eighteenth century a copy of William Hudson's *Flora Anglica* (1762) was presented as a prize to the apprentice who had discovered and investigated the largest number of plants. Although the 'herbor-izing' trips long pre-dated the formal garden establishment, in be-coming the centre for these activities, the Chelsea institution came to have wider significance providing an authoritative place for identi-fying, using and studying plants from the metropolitan region.[4]

The design and usage of the garden was much emulated in other gardens. The Lord of the Manor from 1721, Sir Hans Sloane granted the freehold to the Company on condition that they present the Royal Society with two thousand plants in annual instalments of fifty, which were preserved in their herbarium. A distinguished nat-uralist and collector as we have seen, Sloane was easily the most im-portant patron of the venture, funding improvement of buildings and providing plants from the 1720s. The design and features of the garden are clearly shown in John Haynes's detailed survey under-taken for the Company in 1751 which was intended to provide an 'explanation of the several parts of the garden, shewing where the most conspicuous trees and plants are disposed'. It demonstrates how the garden was laid out broadly according to a symmetrical plan with quarters arranged around a statue of Sloane at the centre erected by the Company in 1733. There were divisions with a series of paths surrounding and bisecting each of the beds with plants, trees and shrubs arranged systematically according to medical usage and pre-Linnaean taxonomies. Potential commercial utility was as much of an incentive for inclusion as medicine, and by the later eighteenth century these included herbs, sugar maple, dragon, mango, orange, lemon and banana trees, and coffee, tea, tobacco and sugar plants from the Americas, the East and West Indies and other parts of the globe. A 110-feet long 'spacious' greenhouse was provided in 1732 by a subscription from Company members on the north side of the garden adjoining Paradise Row with a library of botanical works, a herbarium of dried plants and cabinet of seeds grown in the garden, whilst adjoining this were apartments for the gardener and his family. At each end of the greenhouse two smaller hot houses were provided. Towards the south side of the garden four cedars were planted in 1683, two of which had girths of over twelve feet by the 1790s and dominated prospects from the river such as the engraving in the first edition of Daniel Lyson's *Environs of London* (1795) (Figure 10). The growth rate of these in the English climate surprised the gardeners and patrons providing striking evidence that exotics could be successfully planted although the two others had to be removed in 1771 having been cut too closely due to proximity to the green house.[5]

10 The Chelsea Physic Garden from the River Thames
from D. Lysons, *Environs of London* (1795)

The pleasurable and aesthetic aspects of the garden are also clearly evident in Haynes' depiction of the pond, ornamental and formal planting including the four cedars and along the east and west perimeters, serpentine paths through some of the beds, and particularly the gentlemen and ladies strolling along the riverbank adjoining the garden, observing the passage of small boats. The illustration of the Thames also demonstrates its significance in the positioning of the garden as conduit for patrons, visitors and their guests, gateway to exotic places and productions from across the oceans. Sloane's requirement that at least fifty dried original specimens be sent each year encouraged discovery and importation and almost 4,000 specimens were given between 1722 and 1796, the identification of which has provided a good indication of the sources and global reach of the Chelsea institution. Around 11 per cent of those identified, for instance, hailed from North America.[6]

The success of the Physic Garden encouraged the foundation of local commercial gardens 'much frequented by the nobility and the fashionable world in the spring' such as Colvill's Nursery established in 1786 which had a large collection of exotics. The arrangement of medicinal plants within the Physic Garden was of course regarded as the most important section and a catalogue was published by the Demonstrator Isaac Rand in 1730. This was supplemented in the same year by a general catalogue of all the garden's plants by the gardener Philip Miller (1691–1791), who held the office and resided in the garden for almost half a century from 1722. Haynes's plan and the catalogues were not the only publications to be inspired by the Chelsea gardens, Miller's *Gardener's and Florist's Dictionary* (1724), probably the most important and influential Georgian work on the subject, which was republished many times in successive editions and epitomes was also, in many ways, a product of the institution.[7] For the seventh edition, Miller reluctantly accepted aspects of the Linnaean system including binomial nomenclature, adopting it fully only for the eighth edition of 1768, the final version that appeared in his lifetime. The work lived on and was comprehensively enlarged, revised and updated by Thomas Martyn, Professor of Botany at Cambridge for the ninth edition, producing what was really a new book although he strove to retain the original character.[8]

11 Engraving of the Oxford Botanic [Physic] Garden
by J. and H. S. Storer (1821)

UNIVERSITY BOTANICAL GARDENS

Like the Chelsea Physic Garden, English university botanical gardens owe their existence to the requirements of medical botany, but they too came to have greater significance in Georgian urban public and scientific culture. The Oxford Physic Garden was founded in 1622 by Henry Danvers, Earl of Danby as a 'nursery of simples' and so a Professor of 'Botanicey' could read there showing 'the use and virtue of them to his auditors' and improving 'the Faculty of Medicine'. Situated on meadow ground outside the east gate of the city close to the River Cherwell on what had been an old Jewish cemetery, much soil had to be brought to the garden to raise the ground level and 'prevent the overflowing of the waters'. Loggan's plan from *Oxonia Illustrata* (1675) shows a square garden surrounded by a 'stately free-stone wall' with shrubs growing against the interior on all sides and a small complementary exterior garden beyond the wall to the north. A 'comely gatehouse of polish't stone' or 'great gate' designed by one of Inigo Jones' pupils, was erected next to the east bridge, serving as the main entrance. John Claudius Loudon considered the Oxford garden to be a 'venerable establishment, whilst the 'noble stone archway allowed a 'vista to the other extremity of the garden. He judged the walls of the garden to be about two feet thick and twelve feet high with a 'coved gothic cornice on each side under an elevated gothic coping'. The wall was 'composed of large blocks of smoothly dressed stone' and formed in his view 'the noblest garden wall, speaking architecturally, which we have seen in any country'.[9]

12 The original gate and planting within the Oxford Physic Garden, detail from engraving by J. and H. S. Storer (1821)

The main entrance through the grand gate from the Magdalen Bridge and High Street was to the north but there was another southern gate opposite and two gates to the east and west (Figures 11 and 12). The garden was divided into four quarters intended to represent the four quarters of the globe by a path and gated enclosures with another path running around. A few small trees and bushes, some of which are clipped into shapes are shown adorning the main entrance and exit. Through its design, management and location and as the trees and shrubs grew to maturity, the garden remained a botanically significant institution with a library and museum and a picturesque pleasure ground for the town, although it attracted criticism from Loudon and others for not being large enough. Like the seventeenth and eighteenth-century pleasure gardens and walks of Bath, York, Tunbridge Wells, Exeter and the Chelsea Physic Garden, the Oxford Physic Garden was equally important as an urban pleasure garden and place of resort as a place of scientific education and display. Opposite to the entrance was Magdalen Bridge which led onto the High Street, principal thoroughfare of the city and main entrance from the east, providing a prospect celebrated by William Wordsworth and Sir Walter Scott. Some of what were regarded as the most beautiful walks were situated to the east of Magdalen College around the water meadows, surrounded by the branches of the River Cherwell and associated with Joseph Addison, who partly inspired by them, had composed the first version of his *Pleasures of the Imagination* in 1712.[10]

As at Oxford, the demands of medical education were the principal reason for the foundation of a botanical garden at Cambridge University. Much impetus came from metropolitan medicine and scientific culture and there were close connections with the Chelsea Physic Garden. After several abortive attempts, the Cambridge Botanical Garden (or Physic Garden) was established in 1761 when Dr Richard Walker, Vice-Master of Trinity College provided an estate to establish a foundation. After discussions with Philip Miller of the Chelsea Garden, Walker also bought the grounds of an old monastery in St Edward's Parish for £1600 and five or six tenements in Free School Lane, and revenue came from leasing ground on the north side of the court and a subscription fund. The five-acre site passed to the University in 1762 was 'well-walled round, quite open to the south, conveniently sheltered by the town on the other quarters, with an ancient water course' running through the middle (Figures 13 and 14). When the stoves and greenhouses were ready for the reception of 'tender plants' and the garden 'perfected for the harder sorts', 'trails and experiments' were to be 'regularly made and repeated' to discover the 'virtues' of plants 'for the benefit of mankind' which was the 'principal intent of the garden'. By comparison, 'flowers and fruits' were to be regarded as 'amusements only' although as they did have their 'excellencies and amusements' and were not to be 'totally neglected'. The trials were supposed to concentrate on investigating poisonous plants because their 'great powers in altering an animal body' were thought to indicate their potential to become 'active and valuable medicines', and the results were to be reported to the Royal Society and College of Physicians. Within the mansion house botanical lectures were given, the meetings of governors were held, and there were quarters for the reader Thomas Martyn, and a library including documents relating to the garden, and *hortus siccus* of botanical specimens.[11] Under Martyn, Reader and later Professor of Botany from 1762 until 1825, and particularly after James Donn was appointed curator in 1794, the garden attracted national attention through the publication of successive editions of his *Hortus Cantabrigiensis*. The links with metropolitan and national botany are underscored by the fact that the first curator was Charles Miller (1739–1817) youngest son of Philip Miller of Chelsea whilst Martyn grew up in Chelsea under the patronage of Sloane and scarcely lived at the university, removing entirely after his marriage in 1773 apart from the periods of his lectures, which ceased after 1796 due to lack of interest and ill health.[12]

Although a private University benefaction, as Walker's stipulation makes clear, the Botanic Garden was always regarded as a semi-public venture. It was only through co-operation between the university and the corporation that the garden was established and the latter leased some of the land to the former for 25 guineas per annum, whilst additional money was raised through music festivals and other charitable events in addition to private benefactions.[13] Government of the garden was vested in the hands of the Vice-Chancellor, the masters of Trinity and St. John's colleges, the Provost of King's College and the Regius Professor of Physic. Limited demand for botany amongst college students meant that the trustees of the garden looked to local, national, and especially metropolitan benefactors to supplement income from the estate. A list of benefactors published in 1796 reveals that the largest donations came from clergy, aristocracy and gentry, whilst various medical men subscribed including John Fothergill MD, William Heberden, Robert Taylor MD, William Watson MD and Thomas Hayes MD of Chester. There was also a significant sum of £55.13.4 listed as 'sundry smaller benefactions' from members of the University, the town of Cambridge and other sources.[14]

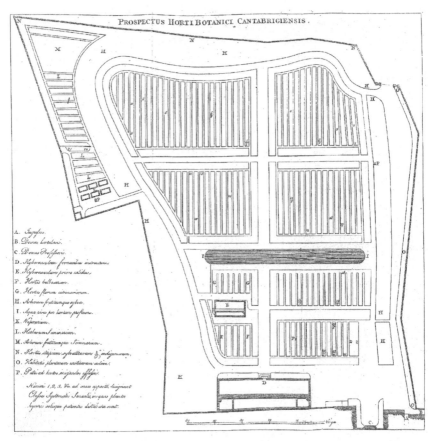

13 Plan of the Cambridge Botanical Garden from T. Martyn, *Mantissa Plantarum Cantabrigiensis* (1772), courtesy of Cambridge University Library

The purpose of the garden was to 'celebrate the wisdom and good-ness of god' and, inspired by the Chelsea Physic Garden the lectures given at Cambridge on medical botany by London physician William Heberden using local plants were supposed to provide practical bene-fits for physic and 'many branches of manufacture and commerce'. By discovering the 'salutary virtues of plants, and their uses for the conveniences of life', the garden would 'promote the public welfare' and help to revive the subject in the university. It would provide a forum for private botanising and collecting which 'for want of public reception' had 'lasted no longer than the collectors'. From 1763 Martyn gave Linnaean botany lectures to students and 'occasional hearers' using the plants and herbarium admitting anyone to these 'without the customary fee' prepared to subscribe ten pounds; however, numbers declined despite the introduction of natural historical subjects into the course.[15] Martyn faced continuous difficulties in obtaining enough money to carry out Walker's objectives and had to make frequent appeals to the university and subscribers for additional funds which were supplemented by the sale of trees and shrubs under Miller.[16]

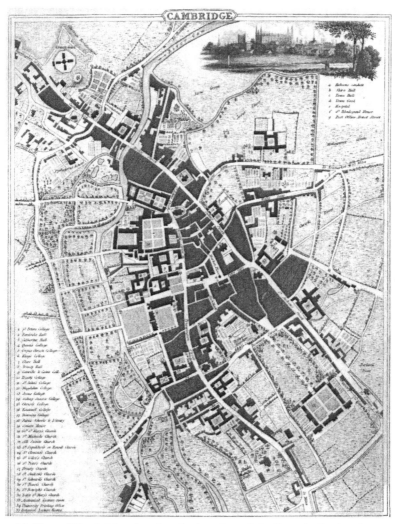

14 Plan of Cambridge showing the location of the Botanic Garden from
R. Harraden, *Cantabrigia Depicta* (1809)

Various changes occurred within the garden and a lecture room was erected on part of the site in 1787 for the Botanical and Jacksonian Professor of Natural and Experimental Philosophy (Figures 15 and 16). From the 1780s, Martyn concentrated on his numerous publications which included the very successful *Letters on the Elements of Botany Addressed to a Lady* (1785) after Rousseau, and a much-augmented and rewritten edition of Miller's *Gardener's Dictionary* published in 1807 after 22 years of work.[17] However, James Donn revitalised the collection and secured the status of the garden through the successive additions of his *Hortus Cantabrigiensis* which superseded an earlier cataogue published by Martyn in 1771. The changing aims of Donn's catalogue and the large increase in number of plants listed illustrate this greater ambition. The fact that 'Botanic Garden' came to be universally favoured rather than 'Physic Garden' demonstrates how botanical study for its own sake came to supplant practical physic from the later eighteenth century. Similarly, whilst medically useful plants were labelled with an abbreviation in the first edition of the *Hortus Cantabrigiensis* (1796) this was removed from the second (1800) as other aspects of taxonomy and botany came to predominate.[18]

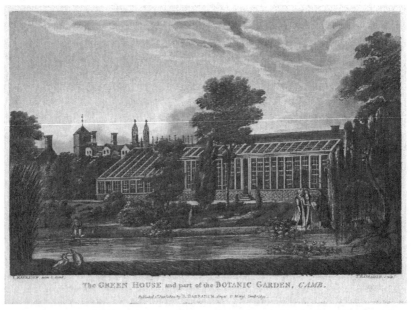

The GREEN HOUSE and part of the BOTANIC GARDEN, *CAMB.*

15 R. Harraden, The green house and part of the Cambridge Botanical Garden (1800), courtesy of the British Library

16 R. Harraden, The lecture rooms in the Cambridge Botanical
Garden (1802), courtesy of the British Library

A detailed plan published in 1771 reveals the disposition of all the
main features of the Cambridge garden, and although there were
some similarities it presents a strikingly different form of arrangement from the four quarters of the globe within the Oxford Physic
Garden, clearly illustrating the impact of Enlightenment taxonomy
(Figure 12). The garden was surrounded by a wall and although not as
prominent as at the Oxford, elegant wrought iron gates were placed
at the main entrance. Exotic plants (O) were positioned along the
interior of the south-west wall and a selection of native plants and
trees, including those growing locally in the county (N) were placed
just inside the south-east wall. Within these a plantation of fruit trees
(H) occupied a belt running around most of the garden except for the
north-west wall, against which the lecture rooms were situated. The
picturesque effect of the trees and shrubs as they grew around the
periphery is evident in the painting of the garden in Ackermann's
History (1815). At the centre was the canal for water plants (I), whilst
water was also supplied from three wells (P) to the north-west, south-
west and north-east. A separate more sheltered section to the south-
east contained the tree and herb nurseries (M and L) and between the

lecture rooms and canal to the north-west were flower beds (G), the hot house (E) and ground for nurturing bulbs (F). A central gravel path running from the entrance to the lecture rooms from north-west to south-east provided a spine for the garden and offered views from the windows and steps along the whole space and across the canal. A series of branching cross-paths between the beds radiated from this to the periphery whilst another path around most of the edge made a full perambulation of the perimeter between fruit trees and exotics and the central beds possible.[19]

Accounts of the Cambridge garden imply that it was immediately arranged according to the Linnaean system from 1761, but this seems to have taken some time to become established, perhaps because like his father at Chelsea, Charles Miller did not immediately accept the Linnaean arrangement. Martyn's 1763 lectures were some of the first on the Linnaean system to be read in England and he was congratulated by Pulteney for having placed 'the science of botany into a new dress by the method and improvements of the illustrious Swede'. At around the same time in Scotland, Dr John Hope (1725–86) Professor of Medicine and Botany at Edinburgh University oversaw the move of the university botanical garden to a site next to Leith Walk, using it to teach the Linnaean system.[20] Although Linnaean botany came to predominate in the Cambridge garden, during the 1760s under Miller a more pragmatic attitude seems to have prevailed in the garden. Martyn's *Plantae Cantabrigienses* (1763) was arranged according to Linnaean classes but in three parallel columns, the first being generic and trivial Linnaean names, the second names based upon his father's *Methodus Plantarum* and the third following terms used in Ray's alphabetical *Catalogus Plantarum*. This demonstrates the extent to which the earlier terminology and taxonomy of Ray and Martyn senior continued to be used and probably reflects the practical physic and botanical arrangements pursued in the garden by Miller. In October 1768 the Cambridge trustees resolved that it was 'absolutely necessary to facilitate study of botany and 'render the garden of general use' that plants 'be arranged and marked according to the system of Linnaeus' and that a catalogue be printed and only when Martyn assumed the role of both curator and professor after Miller's departure in 1770 was this fully accomplished.[21]

138

According to the 1771 plan, the 24 Linnaean classes were positioned in beds that occupied most of the central ground, generally in numerical order, however, there were some significant variations in the numerical and spatial disposition of each determined by gardening practicalities, the functions and objectives of Walker's endowment, taxonomic considerations and to a lesser extent, aesthetics. Labels are not mentioned at this stage but they were certainly in existence by Donn's time as William Kirby, a visiting naturalist, noted how well the collection had been augmented and how accurate the labels were in 1797. The amount of space occupied by each class varied, some being spread around out of numerical order. The latter included the first (Monandria) and second (Diandria) classes, which were distributed in different parts of the garden because they included a large variety of exotic spices and medicinal plants including ginger, cardamom and turmeric. Some classes such as numbers four and fourteen occupied a relatively large amount of space. The fourth Linnaean class (Tetrandria) was a large and highly artificial division containing a variety of plants which probably explains the small amount of space allotted to it, and it is striking that the fourteenth class (Didynamia) occupied a large space of ground in the centre. The latter included many herbaceous plants with medicinal applications central to the garden's role in teaching and improving physic, satisfying the expectations of subscribers. The nineteenth class (Syngenesia) also included some plants with medicinal virtues, but the space allotted to it and its location between the lecture theatre and professor's house and the canal beside the main flower beds was probably determined by the fact that it contained many of the most beautiful flowering plants such as dahlias, chrysanthemums, asters and sunflowers for the pleasure of students and subscribers. Some classes were allowed little room such as the twenty-second (Dioecia) and twenty-third (Polygamia), probably because these were mixed and contained many different kinds of trees such as firs most of which were too large to be accommodated in the garden as well as grasses and other plants. It is also significant that the troublesome twentieth (Gynandria) and twenty-fourth (Cryptogamia) classes were positioned in their own small space between the herb-nursery in the sheltered garden periphery. This was because the former included may delicate exotic plants with what were regarded as anomalous flower structures such as orchids, whilst the latter which included living forms as varied as fungi, lichens, ferns, mosses, liverworts and algae, differed greatly from other Linnaean classes in having either hidden, imperfect or non-existing sexual organs. The uncertainty within Linnaean botany surrounding Cryptogam reproduction, whether the class was anomalous or even whether they were plants at all was reflected in their situation. Whilst the garden provided an ordered representation of the Linnaean system there remained considerable scope for variation with other botanical gardens.[22]

WILLIAM CURTIS AND WILLIAM SALISBURY'S LONDON BOTANICAL GARDENS

As we have seen, under Sloane and Miller, the Chelsea Physic Garden had an important impact upon eighteenth-century botanical culture, particularly in the metropolis. As semi-public institutions with subscribers, patrons and visitors, university botanic gardens also encouraged the development of public botanical culture. Other kinds of Georgian garden such as charitable, commercial and institutional nurseries attracted public attention, furthered Linnaean botany and encouraged and informed botanical and horticultural treatises. In Leicestershire for example, Rev. William Hanbury (1725–78) Rector of Church Langton, who had studied at Oxford and knew the Physic Garden there well, established ten large nurseries covering forty acres spread between four parishes and the neighbouring villages of Tur Langton and Gumley. The profits from these were intended to fund a charity for church improvement, a new school and other schemes (including a library and physic garden) with local gentry serving as trustees. Hanbury organised annual choral festivals to provide additional funding for the charity and nurseries. The latter raised 'many hundred thousand' of 'all curious foreign trees, shrubs, tender exotics, perennials, and annuals' renowned for their 'excellence, luxury and beauty'. Obtaining assistance from 'numerous' correspondents abroad, he tried to procure 'every sort of seed', and many fruit and forest trees and kitchen garden varieties were planted using the techniques of layering, grafting and budding. Hanbury argued that the 'noble science of botany and the delightful art of gardening' could be 'improved and heightened through knowledge of each other, particularly through the application of the Linnaean system'. He used his experience, 'practice and observation' at the nursery gardens to inform his *Complete Body of Planting and Gardening* (1770) which included a detailed exposition of the Linnaean system and lengthy analysis of the 'culture and management' of different sorts of plants and gardens. He intended that his work would encourage the improvement of estates for the glory of the nation, florists eager to win prizes and ladies and gentlemen interested in natural history, whilst revealing the divine wonders of nature.[23]

Some privately owned and managed subscription botanical gardens also facilitated the transformation of the 'science of botany' from being largely 'an appendage to physic' into being part of practical gardening education and a polite pleasurable activity for the wealthy and middling sort. Whilst considerable attention has been focussed upon Kew, it is important to remember that despite Sir Joseph Banks's belief in the pre-eminence of his gardens, other kinds of institutional, subscription and commercial botanical gardens played a greater role in facilitating Georgian plant study nationally than the exclusive royal domain.[24]

One of the most influential was the 'London Botanic Garden' opened by William Curtis at Lambeth in 1779. Metropolitan Enlightenment botanical culture and experience of working in established botanical gardens and creating others had a profound impact upon Curtis's career. He began plant collecting around London as an apothecary and continued as Demonstrator at Chelsea, expanding this work into a systematic attempt at a comprehensive London flora which culminated in the *Catalogue of Plants Growing Wild in the Environs of London* and ultimately the lavishly illustrated *Flora Londinensis* (1777) which was not a financial success. The establishment of Curtis's semi-public commercial London botanical garden was facilitated by the city's status as easily the largest port and the most important political, imperial, economic, industrial and cultural centre in Georgian England, with the largest concentrations of wealthy resident aristocracy, gentry, professionals and middling-sort residents. Although not as dominant as it was to become during the nineteenth century, the vast audiences for science, literature and the arts stimulated by the declining importance of court, print culture and presence of national and royal institutions such as the Botanic Gardens at Kew, Society of Arts and Manufactures and Royal Society, also underscores the city's cultural pre-eminence. Another major factor was the number of gardening, floricultural and botanical clubs and gardens from the Chelsea Physic Garden and royal parks to market gardens, commercial nurseries such as the well-established Brompton Park business and fashionable pleasure gardens, which encouraged a London audience for Curtis's ventures. Cultural differences, diversity and originality were also heightened in the myriad spaces of a city with approaching a million inhabitants and intersections of social ranks and orders in close proximity.[25]

Curtis came to botany through medicine having served an apprenticeship as an apothecary. He subsequently attended St Thomas's hospital and the anatomical lectures given there by Joseph Else, as well as the lectures of Dr George Fordyce, senior physician who encouraged his pupils 'into the fields and meadows near town' to instruct them 'in the principles of the science of botany'. On these occasions, Curtis assisted Fordyce in demonstrating the plants, often doing this unaided, which encouraged him to start giving public subscription courses of lectures and publish in entomology and botany during the 1770s. Curtis led pupils in plant-collecting walks around the London vicinity, after which the plants were demonstrated and discussed over dinner, which led to his appointment as demonstrator.[26] He had served an apprenticeship as an apothecary with his grandfather before becoming a lecturer and Demonstrator of Plants for the Society of Apothecaries at the Chelsea garden, and botanising gradually took over from his medical practice as his principal source of income.

In 1771 in conjunction with Thomas White brother of Rev. Gilbert White, of Selborne, Hampshire, Curtis occupied a very small garden near the Grange Road at the bottom of Bermondsey Street where he first conceived the design for publishing the *Flora Londinensis*.[27] The *Flora Londinensis* was supposed to include all hardy plants growing within a 10-mile radius of London and subsequently 'all the plants which grow wild in Great Britain', the potential being illustrated by a catalogue of plants 'growing wild' in the environs of Settle, Yorkshire undertaken by Curtis at the request of the wealthy metropolitan Quaker physician, author and philanthropist, John

Coakley Lettsom, who had his own domestic botanic garden. The letterpress provided Linnaean and English names, synonyms, detailed description in Latin and English followed by an account of the 'history, peculiarities, oeconomy and uses of the plant' in terms of medicine, agriculture, 'rural oeconomy' and other arts. However, the work was never completed. The cost of production and gradual issue of the work in parts distributed by the publishers and Curtis from his garden, ensured that, despite the degree of accuracy and elegance which helped to make it an 'object of universal admiration', 'the sale of the work never equalled its unrivalled merit'. However, according to 'Kewensis', the size, accuracy and 'masterly exemplification of dissection of flowers', of the *Flora Londinensis* did 'as much for the establishment of the Linnaean system as any work which ever appeared', the plates of mosses alone being so minutely accurate as to 'of themselves initiate any one into the knowledge of that branch of the Cryptogamia'.[28]

The Grange Road garden was soon too small for Curtis's 'extensive ideas' prompting the move to a larger piece of ground in Lambeth Marsh in 1777 which opened to the subscribing public in 1779, bringing together what purported to be the 'largest collection of British plants ever brought together into one place' (Figures 17 and 18). Having taken a partner in the apothecary business, Curtis subsequently 'declined the practice of physic altogether' in order to concentrate upon garden management and publications, the 'precarious profits' of which became his only income. He publicised the new venture calling for subscriptions from physicians, apothecaries, 'students in physic', the 'scientific farmer', botanists, the 'lover of flowers' and 'the public in general' in a pamphlet of 1778 where he described himself as an apothecary and author of the *Flora Londinensis*.[29] The garden was a success reaching what Curtis described as a 'degree of perfection with respect to the objects it embraced' and a 'somewhat more than adequate' subscription, and he hoped it would become a 'permanent' institution in the metropolis for the 'diffusion of useful knowledge'. However, he was compelled to abandon it for various reasons. Searching for a site with 'more favourable soil, and purer air', in 1789 Curtis chose a 'spacious territory' at Fulham Road near Queen's Elm, Brompton, in the parish of Kensington, and from here conducted his botanical operations until his death. The proximity of the Chelsea Physic Garden and commercial nurseries such as the large and successful Brompton Park business once conducted by George London and Henry Wise were an important attraction. Another reason for the selection of Brompton was the fact that most of Curtis's Lambeth patrons and subscribers resided to the west of the city, so would find it easier to visit the new establishment. These included aristocracy and gentry such as the Duke of Gordon, Earl of Carlisle and Earl of Bute, botanists and philosophers such as Joseph Banks, Thomas Martyn and Thomas Gisborne, medical men such as John Ford, William Heberden and his friend Lettsom. Much of the audience for Curtis's gardens was female illustrating the degree to which botany had become part of polite culture. Sixteen per cent of subscribers listed in 1790 were women acting in their own name, but many more would have gained admission as family members or friends. It was conventional for family subscriptions to be listed under male heads of household and other Georgian scientific societies excluded women from becoming members. Significantly, almost a third of the female subscribers were aristocrats, who had greater authority to act in their own name, including the Countess of Aylesford, Countess of Cremorne, Lady Hume and the Countess of Lonsdale.[30]

17 Detail from J. Sowerby, 'William Curtis's botanic garden
at Lambeth Marsh' (before 1787) from M. Galinou ed.,
London's Pride (Museum of London 1990)

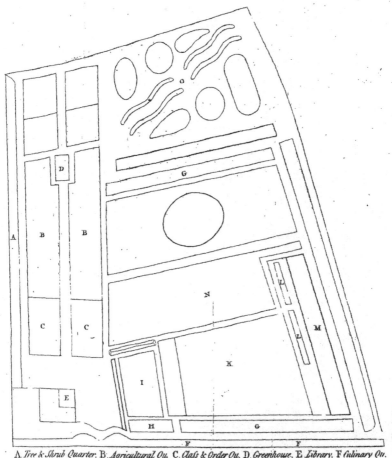

A *Tree & Shrub Quarter.* B. *Agricultural Qu.* C. *Class & Order Qu.* D. *Greenhouse.* E *Library.* F *Culinary Qu.*
G *Medicinal Qu.* H *Poisonous Qu.* I *Wood Qu.* K *Grass Qu.* L *Wall.* M *Water & Bog Qu.* N *General Qu.*
O *Ornamental Qu.*

18 Plan of the London Botanic Garden from W. Curtis,
A Catalogue of ... Plants Cultivated in the London Botanic Garden (1783)

Apart from the important addition of a nursery for the supply and
sale of plants to subscribers and customers, the 'Brompton Botanic
Garden' was conducted upon the same plan by Curtis and later William Salisbury his partner who created another during the early 1800s
opposite to Cadogan Place, Sloane Street in Chelsea, which was run
on a similar basis with planted divisions, a library and botanical lectures. The aim of the botanical gardens was to provide instruction
and 'rational amusement' in the 'beauties of nature', whilst furthering, through botanical study, improvements in food, medicine, agri-

145

culture. One guinea per annum subscription entitled subscribers with one guest to walk the garden, inspect plants and utilise the library, the hours of opening being between six and eight o'clock from April to September inclusive, and between eight and four for the Autumn and Winter. For two guineas, subscribers could bring an unlimited number of guests and receive roots and seeds 'that can be spared' to the amount of their subscription, so long as sufficient notice was given. A charge of two shillings and sixpence admitted other visitors to the garden raising useful additional income, gratifying casual curiosity and, as at Vauxhall and Ranelagh, was intended to 'prevent the intrusion of improper company'. Other regulations demonstrate the problems caused to Curtis by the novelty of public botanical gardens as an institution and differing expectations and types of behaviour within public and private gardens and suggest that he was not able to fully control behaviour within the institution. There was a conflict concerning behaviour norms associated with pleasure gardens and private gardens and those demanded by a botanical garden proprietor, particularly because of the care many of the scarcer more delicate plants required. The lending of library books as at Lambeth had to be discontinued because too many subscribers were walking off with them; as the Lambeth Marsh garden had 'suffered considerably from the depredations of the inconsiderate or designing' who had made 'bouquets of the flowers' and taken 'cuttings of valuable plants' and 'unique specimens', Curtis warned in the regulations at Brompton that practices 'so contrary to its established rules' and 'unfavourable' to its purpose ought to be curtailed. Similarly, he asked that dogs be not brought into the garden and that visitors did not tread on the borders or 'suffer those under their care to do so'.[31]

THE

BOTANICAL MAGAZINE;

O R,

Flower-Garden Diſplayed :

IN WHICH

The moſt Ornamental FOREIGN PLANTS, cultivated in
the Open Ground, the Green-Houſe, and the Stove,
will be accurately repreſented in their natural Colours.

TO WHICH WILL BE ADDED,

Their Names, Claſs, Order, Generic and Specific Charaɕters,
according to the celebrated LINNÆUS; their Places of
Growth, and Times of Flowering:

TOGETHER WITH

THE MOST APPROVED METHODS OF CULTURE.

A   W O R K

Intended for the Uſe of

Such LADIES, GENTLEMEN, and GARDENERS, as wiſh to become
ſcientifically acquainted with the Plants they cultivate.

L O N D O N:

Printed for W. CURTIS, at his BOTANIC-GARDEN,
Lambeth-Marſh; and Sold by all Bookſellers, Stationers, and News-
Carriers, in Town and Country.

M DCC LXXXVII.

19 Title page of the first edition of the *Botanical Magazine* (1787), courtesy
of the Department of Special Collections, Memorial Library,
University of Wisconsin-Madison, USA

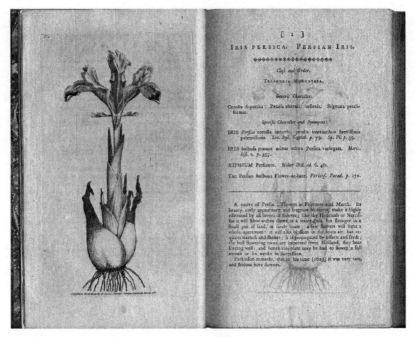

20 Persian Iris from the *Botanical Magazine*, 1 (1787), courtesy of
the Department of Special Collections, Memorial Library,
University of Wisconsin-Madison, USA

Courses of botanical lectures were offered at the Lambeth and
Brompton gardens and plant hunting trips were undertaken, initially
with London medical students on the model of the Society of Apothecaries. Curtis's experience as apothecary, botanist, teacher and gardener encouraged him to compose a series of botanical and gardening works, easily the most successful and enduring of which was the
Botanical Magazine which appeared in 1787 and continued after his
death doing much, as Loudon remarked, to 'diffuse a general taste for
botany' (Figures 19 and 20). Curtis strove to capitalise upon the success of the gardens, lectures and *Botanical Magazine* by publishing a
companion volume offering an elementary introduction to botany
and 'the substance of a course of lectures, chiefly explanatory of the
Linnaean system' read at the Lambeth Marsh garden. His son-in-law
Samuel Curtis published the lectures posthumously from notes in
1805 with illustrations from the *Botanical Magazine*. After an introduction to practical botany with a short bibliography, a lecture on
animal and vegetable bodies (probably added in the 1790s) and a

series of lectures on plant parts and organs, Curtis proceeded to a detailed exposition of Linnaean taxonomy illustrated with plants from his botanical gardens. Although the *Botanical Magazine* was intended as a more popular publication than the *Flora Londinensis*, of which no more than around three hundred copies were published, the former had 'such a captivating appearance' and was 'executed with so much taste and accuracy', that it became a major success. From its 'unvaried continuance in excellence and popularity' and the 'original and excellent observations' interspersed within, the *Botanical Magazine* 'continued to be a mine of wealth' to Curtis until his death and ensured his botanical fame. Through the success of the gardens and his publications, there was 'not a naturalist of any eminence who did not court his acquaintance' including Banks, Jonas Dryander, John Sims, Thomas Frankland, Withering, Lettsom, Martyn and Darwin.[32]

A striking water-colour painting entitled 'William Curtis's botanic garden at Lambeth Marsh' by James Sowerby, which dates from before 1787, shows how the garden appeared when the *Botanical Magazine* first appeared (Figure 18). As we can see from Sowerby's painting and the *Catalogue of ... Plants Cultivated in the London Botanical Garden* (1784), the garden was divided into a series of distinct geometric sections surrounded by trees and shrubs with a circular pound and narrow ornamental oblong canal or pond in which were growing aqueous plants. Open to the sunshine and elements, many of the trees and shrubs look like they have been planted quite recently. Curtis's collection eventually accommodated thousands of species and his garden was divided into physic, poisonous, culinary, agricultural, British ornamental plants and shrubs surrounded by gravel paths with separate beds laid out according to the Linnaean system. The plants were arranged in their 'various stages of growth' with proper names in Latin and English painted upon labels. Lectures were conducted in the garden, presumably outside or in Curtis's house in inclement weather, whilst those who paid a higher subscription were enabled to remove plants and seeds for their own gardens. A library was provided at Lambeth containing the 'more necessary books for students in botany, the most celebrated treatises on the *materia medica*' and various agricultural tomes for perusal and study. The list of books shortly after their removal to Brompton underscores the core purposes of the gardens and demands of subscribers and included studies of taxonomy and practical botany by Linnaeus, Adanson, Martyn, Withering and others, natural historical works, dissertations on medical botany, accounts of voyages and international exploration, national and regional botanical surveys such as Charles Deering's *Catalogue of Plants about Nottingham* (1738) and John Lightfoot's *Flora Scotica* (1777) and White's *Natural History And Antiquities of Selborne* (1789), journals from relevant academies and scientific associations and books on gardening, horticulture, agriculture and arboriculture. The collection confirms the importance of national and international botanical networks and the relationship between Curtis's London gardens and other physic and botanical gardens, quite a few works having been presented by authors. Books inspired by these included William Aiton's *Hortus Kewensis* (1789), the Lichfield Botanical Society's *System of Vegetables* (1782) and *Families of Plants* (1787), and poetical celebrations such as William Mason's *The English Garden* (1783) and Erasmus Darwin's *Loves of the Plants* (1789).[33]

Lambeth Marsh was ideal for a garden in some ways being close to the river and the city with its institutions and potentially large audience for botany, but not enveloped by it, whilst the ready supply of water and boggy ground was congenial for the growth of some plants such as grasses, although inundations proved a problem for others. Like the Chelsea Physic Garden, Vauxhall, Ranelagh and some other metropolitan pleasure gardens, Curtis took advantage of the river to encourage visitors by boat and supply plants and other materials. As the trees, shrubs and hedges grew and became 'ornamental in themselves', shelter was being afforded for more delicate kinds. However, the factors that helped to ensure initial success later proved to be detrimental in other respects. The proximity of the city meant that as it expanded during the 1780s, so like various other pleasure gardens, the Lambeth garden became surrounded by buildings, whilst the 'obscurity of the situation', 'badness of roads' leading to it, increasing city smoke and the 'highly offensive' effluvia rising from the marsh helped to induce the move to Brompton. Unless the wind blew from the south, the smoke from London chimneys and furnaces became so bad that it 'enveloped' all of Curtis's plants, destroying many and injuring others, especially Alpine kinds.[34]

To survive as a successful commercial and scientific operation, Curtis's and Salisbury's London botanical gardens combined some of the features and management of urban subscription walks and pleasure grounds with those of physic gardens, Linnaean botanical gardens and commercial nurseries. Salisbury's six-acre Sloane-Street establishment, for instance, laid out from 1807, aimed to attract the patronage of British and foreign aristocracy, gentry and middling sort as well as gardeners and botanists and was promoted like the pleasure gardens as a 'delightful promenade for company' with two miles of varied paths, 'independent of the scientific arrangements of the different departments'. However, the provision of botanical lectures was continued as well as the seventeen sections including ones for trees and shrubs, exotic trees and shrubs, a department 'to illustrate the Linnaean system' and another to exhibit 'the natural orders of other botanists'. The symbiosis between plant collections, gardens with their botany libraries and the *Botanical Magazine* is evident from Curtis's determination, as he emphasised in the first volume, to provide a 'display of the flower garden of ornamental foreign plants cultivated in the open ground, the greenhouse and the stove'. As the alternative title of 'Flower Garden Displayed' suggests, the *Botanical Magazine* provided an imaginary walk through a botanic garden and Curtis emphasised the fact that the hand-coloured engravings representing 'natural colours' were based upon living specimens, especially those in his botanic garden. The text accompanying the 'ornamental foreign plants' included served, like the labels in the ground in his botanical garden, to delineate the name, class, order, generic and specific character according to the Linnaean system so that 'ladies, gentlemen and gardeners' would become 'scientifically acquainted' with the plants they cultivated. Also included were the 'places of growth' and whether plants were appropriate for open ground, green houses and stoves and with times of flowering, 'the most approved methods of culture'. The *Magazine*'s most striking feature was the high quality and extremely detailed hand-coloured plates, with a text that provided Linnaean name, pre-Linnaean synonyms, provenance, periods of flowering and cultivation instructions. Despite – and perhaps because of – its expensive cost, the *Botanical Magazine* succeeded because it served as a virtual tour through Curtis's gardens, with letterpress redolent of the lectures he gave surrounded by his precious plants.[35]

21 View of Liverpool in 1650 from S. Smiles,
Lives of the Engineers: Boulton and Watt (1904)

22 Duke's Dock, Liverpool from S. Smiles,
Lives of the Engineers: Boulton and Watt (1904)

THE LIVERPOOL BOTANICAL GARDEN

Established in 1800 and opened in 1802, the Liverpool Botanical garden served as the most important early nineteenth-century model for kindred institutions, and the wealthy and cultured banker, scholar and botanist William Roscoe's opening address provided an influential manifesto for the public Enlightenment botanic garden. Like the Oxford, Cambridge, Chelsea and London botanical gardens, it demonstrated the importance of patronage from corporations, merchants and local citizenry in ensuring the success of these ventures, which helped to determine design, usage and public character. Scientific culture and education were greatly encouraged in ports such as Hull, Bristol and Liverpool by the practical requirements of navigation and business, the opportunities presented by international trade and the partly contradictory need to project politeness and refined culture (Figures 21 and 22). As we have noted, one manifestation was the range of literary and scientific associations which included a debating society, a Mathematical Society, the Academy of Arts (1767), the Liverpool Philosophical Society (1779) and another philosophical society in 1784. Such institutions also provided a means for the middling sort and elite to present a united public culture in the face of religious and political differences including disagreements concerning slavery, a major contributor to merchant wealth in the town. Science and international connections were therefore unusually prominent in local urban culture, helping to explain why a subscription botanical garden was established in Liverpool before most other English towns.[36]

About ten acres of land were purchased for the Liverpool scheme, over half being utilised for the garden and the remainder being sold, recouping most of the original purchase price due to a rapid increase in land prices (Figures 23 and 24). With Roscoe's encouragement and to aid the general improvement of the town, the corporation facilitated the venture by making a free grant to the proprietors of the reversionary interest of the garden and buildings, allowing the land to be held by a renewable lease on condition that it was always used for the original purpose. Like his friend James Edward Smith, Roscoe articulated the intentions and hopes of the promoters, emphasising the importance of the gardens for medicine, practical botany, natural

153

theology, polite culture and commercial economy. The latter was, of course, regarded as very important in a town of international trade and commerce, and Roscoe was careful to reassure those unused to wandering too far from warehouses and balance sheets that if properly managed the gardens would become a self-funding business. The 'prevailing taste for botanical studies' and 'liberality displayed by Liverpool inhabitants in the 'encouragement of scientific pursuits' revealed the gardens to be 'a desirable and attainable object'. Roscoe noted that as many of the 'great repositories' of plants were in nurseries in the London area, provincial botanical gardens provided an opportunity for the benefits to be conferred upon inhabitants of other great towns. He associated the Botanic Garden with local pleasure gardens and public walks, suggesting that as so much 'expense and attention' had been 'bestowed by many respectable individuals' in supporting these, so a garden 'properly laid out, and supplied with every beautiful production of vegetable nature' yet 'enjoyed at the small expense of an annual subscription' would prove equally attractive. There was little doubt of the advantages to be derived from botany for a commercial town such as Liverpool and for Britain in general which included applications to 'agriculture, gardening, medicine, and other arts, essential to the comfort and even support of life'. Botany provided a form of elegant amusement, but also an activity that developed the understanding and promoted 'the health and vigour of the bodily frame' whilst plants and seeds could be sold and distributed around the population and professionals such as nurserymen would be able to observe them before sale or purchase.[37]

154

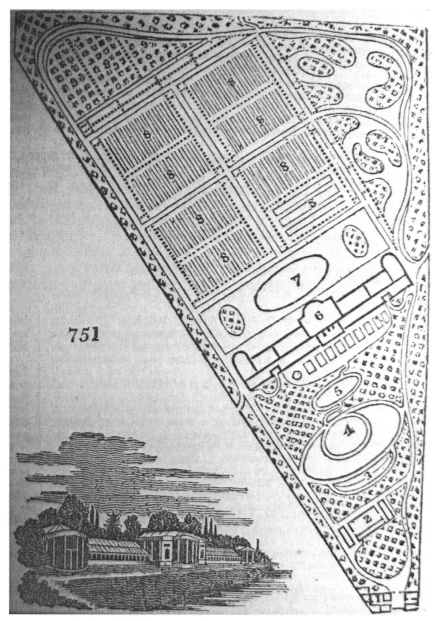

23 Plan of the Liverpool Botanic Garden from J. C. Louden,
Encyclopaedia of Gardening, second edition (1824)

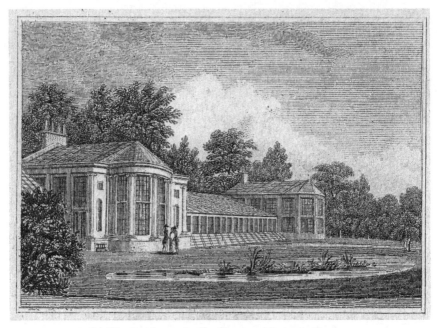

24 The hot house in the Liverpool Botanic Garden
from *Marshall's Select Views in Great Britain* (1824)

Whilst Roscoe was a devotee of taxonomy it was the qualities of trees and plants as living entities that appealed to him and he suggested that even the cultivation of the fine arts ought to 'yield the superiority to the study of nature' which demonstrated how artists derived their concepts of beauty and proportion. Only careful study of vegetable anatomy and physiology enabled appreciation of the 'wonderful conformation' of parts combining the 'united purposes of elegance and utility' in nature. Like his friend Smith, Roscoe regarded the taxonomy of 'that great father of the science, the immortal Linnaeus', as preceding 'every other kind' of plant knowledge and it was only in a botanical garden that all dimensions of botany could be studied.[38] As the Chelsea apothecaries had recognised, whilst dried herbariums and 'the most beautiful and accurate' drawings were valuable, these fell 'infinitely short' of the 'delicacy and minuteness of parts' on which 'scientific distinctions' depended, which could only be studied using living plants. Private collectors tended to choose their own 'favourite class of productions' whilst motivated by profit, commercial nurseries selected plants that best repaid 'the attention of the cultivator', thus the achievements of many 'long and dangerous voyages' would be wasted as collections died or were dispersed. Only 'public institutions' offered collections comprehending 'every known vegetable production' and provided the 'copiousness or permanency' to preserve them over the years which was 'indispensably necessary to their perfection'. Hence, a department within the Liverpool garden was devoted to 'trees, shrubs, and hardy herbaceous plants of every description' forming a 'general collection' of those that could be 'obtained from every part of the world' and were sufficiently hardy to bear the English winter.[39]

The plan of the Liverpool garden demonstrates how Roscoe and the committee strove to satisfy the demands of public walk and practical botany (Figure 20). The five-acre triangular-shaped site was enclosed within a substantial stone wall, entry being obtained through two lodges which were constructed to house the residence of the curator, committee meetings and an intended herbarium and library. By 1808 some 3,000 dried plants had been obtained from merchants, nurserymen, collectors and other gardens including many collected by the German naturalist Johann Reinhold Forster (who had once taught at the Warrington Academy) during his South Sea voyages with Cook. Seeds and specimens were sent by – and exchanged with – other botanical gardens including those of Kew, Oxford, Cambridge and Dublin. As Richards the curator emphasised, the gardens were able to take advantage of the fact that they were situated within a short distance from a 'large sea port ... from which communication is direct and frequent to the most remote parts of the earth' and Liverpool merchants and sea captains were thanked by the proprietors 'for their numerous and valuable presents of plants and seeds to the Botanic Garden'. A walk surrounded the garden which led successively to the stove and places for rock and bog plants, a greenhouse ground (5), conservatory (6), aquarium (7), zone for herbaceous plants (8) and gramineum (9). The large conservatory, depicted on the frontispiece of the 1808 catalogue, was described as an 'elegant and spacious range of buildings, 240 feet long, and in the centre 24 feet high'. It was divided into five compartments which were heated to different temperatures so that plants from different climates from 'every part of the world' could be accommodated. There was also a small conservatory with bark pits and an aquarium intended to accommodate a single specimen of 'every rare and tender aquatic' and a 'large compartment with small frames and wooden covers' for the preservation of tender herbaceous plants requiring winter shelter.[40]

The systematic manner in which the Liverpool gardens were laid out was praised by the botanist Peter William Watson who noted that plants were positioned in 'geometric and scientific order, and in line, space, group ... with appropriate neat labels to each, and with a zeal to fill up the blanks', which, as we shall see, inspired the design of the Hull Botanical Gardens.[41] Having met Roscoe, Benjamin Silliman, an American merchant, was taken on a tour of the Botanical Garden in 1806 and he considered that under patronage of Roscoe and Currie, it had 'flourished rapidly' and become 'a fine establishment'. He noted that the ground was 'well adapted to the purpose' whilst the garden included 'a pond and a portion of marshy land in the middle' for 'such plants as require a wet soil or constant immersion'. The hot houses were 'extensive and handsome' exhibiting 'a great variety of exotics' and entire garden was 'a place of great beauty'.[42] The gardens with their hot houses also had an important impact upon the botanical work of Roscoe, informing his defence of Lin-naean botany and enhancing his reputation as a botanist. Defending the Linnaean system against advocates of Jussieuean taxonomy, particularly the French, Roscoe argued that a genuine natural system would only be obtained through 'a long and intimate acquaintance' with the changing physical characteristics and products of plants growing in gardens and their native places.[43] In his work on the Mon-andrian order of tropical plants, which included 112 hand-coloured lithographed plates, Roscoe employed his experience of botany and the Liverpool garden to attempt a more 'natural' system of classifi-cation upon a group that had hitherto been held as an artificial Lin-naean class defined by the flower's solitary anther alone. Within the work, Roscoe offered a virtual tour through the order providing binomial names and an abbreviated technical description followed by 1 or 2 pages of text for each specimen and concluding with observa-tions concerning the source of the plant illustrated, its origins, previ-ous descriptions and references to the numbered dissections on each plate.[44]

HULL BOTANICAL GARDENS

The foundation of the Hull Botanic Gardens in 1812 was encouraged by the Liverpool venture and Roscoe's address, by significant growth in maritime trade and the development of public urban culture signified by the opening of a subscription library (1775), zoological gardens and the Hull Literary and Philosophical Society. The example of Kew and other university and British and Irish subscription botanical gardens also provided much inspiration and fostered inter-urban rivalry. As at Liverpool, the Hull Botanic Gardens provided a forum for polite sociability for the merchant, gentry and middling-sort elite and was intended to offer economic, medical and botanical benefits. The impact of the Liverpool institution is evident from the role of curator John Shepherd in designing the Hull gardens whilst the Scotsman William Donn (1787–1827) nephew of James Donn curator at the Cambridge Garden was appointed as first Curator. Donn trained at Dr Isaac Swainson's Heath Lane Lodge botanical gardens at Twickenham near London which were designed by Kensington nurseryman Daniel Grimwood during the 1790s and subsequently managed by Arthur Biggs (later curator of the Cambridge Botanic Garden), before being employed under Thomas Hoy, head gardener for the Duke of Northumberland at Syon House. Donn's botanical and gardening knowledge and experience, which was praised by Loudon, and the assistance provided by Banks and the Kew gardens clearly illustrate the importance of private, commercial and public botanical and nursery garden networks.

Promoters of the Hull gardens emphasised the need for new town walks and reassured potential subscribers that such were the 'attractions' of the Liverpool Botanic Garden 'as a place of fashionable resort' that previous to recent mercantile difficulties caused by the French conflict, shares had been selling at a hundred per cent premium. William Spence contended that 'one of the chief desiderata of Hull' was 'an umbrageous walk' and whilst the Humber banks, the piers and Spring Bank were all 'favourable', unlike the Botanical Garden, because of the lack of trees and shrubs, they did not offer sufficient shelter from the summer sun or elements. [45] In addition to providing the usual medical benefits, the study of local and imported trees and shrubs was thought to foster the growth of new varieties of plants and fruit trees for agriculture and horticulture. It was also hoped that the garden would be a 'repository' for the importation, study and distribution of exotics by town merchants such as tea trees, coffee plants and olive trees whilst neighbourhood gentry would be able to examine the 'effects of different shrubs and plants' before making selections for their plantations or greenhouses and gardeners would be able to obtain 'correct names for their vegetable treasures', so often 'miscalled' by those who supplied them.[46] Local botanists could study the systematic collections and understand the different forms of arrangement whilst 'wives and daughters having now the means' of following a study which had become 'one of the most fashionable', would be led to cultivate it.[47] Beyond the needs of economic, medical and botanical utility and refined sociability, the gardens would also help to celebrate the successes of British science at a time of war, whilst in terms of natural theology, systematic displays celebrated the handiwork of the 'supreme and intelligent author' of creation.[48]

Watson considered support for the gardens to be 'respectable' for a 'trading town of 30,000 inhabitants', although he regretted that numbers remained inadequate for the 'expense of an extensive establishment of its nature'. By 1824 there were 498 shareholders (213 at fives guineas and 285 at six pounds per share), whilst the number of subscribers at one guinea and a half per annum was 277, with 7 at 2 guineas.[49] Leading promoters of the institution included botanists, medical men, clergy, a few aristocrats and gentry, and one or two merchant and naval officers. Author of various medical and horticultural works, John Alderson (c.1757–1829), one of the main supporters, was physician at the Hull General Infirmary and served as president of the Hull Subscription Library from 1801 and the literary and philosophical society in 1822.[50] The botanists included William Spence (c.1782–1860), Peter William Watson (1761–1830) and Adrian Hardy Haworth (1768–1833). Watson made a living from trade before pursuing natural history and chemistry and becoming a skilful landscape painter. Spence had various commercial and improving interests but pursued horticulture, agriculture and botany, serving as president of the Holderness Agricultural Society, becoming a member of the Linnaean Society in 1806, a Fellow of the Royal Society and a prominent supporter of the London Entomological Society. Spence also published many other articles on natural history and with Rev. William Kirby the four-volume *Introduction to Entomology* (1815–1826).[51] Like Spence, Haworth and Watson moved in national, metropolitan and local botanical circles and were able to take advantage of these to develop the Hull gardens. Haworth was closely connected to the town's trading community, his father Benjamin having been a successful merchant and landowner who resided at Hullbank or Haworth Hall nearby.[52] Haworth and his family moved to Little Chelsea in London during the 1790s, a district still on the periphery of the spreading city where, as we have seen, various market garden, nursery and florist businesses were located. Like his friend Spence, Haworth helped to found the Aurelian and Entomological societies and was elected a Fellow of the Linnean Society in 1798, making full use of the herbarium for his work on succulents in addition to many London gardens and nursery collections including Kew.[53] Haworth took a close interest in regional herborising, forming a private botanical garden, library, natural history museum and herbarium which informed his *Lepidoptera Britannica* (1803–28), which claimed to be the first authoritative national study of English butterflies and moths.[54]

The Hull Botanic Garden was situated about a mile from the town centre on the Anlaby Road at the bottom of a road later called Linnaeus Street. As laid out by Shepherd and the committee, the principal walks surrounding and intersecting the garden were 8 or 9 feet broad and formed a total length of almost three quarters of a mile. Compartments were set aside for bog, alpine and greenhouse plants, there was a thirty-yard long pond for the growth of aquatics and a twelve-foot high mound on the south-west corner which admitted though surrounded by trees, 'an extensive view of the Humber, the Lincolnshire coast and the Wolds'. At the entrance were two lodges one of which was the curator's residence and the other used as a botanic library for subscribers and committee members. Sufficient room was left for 'one specimen at least of every tree, shrub, and hardy plant in the kingdom' as well as for a large number of exotics which were very important given that as Loudon emphasised the country round was 'almost without trees'. According to Spence the two-acre belt of trees provided both picturesque beauty and the shelter necessary for promenading subscribers and their families. Soon after the garden was created the extensive nursery stock of the late

John Philipson of Cottingham was purchased, and other plants were procured by Donn from London nurseries, the Cambridge Botanic Garden through his uncle and other individuals whilst Hull merchants instructed their agents abroad to collect and return seeds. The importance of local flora is evident from the fact that Watson provided 'many indigenous plants' collected 'at considerable expense and labour', traversing 'the whole East Riding of Yorkshire' in his gig 'with proper apparatus for cutting up roots, collecting seeds'.[55]

During the first three or four years, according to Watson, the number of hardy plants 'increased rapidly' and the garden 'bid fair to possess those scientific features which constitute the real intrinsic value of such an establishment', whilst almost £5,000 was expended on plants and other features by 1816. At the instigation of Watson, Hardy and other promoters assisted by Shepherd the Liverpool curator, the perennial section was not laid out according to the Linnaean system but according to a more natural arrangement. Initially this followed the 'valuable and elegant synopsis' of the Dutch/South-African botanist Christiaan Hendrik Persoon (1761–1836), whose *Synopsis Plantarum* detailed over 20,000 plant species, but whose main efforts were devoted to systematic mycology. Each plant was placed within its own generous space of 3 per row separated by half a foot 'to leave room for all expected acquisitions' with generic label sticks for each space paged 'to facilitate immediate reference' to Persoon's *Synopsis Plantarum*. The rather regimented beds were two and a half feet wide with smaller intermediate beds of one and a half feet between them and contained 22 plants each with number stakes to be referred to on a plan of the garden that was hung up in the committee-room. Subsequently the committee decided to remove the Persoon arrangement and replace this with the 'certainly more enlightened [generic] system of Jussieu', work that was planned and undertaken by Watson. This was subsequently changed into an arrangement based upon the curator, William Donn's catalogue although regarded as scientifically inferior to the Persoon and Jussieuean systems by Watson.[56]

These changes help to explain the continuing allure of the Linnaean system and the problems experienced at botanical gardens realising planted arrangements of competing artificial and natural systems in the face of numerous exotic imports. In 1767 Thomas

Martyn complained to Richard Pulteney that botanical 'system mad-
ness' was 'epidemical', the sixty new systems detailed in [Michel]
Adanson's *Familles des Plantes* (1763) not being 'worth one farthing',
but such pronouncements were of little use to curators and gardeners
having to satisfy the needs of subscribers and patrons for clearly
ordered – yet aesthetically pleasing – collections. As a result, although
in Watson's view the committee had always been 'well intentioned',
they had insufficient expertise and knowledge to maintain informed
taxonomic displays whereas the Hull gardens required 'ardent, constant, and scientific attentions' to 'render them of importance to the
sciences', which could only be provided by botanists. More was
needed than just maintaining the garden as 'neat, clean, and pretty'
which though it had its place was 'mere mechanic work' and Watson
noted that the venture had inspired no new 'botanic genius' or
explorer such as an Alexander von Humboldt, Carl Peter Thunberg or
James Cook.[57]

CONCLUSION

The Liverpool and Hull gardens inspired the formation of botanical
gardens elsewhere. At Whitby, Yorkshire a botanical garden was
founded in 1812 in association with the Willison nursery company.
Inspired primarily by the Hull Botanic Garden with whom they exchanged plants, the committee was able to publish a catalogue of
plants with rules and regulations of the institution in 1814.[58] Despite
the fact that the Whitby garden experienced difficulties, semi-public
botanical gardens subsequently appeared in numerous British towns
including Sheffield and Birmingham. British gardens were also able to
take their position in a network of colonial botanical gardens which
came to centre upon Kew through the actions of Sir Joseph Banks. As
Aylmer Bourke Lambert put it, the greater 'intercourse with foreign
parts, and consequently the 'facilities which scientific travellers find
in visiting ... remote regions', had 'tended infinitely to enlarge ...
every branch of natural history', providing a 'vast increase of new
species' and a 'rich fund of interesting facts and observations'. This

required considerable co-operation and facilitated the international exchange of specimens and information.[59]

The experience of designing, managing and using these gardens impacted upon the theoretical spaces of botany, notably through taxonomic reappraisals. As we have seen, Enlightenment botanical gardens at Lichfield, Cambridge, Liverpool and Hull inspired various botanical publications which, in turn, impacted upon the form and usage of the gardens. Darwin's Lichfield garden helped to prompt the Lichfield Botanical Society's translation of the works of Linnaeus into English and both of these subsequently inspired the composition of the *Botanic Garden* (1791). More usually however, the first publications inspired by institutional gardens were prospectuses and calls for subscriptions and patronage presenting idealised views seldom fully realised in practice. Typically as at London, Liverpool, Cambridge and Hull, catalogues of plants were published such as Donn's *Hortus Cantabrigiensis* which promoted idealised views of the gardens. Donn's catalogue came to be primarily used by those who never visited the gardens and if they did, would only interpret and experience it through a more perfect, virtual idealised form. Similarly, the lectures associated with Curtis's London botanical gardens inspired the *Flora Londinensis* and the tremendously successful *Botanical Magazine*.

The popularity of botanical gardens in Liverpool, London and elsewhere prompted Loudon to call for a new national botanical garden at the Linnean Society in 1811. He contended that the English gardens were inferior to what might be expected and the 'wealth and spirit of the nation', suggesting that 'even ... some of the provincial gardens' such as the collection of plants at Cambridge and the 'plan and general arrangement of Liverpool Garden' were superior to Kew. In the paper and *Hints on the Formation of Gardens* (1812), dedicated to the Prince Regent, Loudon argued that such a garden would be a 'living museum' with representatives of every species and variety of the vegetable kingdom and feature examples of classical, old and modern English and various other historical and national gardening styles, with a school for gardeners, library and lecture hall and grand hothouse. The hothouses and greenhouses might form two concentric rings connected by radial wings, the outer ring of which would be large enough for a carriage drive (Figure 25). Loudon's plan has been described as an 'extravagant vision' but it is clear that he was inspired by university and subscription botanical gardens, commercial pleasure gardens and the achievements of royal and aristocratic gardens at Kew, Woburn, Whiteknights and other places to project a new national botanical institution. His suggestions were intended to generate interest from wealthy benefactors in the venture and were certainly no more extravagant than many Georgian pleasure gardens and quite in accord with Curtis and Salisbury's ventures. Shortly before, William Fordyce the surveyor of woods and forests had urged that Regent's Park being developed by John Nash, would be ideal for new public gardens and in 1812, supported by Nash and responding to demands from Loudon and others, Salisbury proposed a new national botanical garden to incorporate the Kew collections. These calls were encouraged by the temporary Whig optimism following the establishment of the Regency after George III's final mental deterioration from 1810, which prompted renewed calls for peace, political reform and a government of national unity. A dinner was held to launch the venture chaired by the Duke of Sussex and a prospectus was printed seeking subscriptions for the six and a half acre garden with library, hothouses but the venture failed to progress partly through opposition from Banks who complained that botany was best improved away 'from the crowded populations of the metropolis' to prevent interest from those motivated by mere 'idle curiosity'. Yet as Thomas Martyn, Curtis, Salisbury and Loudon recognised, public curiosity could be harnessed to further botanical science by creating wonderful new institutions for promoting the subject to wider audiences.[60]

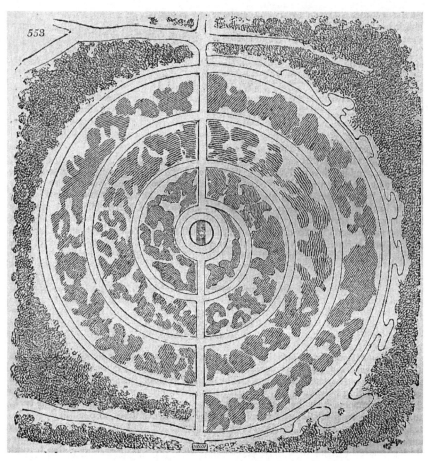

25 Design for spiral botanical garden from J. C. Loudon
Encyclopaedia of Gardening, second edition (1824)

6 'Enlightened and Animated Circles': County Towns and Counties

INTRODUCTION

This chapter explores scientific culture in English counties and county towns, which it contends was an important manifestation of the Georgian urban renaissance as well as the British Enlightenment. County designations and identities helped to inspire – and were simultaneously forged through – the provincial English sciences, as the surveying of new county maps and publication of county floras demonstrate so well. Case studies of Leicester, York, Norwich and Derby underscore the importance of counties in the production and consumption of natural history and natural philosophy, and the role of their capitals in promoting these endeavours.[1] County town cultures are compared with those of other types of urban centre and it is argued that varied socio-economic and geospatial features helped to determine the promotion, audiences and consumption of philosophical culture, although typological characters cannot be defined too rigidly. As Livingstone and Withers have argued, the study of scientific culture allows us to reassess the relationships between metropolis and province, within and between regions and between county towns and their rural hinterlands. This is apparent when we consider the role of national networks in relation to the production, consumption and circulation of scientific ideas within and between English county towns.[2]

The juridico-political and ecclesiastical status of county towns allowed them to continue to 'punch above their weight' and promote distinctive public cultures, which remained the case well into the nineteenth century despite the accelerated eclipse of many county towns in demographic and economic terms. The relative decline of county towns compared to manufacturing and industrial centres is apparent in population terms, with 11 out of the largest 25 English provincial towns in 1700 being county towns, a figure that had fallen to 8 by the time of the first census of 1801 and 6 by 1841.[3] One of the most important reasons for the cultural status of county towns was, of course, the residence and patronage of aristocracy and gentry. However, it was the related role of professionals such as medical men, lawyers and clergy as the lynchpins of provincial natural history, antiquarianism and natural philosophy that became more decisive. Only the gathering pace of professionalisation, metropolitan institutionalisation, specialisation and other developments within the sciences and society during the nineteenth century threatened this county-focussed and largely professionally inspired county intellectual culture.

Given their importance as audience and practitioners in the creation of public science from the seventeenth century, professionals were well versed in the procedures and practices of natural history and natural philosophy most of which were modelled on those of medicine, law and the church. Hence their historic importance, status and administrative functions and their concentration within county towns meant that they often played a crucial role in provincial scientific culture. Although sometimes strained by professional disagreements and religious and political differences, the education and status of urban professionals gave them the authority to make pronouncements on scientific matters which were respected locally. They had the education and authority to enter into minute disputes on scholarly matters and authenticate, verify facts and controversial or contingent judgements, for example on religious, antiquarian or scientific controversies. Their procedures and methods of enquiry, verification and disputation helped to shape and define the sciences.[4]

A number of factors have been proffered to explain the diversity of Georgian urban cultural expression and the reception of scientific ideas, many of which emphasise socio-economic differences. Thackray for instance, with the case of Manchester in mind, suggested a number of determinants. These include varied population growth, the economic success of merchants and manufacturers in some centres, the relative isolation of provincial towns prior to the coming of the railways, and the possible congruence of scientific culture with certain religious and political values such as Dissent and radicalism.[5] Other explanations for this geographical differentiation have been suggested by Emerson and Inkster. Using the experience of Scottish towns, Emerson tried to ground Enlightenment culture in its socio-urban roots by defining four categories of town according to their levels of institutional complexity and arguing that characteristics of this culture were partly determined by these individuating features. Inkster has emphasised the centrality of factors such as the practical utility of science which was important in industrial centres, the rivalry between different social groups – not necessarily class groups – and the varied socio-economic structure of provincial centres.[6] A relationship between population growth and the vibrancy of scientific associational culture has been posited, although how this is defined is problematic. The difficulty is that such variables cannot be considered in isolation, as they may have acted in combination or against each other. Apportioning weight to each in a functional causal manner is difficult, whilst 'culture' cannot simply be regarded as an epiphenominal result of 'economic' factors, being part of an intertwined nexus of reflexive phenomena. What causal weight, for instance, should be assigned to the historical status of county towns in determining their status by the early nineteenth century? A further problem is that, partly as a reaction against earlier work that posited a strong association between British scientific culture and industrialisation, insufficient weight has been given to the varied intellectual character and activities of provincial philosophy. Analyses of the intellectual content, character and practices of provincial scientific cultures enhance understanding of their varied geographical character and the differential reception of scientific ideas and practices. They also show how provincial identities were negotiated with – and sometimes conflicted with – national and metropolitan values and identities.[7]

Networks of early-modern natural history, natural philosophy and antiquarianism, commercialism, the public sphere and print culture afforded opportunities for the middling sort and gentry to transcend localism and provinciality and adopt aspects of metropolitan culture. As we shall see, knowledge transmission and cultural emulation was not unidirectional between London and provinces, but contingent, negotiated and contested reflecting the reality of provincial and metropolitan identities as fluid, complex and interlocking rather than fixed and determined. Industrial, philosophical and cultural innovation, of course, flowed from the provinces and county-focussed intellectual endeavours helped to stimulate metropolitan science in countless ways from immigration to importation, audience creation and knowledge consumption. Varied cultural forces transcended elite groups, individuals often having multiple allegiances and fractured identities, particularly with respect to politics, religion, social rank and status. Natural history, for instance, when subsumed in the county chorographical tradition, provided producers and consumers of knowledge with opportunities to assert and promote different provincial identities, for instance through the production, consumption and usage of maps and guides, or membership of local gentlemanly, philosophical or agricultural societies. The international character of scholarship facilitated by religion, neo-classicism, travel, communication, learned societies and journals, allowed *savants* and audiences to counter the metropolitan pull by embracing the international republic of letters.[8] Naturalists sent specimens from the North American colonies and English provincials provided materials for London collectors, institutions and audiences. International scientific cooperation was able to continue, to some degree, throughout the eighteenth and early nineteenth centuries, despite incessant rivalries between European powers, especially the French and British empires. As we shall see, continental philosophers and travellers were welcomed to the English counties by local philosophers who guided them around natural historical and industrial sites of interest such as those in Derbyshire and Yorkshire, whilst botanical and mineralogical specimens and publications circulated internationally.

COUNTY TOWNS, COUNTIES AND SCIENTIFIC CULTURE

In a seminal paper, Everitt elucidated some of the differentiating features and principal functions of county towns as regional foci for communities uniting country and town, and these help to explain the characteristics of their cultural life.[9] He pointed to the occupational diversity in county towns, their role as 'inland *entrepots*', their position as professional and entrepreneurial centres, their status as residential bases for the new urban gentry, their not-unrelated development as centres of leisure, and their role as centres of craftsmanship and nurseries of skill. The position of county towns as economic centres was an important factor in explaining the vibrancy of their literary and philosophical culture, however, other towns that were trading, manufacturing and industrial centres enjoyed equal or greater economic success. What is perhaps more interesting is the cultural importance that county towns had as a result of their juridico-political status and the manner in which this status helped to determine the character and content of their scientific culture and reinforce its distinctiveness and provinciality.

It is unnecessary to explicate in detail all the juridico-political functions that county towns and the institutions that were associated with these had, such as the county courts or gaols and their physical manifestations. Such functions help to explain why these towns tended to be well served by superior transport communications by the 1770s, particularly turnpike roads, which in turn had an important cultural impact, facilitating the regular appearance of itinerant lecturers and other travelling performers for example. They were also important in helping to determine the strength, relative popularity and character of urban cultural institutions because of the relationship between social occasions such as assemblies and theatrical performances, their functions, and their positioning within assize week or borough and county elections. County towns also tended to be centres for charitable foundations, which as part of general urban improvement, were the occasion for some of the greatest social occasions of provincial towns. These included the charitable music festivals in support of hospitals, which were patronised, as Borsay and Clark have shown, by distinctively urban gentry.[10] Furthermore and again related to their juridico-political and other functions, county towns often had a more varied occupational structure than other towns. Many lawyers, clergy and medical men, for instance, lived in them to serve the urban gentry, middling sort and institutions such as courts, hospitals and churches. The significance of such professionals in local scientific culture is evident from the number of clerical naturalists, meteorologists and electricians. Encouraged by commentators such as Thomas Gisborne, they considered that scientific study would improve their education, knowledge of natural theology, whilst providing valuable social contacts and enhancing their status and professionality.[11]

Benefactors of schools, libraries and other public institutions directed their appeals to such professionals. Shapin and Schaffer have argued that because of trust in their education and gentlemanly status, such men were invested with the authority to authenticate information in the republic of letters, whilst clergy could vouch for worth in natural theological terms.[12] As Jankovic has argued, 'with an emphasis on social accreditation and the geographical expansion of fact-gathering' eighteenth-century natural historians such as Rev. William Borlase in Cornwall were regarded as trustworthy authorities.[13] Informed by the intricacies of theological and philosophical argument, recognition of factual contingency encouraged the development of a scholarly community operating in the public and private spheres. The intellectual response to the growth of uncertainty was a greater reliance upon the professional status of clergy, medical men and lawyers whose methods of enquiry, knowledge dissemination and factual verification were adopted. Similarities between the methods adopted by natural philosophers, historians and antiquarians and the inquisitorial aspects of the legal system, the study of the body and formal autopsies, help to explain their significance in the sciences.[14]

County towns were also, of course, cultural and educational centres. After a study of Georgian provincial newspaper advertisements Plumb discovered that, perhaps surprisingly, 'schools were to be found more frequently in the old county towns, and the surrounding districts, than in the new manufacturing towns'. Few schools were advertised in the *Leeds Mercury*, yet the Northampton, York, Norwich and Ipswich newspapers were full of school notices in the eighteenth century.[15] Likewise, some of the earliest subscription and circulating libraries and provincial newspapers tended to be established in old county towns such as Huntingdon, Norwich and Newcastle.[16] Certain other factors as intellectual and commercial manifestations of the provincial urban flowering, served to enhance the image of county towns in the face of the relative economic decline. Changes in geographical education partly resulting from humanistic antiquarianism and scholarship emphasised the individuating characteristics of counties. This emphasised unusual natural phenomena and landscape wonders, heraldry, family genealogies and surveys of antiquities and estates, which inscribed new – county framed – meaning onto space and the landscape. This scholarship was supported by aristocracy, gentry and professionals who had their names celebrated in the subscription lists, maps and illustrations, whilst the gentry and middling sorts adorned their walls with cartographic and visual representations of their localities.[17] Similarly, there was an increase in the number of county histories, whilst regional social and political affairs began to feature in the national and provincial press. Systematic mapping of counties and towns became more common to satisfy the legal, commercial and industrial requirements for accurate surveys for enclosure, improvement acts, road building, canal formation or other purposes. Supported by local gentry and the Society of Arts, county towns were given special prominence in county maps. Geographical study concentrated less upon antiquity, the Bible and classical authors and more upon surveying, commerce, military education, mathematics and astronomy. Basic statistical information, antiquarian and chorographical data became part of the cannon of geographical knowledge including descriptions of county towns, their buildings and institutions in gazetteers, atlases and geographical textbooks.[18] Standard textbooks such as Richard Phillips' *Grammar of British Geography* carried descriptions of county towns with engravings and names printed in bold type on maps (Figure 26). Likewise, children's geographical puzzle games such as Abbe Gaultier's *Complete Course of Geography*, rewarded players with tokens or counters on the basis that they learned to recite verbatim answers such as 'Newcastle, the capital of Northumberland, which is one of the fifty-two counties of England, and one of the six to the north'.[19]

26 Counties and county towns of England and Wales from Rev. J. Goldsmith [Sir Richard Phillips], *A Grammar of British Geography* (1816)

Georgian town histories and travel books also emphasised the historical, ecclesiastical and juridico-political functions of county towns as administrative and cultural centres. Whilst Goldsmith's *British Geography* carried some information about manufactures, industry and manufacturing towns there was a distinct bias towards history and antiquarianism. The antiquity of county towns was usually emphasised, whilst pupils were asked many questions about the civil and ecclesiastical status of these centres. Such approaches predisposed them to value county towns which had more of these functions and an apparently venerable pedigree.[20] Urban status was generally more directly related to civic identity in county towns than in other urban centres. This is evident in the promotion of the former for traders and tourists alike in urban histories, gazetteers and town guides, and the emphasis placed upon urban institutional culture and improvement, frequently stimulated by inter-urban rivalry.[21] As we shall see, this helps to explain the relative strength of antiquarianism in county towns such as York when compared with the greater rhetorical emphasis upon utilitarian applications of science in places such as Manchester, Liverpool and Birmingham.[22]

The importance of region, locality and county in moulding the character of English scientific culture is evident in studies of local meteorology which formed an important part of the chorographic tradition. Counties and county identities were reforged, reinvigorated and shaped in new ways as natural historical, antiquarian, literary and chorographical endeavours established boundaries, features and cultural itineraries through mapping, observation, scholarship, publication and associational activity. Although the role of mathematicians, teachers of mathematics and mechanics needs to be acknowledged, eighteenth-century meteorology prized the local and distinctive nature of phenomena above the instrumental and general. Jankovic has convincingly argued that it was grounded in narrative studies of unusual events and therefore challenges the perception of Enlightenment science as a unified whole. Studies of unusual natural occurrences such as lightning strikes, fiery meteors and earthquakes, full of detailed description and careful observations, were undertaken and presented in the *Philosophical Transactions* and other works. Such enquiries help to explain the preconceptions and concerns of provincial natural philosophy until the early nineteenth century.[23] With the formation of modern instrumental meteorology perceiving nature as a vast laboratory interest in incidental and localised observations declined, however, county and place continued to inform provincial scientific study through geological and natural history societies and other means.[24]

SCIENTIFIC CULTURE IN LEICESTERSHIRE, YORKSHIRE, NORFOLK AND DERBYSHIRE

Literary and scientific associations have long been accorded an important role in the history of English science and provide an indication of the geographies of provincial scientific activity. The earliest such English associations were the Oxford and Cambridge philosophical societies and the gentlemen's clubs of the seventeenth and early-eighteenth centuries in Spalding, Peterborough and other towns, mostly in the eastern counties. Some were founded in commercial centres such as Liverpool and Manchester or resort towns such as Bath, but many appeared in county towns. By the 1770s philosophical societies had appeared in Coventry, Northampton, Newcastle, Derby, York and Norwich and other centres, often stimulated by courses given by itinerant philosophical lecturers. Although sometimes only ephemeral foundations these organisations were often closely related to more enduring informal literary and scientific circles.[25]

The relationship between the ecclesiastical and juridico-political status of county towns and the strength and character of public culture is evident from prosopographical analysis of organisations such as philosophical societies and later mechanics' institutes, museums and natural history societies, which tended to be based in county towns.[26] Such work confirms the importance of urban-based professionals. County towns tended to have a broader range of occupations than other settlements in the eighteenth century, including numbers of professionals and luxury traders, competing to mould and satisfy the tastes of the urban gentry and middling sort.[27] Other centres in contrast, such as ordinary market towns, were less likely to have the size or status to support a successful urban associational culture. Some manufacturing, industrial and trading centres, which did not have county town status and therefore did not tend to be polite or fashionable centres and usually had less historic significance, and weaker scientific institutions. At Leeds, despite the residence of natural philosophers such as Thoresby, Priestley and William Hey (1736–1819), the three attempts to found philosophical societies in the eighteenth century all met with failure until the civic-oriented Leeds Literary and Philosophical Society was instituted in 1819, although a subscription library was successfully instituted. At Bradford, as Morrell has shown, scientific societies were always weak and in both places industrial utility dominated over local natural history. The middling sorts in manufacturing centres frequently remarked upon the gulf between trade and politeness. Edward Baines noted in 1817 that 'Philosophical researches are not much cultivated in Leeds' except for those interested in the utility of science for commerce and manufacture, and this helps to explain the importance of geology in the mining regions of nineteenth-century Yorkshire.[28] At Sheffield and Birmingham because of the nature of local industry and employment patterns, scientific culture tended to make rhetorical appeal to practicality and be utilitarian in character.[29] Similarly in ports such as Hull, Bristol and Liverpool mathematical, commercial and naval academies and tutors competed to offer subjects regarded of commercial utility such as mathematics, navigation, geography, astronomy and modern languages. This in turn shaped local scientific cultures because of the overall demand for practical subjects, the pivotal role of schoolmasters as supporters of scientific institutions and the associations many of these had with Dissenting business networks. As Georgian Liverpool shows, utility, urban improvement and town image were causes on which urban elites could co-operate despite political and religious divisions concerning slavery, parliamentary reform and other matters. Trade, colonial expansion and religious missionary movements also shaped the character of English science in general and that of ports such as Liverpool.[30]

Partly as a function of occupational diversity and the strength of urban professional groups, some county towns tended to have quite large and influential groups of Dissenters, who sometimes, as at Nottingham, Derby and Norwich, took a prominent part in local government.[31] Emphasis upon the theological and civic value of science usually enabled county intellectual culture to embrace Anglicans and Tories as well as Dissenters and reformers. Even where Anglican Tories remained dominant in oligarchic corporations as at Leicester, Nonconformists took part in other aspects of public culture such as charities or musical concerts. Here, the Dissenting community was strong and even the Tory historian John Throsby had to admit 'genteel and numerous', containing 'several of the first families of the town'. The Leicester locality supported a scholarly community that included the medical men James Vaughan, Thomas Arnold (1742–1816) and Robert Bree, the astronomer and mathematician Rev. William Ludlam FRS (1717–88), the radical bookseller Richard Phillips, and the naturalist and agricultural writer Richard Weston (1733–1806).[32] There was also a circle of literary women that included Mary Linwood the composer, author and artist, Elizabeth Coltman, the poet Mary Steele and Susannah Watts, the tutor, poet and political pamphleteer.[33] Literary culture took a number of institutional manifestations by the end of the century. The Leicestershire Agricultural Society, which met in Leicester, supported the application of science to agriculture, and the work of the local Dissenting animal breeder Robert Bakewell, who had his own anatomical museum. Phillips initiated a literary society in 1790 with a subscription library and reading room under the presidency of Arnold whilst helping to found the Adelphi Society which engaged in scientific experiments, although this was suppressed by the corporation during the reaction of the 1790s.[34] Under pressure from the Tory oligarchy, polite urban culture adopted other forms until the 1830s. A music society, for instance, was formed in 1789 receiving support from both Dissenters and Anglicans, but by the 1830s a medical society and two more philosophical societies had been founded and a mechanics' institute. Of these, the Leicester Literary and Philosophical Society with its library and museum laid the deepest roots. Although it experienced initial difficulties, the problem of a severe lack of attendance was rectified by the controversial acceptance of female attendance at meetings and the promotion of courses of lectures. Reflecting Leicester's county status, the Philosophical Society took a strong interest in the geology and natural history of Leicestershire and its collection was the basis of the town museum.[35]

The prominence of county clergy and gentry in polite county town culture is equally evident in York where a philosophical society was established by the 1740s supported by a community of natural philosophers, mathematicians and scholars including the medical men and antiquarians John Burton (1710–71) and Francis Drake (1696–1771), the author of *Eboracum* (1736).[36] This might not be regarded as very remarkable given the status of York as fashionable centre, but by the end of the century the city was stagnating demographically and economically, although it remained an important agricultural centre.[37] As at Leicester, county gentry supported at least three agricultural improvement societies. One met at York from the 1760s, whilst a commercial circulating library (1781), Yorkshire Law Society (1786) and medical society (1781) catered for the needs of prosperous traders and professionals.[38] The antiquity and status of the city as a regional capital helped to justify the claim that York rivalled the metropolis as the second most important centre for freemasonry in the country which was supported by local intellectuals such as Drake, who served as grand master. In 1794 an attempt was made to found another philosophical society which would publish original work on science and literature and offer prizes, though this meta-morphosed into the York Subscription Library.[39] The Yorkshire Philosophical Society (1822) became one of the most prestigious in the country. The principal reason for its inception was the desire to celebrate and study the geology and antiquities of Yorkshire, but the Society was instrumental in the establishment of the British Association for the Advancement of Science which held its first meeting as guests of the Yorkshire philosophers. The express purpose of the British Association was to promote provincial scientific activity through periodic meetings in important centres such as county towns.[40]

The prominence of York society and clergy and the study of county natural history in the Yorkshire Philosophical Society are clear from the activities of original supporters. Those with geological interests included Rev. William Buckland, the Rev. William Venables Vernon (1789–1871) a canon of York minister, the Unitarian minister Rev. Charles Wellbeloved, William Salmond (1769–1838), a retired military man, Anthony Thorpe (1759–1829), a lawyer and antiquarian, the Whig leader and inventor Sir George Cayley (1773–1857), and John Phillips, later president of the British Association. Medical men, clergy, professionals and traders were prominent with various members of the aristocracy, gentry and clergy being enrolled as vice-presidents and patrons. The Yorkshire Philosophical Society acquired hundreds of members in its first twenty years, a library, gardens and a museum, whilst in keeping with its promotion of civic identity and Yorkshire antiquarianism it oversaw the excavation and care of the ruins of St Mary's Abbey. Its politico-religious balance contrasts with the Mechanics' Institute (1827), founded by a group of Dissenters and Whigs. These included Wellbeloved, the principal of Manchester College in York, the Rev. J. Turner, another Unitarian minister, Cayley, and William Hargrove the proprietor of the city's two Whig newspapers, the *Courant* and the *Herald*. However support was still forthcoming from some Anglican clergy and aristocrats such as Archdeacon Wrangham and Lord Milton.

Principally inspired by fascination with county natural history and geology, the Yorkshire Philosophical Society and the British Association for the Advancement of Science were promoted by natural philosophers in a pre-industrial county town experiencing economic stagnation which had been eclipsed in its own county in trading and population terms by Leeds and even Hull. County towns could fight back in the name of civic pride consciously and deliberately using their status, antiquity and historical and cultural importance to determine the character of their associational culture. As the *Yorkshire Observer* candidly admitted in 1823:

> To make [York] commercial, is impracticable; to make it manufacturing is not desirable; to make it considerably more extensive is not necessary. What then remains to be done to restore it to a large portion of its ancient character and magnificence? There remains to confer upon it the distinguished rank of a literary City, and as the centre of scientific attraction.[41]

Norwich too provides an excellent example of a thriving and varied urban literary culture in a county town and regional centre after 1750 (Figure 27). As with York, this is all the more interesting in the face of the town's population stagnating between 1750 and 1810, and relative economic decline until the middle of the nineteenth century.[42] Like Gloucester, Hereford and Worcester, Norwich supported regular music festivals, subscription concerts and societies from the early eighteenth century. Such activities became, as we have seen, an important feature of county town life in the nineteenth century.[43] Norwich had a rich public culture by the end of the seventeenth century centred upon its coffee houses, taverns and libraries.[44] The city boasted of the Old City Library of 1608 – the first independent provincial municipal library in England – and in its later incarnation from 1656 as the 'Public Library of Norwich', apparently the first subscription library in the provinces. A second public subscription library was founded in 1784, which later merged with the Norfolk and Norwich Literary Institution (1822) in 1886.[45] A lively society known as the Tusculan debated religious, political and aesthetic questions when the reform societies and radical reputation of the city were at their height during the 1790s.[46] Norwich also became renowned as a centre for the fine arts producing its own 'Norwich School' with a special county and regional identity, regarded by art historians as of unique importance in the English provinces, whilst the landscape gardener Humphry Repton (1752–1818) found opportunities to flourish in Norwich society and associational culture.[47] Another county-focussed association gained inspiration from the locality and helped to assert county identities. The Norfolk and Suffolk Institution for Promoting the Fine Arts (1803–33), encouraged local artists by exhibiting annually around assize week, whilst classic works were borrowed from local gentry for display, cementing the alliance between county inspiration, audience and patronage.[48]

The city received visits from itinerant lecturers by the 1730s and 1740s, which stimulated the foundation of a Norwich Natural History Society in 1746. One of the founders, William Arderon FRS (1703–67) described how they were inspired to conduct electrical experiments through attendance at such a lecture.[49] A variety of later societies were founded attracting support from across the county, all of which received some inspiration from the unique Norfolk landscape, flora and fauna. These included the Speculative Society (1790), the Norwich Philosophical Society (1812), Norfolk and Norwich Literary Institution (1822), Norfolk and Norwich Museum (1825), Norwich Mechanics' Institute (1825) and the Norfolk and Norwich Horticultural Society (1829). Some of these had hundreds of members and staged exhibitions and events that attracted a large public audience. The Norwich Public Library for instance, by the 1840s, had a collection of 20,000 volumes and 500 proprietary members, and was visited by hundreds of annual subscribers, whilst the Literary Institution had 270 members and library of 11,000 volumes.[50]

One constant of Norwich scientific culture for almost a century was the special attention given to Norfolk botany and natural history and the part that this played in promoting county identity. Founded around 1760, the Norwich Botanical Society became the leading such English association. One of its members was Hugh Rose, who published a translation of Linnaeus, another was Charles Bryant author of the *Flora Diaetetica, or History of Esculent Plants* (1783) and a dictionary of trees, shrubs and plants (1790) also arranged according to the Linnaean system. James Edward Smith, who lectured in the city, as we have seen, went on to obtain the herbarium of Linnaeus and found the Linnean Society, becoming the most influential figure in British botany after the death of Joseph Banks in 1820, before returning to Norwich with his collections.[51] As at Derby and York, some of these scientific groups had a Masonic flavour, and may have grown out of freemasonry which by the 1720s was flourishing in Norwich. One example is the Society of United Friars (1785–1828), which, was, in many ways, another literary and philosophical institution, but whose members underwent an initiation ceremony before being placed into religious 'orders'. In addition to acting as a debating society and supporting various local improvement and charitable ventures, the Friars were keenly interested in agriculture and natural philosophy, building a library, conducting scientific experiments and hosting lectures from itinerants such as Adam Walker. The Friars' varied and politico-religiously diverse membership included shopkeepers, professionals, Anglican and Dissenting clergy, artists, musicians and cartographers. These included the Unitarian clergyman George Morgan, medical men such as Edward Colman and James Martineau, Repton, Richard Taylor the land surveyor, geologist and antiquary, and the publisher and botanist Simon Wilkin, who also created the Norfolk Entomological Society (1811).[52]

Norwich Dissenters held prominent positions in local society and government, and were also instrumental in supporting literary and scientific activity, for example, the foundation of the Norwich public subscription library in 1784. As at Leicester and Derby, the strength of the Dissenters and their theological and educational ideologies in the context of county town society created opportunities for middle-class women.[53] Morgan, minister of the grandiose Octagon Chapel and a friend of the Norwich instrument maker and scientific writer Abraham Brooke, was an experimenter who published *Lectures on Electricity* in 1794, and various pieces in the *Analytical Review*.[54] Later ministers at the Chapel including William Enfield (1740–97) – who acted as one of the earliest presidents of the Public Library – and Pendlebury Houghton (1758–1824) were also natural philosophers, whilst Enfield collaborated with Amelia Opie (1769–1853) and others on *The Cabinet* periodical. Norwich literary culture and Dissenting networks provided the support that launched the careers of Opie and Harriet Martineau (1802–76), who were both members of prosperous families with strong Dissenting connections. Likewise, Susannah Taylor's Unitarian household has been described as 'a central place in the brilliant literary and political society of Norwich' and a fertile context for female education.[55]

The case of Derby demonstrates the importance of county traditions of natural history and antiquities in shaping the character of provincial scientific culture, whilst showing the role of industry in encouraging a utilitarian approach to the study and promotion of the sciences. Like Norwich, Derby underscores how county town occupational diversity and the size of the middling sort encouraged Dissenters in public life. The county character of scientific culture in Derby was as pronounced as that of Norwich, being shaped and challenged by the landscape, geology, natural history, industry and traditions of Derbyshire. Derby and Derbyshire physicians took a keen interest in the geology of the Peak and the healing properties of the springs (Figures 27 and 28). As at Norwich and Nottingham, by the mid-eighteenth century the town was receiving visits from itinerant lecturers who taught mathematical and scientific subjects alongside tutors from private academies and a local Dissenting academy which offered similar subjects. Unlike coal mining, lead mining tended to stimulate geological study because of the perceived importance of understanding the geological context of veins, whilst a thriving trade in ornamental minerals and fossils from the Peak was active in Derby by the later eighteenth century. As John Farey's richly delineated *General View of the Agriculture and Minerals of Derbyshire* (1811–17) demonstrates, local engineers and mine owners acquired a good knowledge of local geology. The nature of Derbyshire industry meant that the interest in county natural history and geology also had a utilitarian flavour. The chorographical tradition was also pronounced from the seventeenth and eighteenth centuries, embodied in the seven wonders of the Peak tradition, only one of which was a building, the rest having natural historical significance. They were celebrated in Thomas Hobbes' *De Mirabilibus Pecci* (1636), Izaak Walton's *Compleat Angler* (1653) and his friend Cotton's *The Wonders of the*

Peak. In the eighteenth century, county natural history was described in the Rev. James Pilkington's *History of Derbyshire* (1789), whilst the many tourist guides, gazetteers and maps from the early nineteenth century all alike gave prominence to the subject.[56] William Adam's *Gem of the Peak* (1838) for example, included detailed accounts of the geology, mineralogy and botany of Derbyshire.[57]

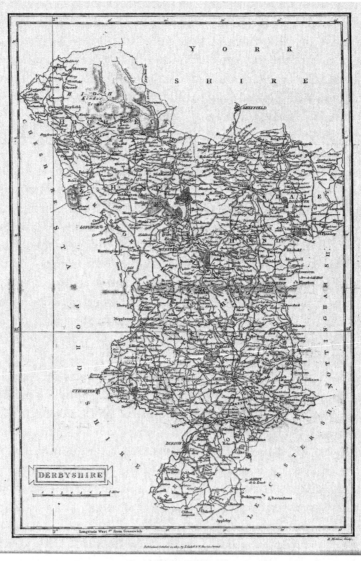

27 Map of Derbyshire from D. and S. Lysons,
Magna Britannia, Derbyshire (1817)

28 Rocks in Middleton Dale, Derbyshire from D. and S. Lysons,
Magna Britannia, Derbyshire (1817)

The circle around the artist Joseph Wright and the clockmaker
and engineer John Whitehurst celebrated the landscape, scenery and
industry of Derbyshire and entertained distinguished philosophers
such as Benjamin Franklin and James Ferguson. They supported the
efforts of the cartographer Peter Perez Burdett to produce the first
comprehensive survey of the county which symbolically used All
Saints Church in Derby as the base for all distance measurements and
relied upon the support of local subscriptions from gentry, miners
and philosophers. In return, Burdett included depictions of large
houses and family names in addition to industrial features such as
mines, antiquarian, historical and natural historical information.
Wright's paintings portrayed the landscape, scenery, meteorology, industry and history of Derbyshire and are, of course, some of the most
famous encapsulations of English Enlightenment provincial scientific
culture. His portrait of Whitehurst for example, managed to celebrate
the county landscape and the geological studies it inspired in his *Enquiry into the Original State and Formation of the Earth* (1778). This included an appendix featuring a stratigraphical depiction of a section

of the strata at High Tor, Matlock, which was shown in Wright's portrait.

As we have seen, the residence at Derby of the physician, natural philosopher and poet Erasmus Darwin enriched the town's scientific culture and encouraged the foundation of the Derby Philosophical Society (1783). Darwin celebrated the activities of the Derby philosophers as well as the county landscape and natural history in his epic poems, *The Botanic Garden* (1791) and *The Temple of Nature* (1803), and in the prose work, *Phytologia* (1800), which included much information concerning Derbyshire flora, fauna, horticulture, geology and topography. The exposed limestone cliffs with their embedded sea creatures (the 'might monuments of past delight'), igneous toadstone deposits and warm springs, for instance, were interpreted as providing evidence for the oceanic origins and evolutionary processes of life. Many of the ideas and examples drawn upon so vividly in these works were gained from Darwin's experiences as medical practitioner touring the region to see patients in his large coach, but also his involvement in local politics, charitable activities and industry as well as natural philosophy and natural history. From these experiences and Lunar Society connections also stemmed Darwin's rather uncritical adulation of industry and progress and partial blindness towards the oppression of life in the manufactories. Darwin and philosophical friends such as the members of the Strutt family of textile manufacturers saw Derby as the focal point for their scholarly sociability, contending that intellectual Enlightenment would facilitate urban improvement. Darwin used his medical and philosophical authority, for instance, to argue in a broadside that urban improvements would encourage more from the country to spend time in the county town which would become 'a kind of metropolis ... as York, Shrewsbury, Lincoln, etc.'[58] Other members of the Derby Philosophical Society celebrated their provinciality and for this and their political views they were mocked in the *Anti-Jacobin Review* during the 1790s.[59] The writer, botanist and poet, Brooke Boothby, called his county philosophical friends 'not the least learned and reflecting men in this kingdom' and noted dryly that they were 'not confined to courts and capitals'. Rev. Thomas Gisborne, another founder member, advocated the creation of philosophical societies in all major towns so that they would become intellectual centres for the middle classes to engage in learned discussion and the dissemination of ideas and innovations.[60]

Subsequent Derby scientific culture was represented by a succession of associations including the Derby Literary and Philosophical Society (1808), Derby Mechanics Institution (1825) and Derbyshire County Museum and Natural History Society (1835). Primarily the inspiration of Darwin, the Philosophical Society took a keen interest in the landscape, geology and natural history of Derbyshire. They supported Pilkington's *History of Derbyshire*, White Watson's *Delineation of the Strata of Derbyshire* (1811), Farey's *Geology of Derbyshire* and other such publications, whilst county studies formed an important part of the activities undertaken at the Mechanics institute and Natural History Society. In some respects, these culminated in two large exhibitions promoted by the two organisations which reveal the size of the audience for science. Visited by tens of thousands and celebrated in local and provincial newspapers, they featured artistic depictions of the Derbyshire landscape with natural historical and geological specimens and antiquities. However, perhaps the most important intellectual stimulation provided by the county landscape, geology, natural history and industry within Derby scientific culture was the role that they played in the development of the evolutionary philosophies of Erasmus Darwin and Herbert Spencer.[61]

CONCLUSION

Geographical approaches to the study of Georgian scientific culture should include close attention to the role of counties at many different levels in shaping the promotion, character and reception of ideas and practices. Although more likely to be elite-oriented than manufacturing or industrial centres, county town scientific cultures tended to be more politically inclusive than the former. As the foci of county cultures, county towns demonstrate the effectiveness of urban typologies in such a geographical approach to the analysis of Georgian intellectual endeavours. As such, the rich cultures of county natural history and natural philosophy support Borsay's criticism of Porter's contention that the prime aim of provincial Enlightenment culture was to 'assimilate metropolitan culture and values', being more imitative than innovatory. In fact, what Merton called the universalistic and communistic aspirations of natural knowledge as represented by the collective of Enlightenment scholars, when combined with the general economic and cultural urban renaissance, facilitated the creation of provincial county cultures and identities that were far from being merely passive imitations of metropolitan values and fashion. It is therefore not surprising that reformers and writers such as John Thelwall found refuge, support and patronage in the philosophical circles of Norwich, Derby and Nottingham during the reaction of the later 1790s. Looking back shortly afterwards from a relatively isolated Welsh village, Thelwall complained of how he had shrunk from 'theatres and halls of assembly', from the 'friendly, the enlightened, the animated circles of Norwich' and from the 'elegant and highly intellectual society of Derby'.[62]

Different types of towns had scientific cultures with varied characteristics, encouraged and shaped by the assertion of provincial identities. The juridico-political, ecclesiastical, historical and other characteristics and functions of county towns helped to determine the strength and character of their public urban culture and the role that they played in shaping county identities. With their varied occupational structures, their role as nexuses of polite culture, as centres for county gentry, and their relatively large concentration of professionals, county town scientific cultures tended to be less utilitarian than those of manufacturing centres like Leeds, Sheffield and Birmingham. Medical men, for instance, were often instrumental in the foundation of literary and philosophical societies and tended to be most interested in endeavours that fosted and complemented their professional interests such as natural history, galvanism and chemistry.

County town status continued to differentiate provincial scientific cultures during the nineteenth century as analysis of the 1851 census data on literary and scientific institutions suggests. County-focussed scientific culture continued to ensure that they former 'punched above their weight' in terms of middle-class public institutional culture despite relative economic eclipse by manufacturing and industrial towns. This continuing relative importance seems to be confirmed by the fact that 26 or 52 per cent of the annual meetings of the British Association between 1831 and 1900 took place in county towns such as Nottingham, Ipswich, Norwich, Exeter and York, whilst 22 or 43 per cent took place in ports or large manufacturing centres. A similar pattern is evident when the annual meetings of the Archaeological Association and the (Royal) Archaeological Institute are considered, the latter meeting at Warwick, Dorchester, Lancaster and Leicester during the 1860s. The argument also receives some support from other analyses of the 1851 data. In his consideration of levels of illiteracy and schooling in provincial towns, Stephens found that there was a general tendency for larger northern and industrial centres to have poorer educational provision than their predominantly southern market counterparts. He found that the agricultural areas surrounding industrial towns tended to have higher levels of literacy and better educational provision, whilst market towns were often 'a haven of culture and civilised behaviour' in their respective regions. Stephens's emphasis upon the continuing importance of the economic variety and cultural vitality of market towns and agricultural centres provides some support for the approach elucidated here.[63] Although analysis of the 1851 census data is incomplete, the evidence is that county towns status continued to count and impact upon the strength and character of urban institutional culture.

7 Urban Scientific Culture: Science and Politics in Nottingham[1]

INTRODUCTION

Natural philosophy was, as we have seen, celebrated in the European Enlightenment as one of the principal models and inspirations for cultural and intellectual activities and the development of the public sphere. Taking one English provincial town as a case study, this chapter contends that the sciences were as as important in Nottingham cultural life as they were in other centres such as Manchester, Norwich, Birmingham and Derby. It examines the activities of Nottingham philosophers and visiting lecturers, assessing the importance of the sciences in the curricula of tutors and educational institutions. It explores the role of professionals such as clergy, lawyers and medical men as publicists and practitioners of Nottingham science, exploring how local topography, society, economy and culture shaped the character of local urban scientific culture. Although there were occasional political and religious strains, scientific ideas enjoyed a wide appeal in Nottingham society encouraged by natural theology, progressive reformism and Enlightenment improvement. Whilst broader progressive Enlightenment ideas had a major impact nationally, local differences and identities played a major role in shaping the character of provincial urban scientific cultures. As an agricultural and industrial county town and regional centre with a relatively independent urban corporation and large population of dissenters, Nottingham demonstrates how the sciences intersected with local economies, politics and society.[2]

29 The vicinity of Nottingham from F. C. Laird, *Topographical and Historical Description of the County of Nottingham* (1820)

NOTTINGHAM: COUNTY TOWN AND REGIONAL CENTRE

Nottingham had vital links with the north and east coast through river, road and later canal networks, and was also an important staging post on the routes between north and south (Figure 29). Markets, fairs and races such as the great annual Goose Fair drew in crowds from across the region.[3] With a rising population of 10,720 by 1739, 17,700 by 1779 and 28,861 by 1801, Nottingham was a fashionable regional leisure centre for gentry and professionals attracting hundreds of visitors seasonally, drawing in the gentry and aristocracy and nurturing its own residential urban gentry.[4] They stimulated and were served by a large variety of entertainment and services including assemblies, sports, the theatre, music clubs and societies, book clubs, coffee houses, circulating libraries, museums and exhibitions.[5] Middling-sort consumer demand helped to encourage the development of the town as a craft and market centre for the distribution of mundane and luxury goods. The dominant textile industries were an important result of internal and external demand and these placed a premium on technological and entrepreneurial innovation and mental capital which stimulated educational change.[6]

Nottingham Castle, Sherwood Forest, the legends of Robin Hood and other celebrated aspects of county history began to be employed to define civic identity and reinforce urban status, encouraged by antiquarian studies which helped promote commerce.[7] As 'a town of note the age of 900 years', a 'considerable borough' and mayoral centre, Nottingham's historical status was an important manifestation of the town's identity, employed to counter religious and political divisions.[8] In the face of cultural and economic rivalry from Leicester and Derby, Nottingham histories, guides and gazetteers asserted in their rhetoric the geographical superiority, venerable antiquity, status and commercial strength of what they held to be one of England's most important towns.[9] The surrounding country estates and number of orchards, gardens and open spaces within Georgian Nottingham dominated prospects of the town and impressed travellers. Charles Deering a German physician who moved to the town and composed its first major history, argued that no naturalist could have suggested a superior location nor one better designed to promote health and longevity. He asserted that it was 'hard to judge whether art or nature' had the greatest share' in the 'surprisingly grand and magnificent' view from the south which put 'even a person the most acquainted with all the parts of England to a stand, to name its equal'.[10]

Gentry and landowners, who supported the publication of maps and historical works – even if they did not live in Nottingham – often had mansion houses there or at least came into the city on special occasions such as the assizes or races. The most important were the

192

dukes of Newcastle who held Nottingham Castle which was transformed into a grand neo-classical mansion dominating the town from its rock in place of the medieval edifice pulled down during the 1660s.[11] As the natural history sections in Deering's *Nottinghamia* and the information contained in county maps demonstrates scientific study featured prominently as part of humanistic scholarship.[12] Maps helped to resolve boundary and other legal disputes, celebrating families and the extent of estates, satisfying heraldic pride, demonstrating apparent antiquity, defining enclosure and engineering measures such as river improvements, canals, roads, railways. They also celebrated and contributed towards estate improvement in a county with a concentration of ducal seats. Nottinghamshire aristocrats hired landscape gardeners and architects such as Humphry Repton and Thomas Wright to design improvements. They also supported an Agricultural Society for Nottinghamshire and the West Riding which usually met at Bawtry, advertised its meetings in the newspapers, and offered prizes for agricultural innovations, estate improvement, arboriculture and other activities similar to those encouraged by the Society of Arts.[13] Nottinghamshire newspapers frequently carried detailed agricultural, horticultural and meteorological reports, particularly concerning unusual occurrences such as diseases, fire balls and shooting stars. Like other aspects of local chorography, they tended to emphasise the local and distinctive above the instrumental and general.[14]

Supported by county gentry and aristocracy, the juridico-political status of Georgian Nottingham provided a major stimulus to its intellectual culture. The economic importance and civil and administrative status of the city help to explain the quality of its transport communications and the vitality of its public culture. Cultural institutions and events were often shaped by official occasions on the civic calendar such as the assizes or borough and county elections.[15] Nottingham was also the site for most major county charitable institutions such as the General Hospital and the Lunatic Asylum. Religious, civil and legal institutions and the gentry, middling sort and aristocracy who supported them were served by resident professionals who, as we have seen, were prominent in urban scientific culture.[16] The Rev. William Standfast for instance, endowed a subscription library with a large number of medical and natural philosophical texts to the citizens of Nottingham in 1744 'for the use of the clergy, lawyers, phicitians [*sic*], and other persons of a liberal and learned education' living in the vicinity.[17]

30 View of Nottingham from the North Road from F. C. Laird, *Topographical and Historical Description of the County of Nottingham* (1820

Inspired by the education that they had received, often in Scottish or continental universities, physicians and surgeons took a particular interest in chemistry, electricity, physiology and botany, often using these as a basis for wider philosophical studies. It is noteworthy, for example, that the first attempt at a comprehensive Nottinghamshire flora was undertaken by Deering with his antiquarian and historical studies. Making a living from private practice and the opportunities and status provided by attachment to the General Hospital, Lunatic Asylum and other Nottingham institutions, medical men such as the Nottingham physicians John Storer and Snowden White and clergymen such as Rev. White Almond were prominent supporters of scientific culture for decades. Formal scientific institutions often grew out of informally constituted coteries whilst clubs not avowedly scientific in purpose such as debating societies, Masonic lodges or book clubs sometimes promoted the sciences. The Nottingham Book Society, for example, although a circulating subscription library, received special thanks from the aeronaut Cracknell for promoting his 'grand philosophical' balloon launch in July 1785.[18]

The careers of Storer and White illustrate the importance of medical professionals in Nottingham intellectual cultures. Originally serving as an officer on the army medical staff in Holland, Storer 'soon gained the confidence of the leading families of the town and county' and the medical profession after moving to Nottingham in 1781. He played a 'leading and active part in the founding and conducting' of the General Hospital, Lunatic Asylum and Vaccine Institution. Storer and White were both were members of a book club at the White Lion and helped to promote the Nottingham Subscription Library (1816) and scientific endeavours associated with it, the former serving as first president.[19] White and Storer were also members of the Derby Philosophical Society which served as a regional scientific society. Although they borrowed many medical texts from its library including Andrew Duncan's *Medical and Philosophical Commentaries*, Thomas Kirkland's work on the improvement of surgery (1783) and Thomas Fowler's study of the effects of tobacco and arsenic (1786), they read others with few obvious medical applications. White borrowed Adam Smith's *Wealth of Nations* (1776), Roger Long's *Astronomy* (1784) and John Whitehurst's *Inquiry into the Original State and Formation of the Earth* (1778) for example, whilst Storer also borrowed Whitehurst's works along with Anders Sparrman's *Voyage au Cap le Bonne-Esperance* (1787) and Charles Hutton's *Tracts: Mathematical and Philosophical* (1786). They also read books on botany, electricity and chemistry such as Carl Linnaeus' *Family of Plants* translated by Darwin and the Lichfield Botanical Society and Richard Watson's *Chemical Essays* (1782).[20]

NEWSPAPERS, SCIENTIFIC BOOKS AND INSTRUMENTS

The degree of interest in the sciences in Nottingham is evident from the number of scientific works promoted in the town, the activities of tutors and schools and the frequency and popularity of lectures on natural philosophy. The sciences were encouraged part of practical mechanics, genteel learning and the developing public sphere whilst being encouraged by the Dissenting ideal of personal education and improvement. As we have seen although the Newtonians strove to dominate Georgian natural philosophy supported by the Royal Society, their precepts remained contentious, contingent and incomplete as the queries closing Newton's *Optics* emphasised. Boundaries between different paradigms and conceptions of rationality were fluid and negotiated in a variety of social and geographical contexts.[21] Natural philosophy and astronomy remained closely intertwined with astrology which was condemned as irrational but served as the inspiration for many mathematical and astronomical inquiries as the career of Thomas Simpson demonstrates.[22] On the arrest of Robert Andrews a Grantham astrologer, the *Nottingham Journal* commented that the fact that 'such characters could exist in an age so enlighten'd as the present, appears somewhat extraordinary and must occasion many risible countenances' as being 'inconsistent with reason' and 'ridiculous in nature'.[23] However, the newspaper also advertised works of astrology such as William Partridge's *Book of Fate: or Universal Fortune Teller* (1770). It also supported the work of local astronomers and astrologers such as the Bingham schoolmaster Daniel Stafford, who compiled almanacs for the Company of Stationers after the death of Robert White, another local mathematician and astronomer.[24] Serious polite rationality combined with family amusement, humour and entertaining pursuits like drinking and socialising in taverns and theatres. Whiffin's museum of natural curiosities (1782), the microscopic exhibition at the Feather's Tavern (1787), the fabulous 'world in miniature' (1775), and lectures by Sieur Herman Boaz and his 'miraculous' and 'wonderful' display of the 'occult sciences' (1780), offered popular introductions to some aspects of the sciences. The mechanical world in miniature, for example, claimed to distil all the 'beauties of art and the elegancies of architecture, sculpture, painting, music, and astronomy' using 'a vast variety of figures, moving in the liveliest manner, where the imagination is ever delighted.'[25]

Nottinghamshire craftsmen such as the watch, clock and instrument maker William Mordan advertised the sale of household scientific and geographical instruments.[26] One shipment advertised as 'just arrived from London' and sold at Thurland Hall, Nottingham in 1781 included maps, drawing instruments, telescopes, prospect glasses, microscopes and 'dissected maps to teach geography'.[27] Rocks, minerals and fossils were sold as ornaments and as part of furniture items for incorporation into domestic fire places and chimney pieces at the marble and petrifaction works under Radcliffe and later the Strettons.[28] Scientific books and instruments were often promoted and sold by Nottingham newspaper proprietors, especially Samuel Creswell and George Burbage. Examples include the promotion of the new edition of Chamber's *Cyclopaedia* edited by Abraham Rees in 1778 and geographical works such as Thomas Bankes' *System of Universal Geography* (1790), accounts of James Cook's voyages and geographical magazines sold with globes included *gratis*.[29] Considerable correspondence on mathematical and scientific subjects also appeared in the Nottingham newspapers with frequent accounts of astronomical and meteorological phenomena. The astronomer and mathematician George Burroughs of Sedgebrook for example, sent detailed letters to the *Nottingham Journal* on astronomical subjects such as eclipses during the 1780s often illustrated with diagrams encouraging others to observe.[30] The *Journal* did more than print letters it also gathered information, assessed the authority, validity and accuracy of reports and offered explanations utilising the latest philosophical theories.[31] Considerable knowledge of the sciences was assumed, one lengthy and detailed description of a lunar eclipse in 1776 commenting that as 'the first principles of astronomy and geography are now so well known, it seems unnecessary to give a particular description'.[32] The Nottingham newspapers also carried supportive accounts of the meetings of philosophical clubs in the region during the 1780s and 1790s such as the Chesterfield Debating Society, which helped to stimulate local interest. One correspondent praised the 'rational method of spending an evening' of 'instruction and amusement' in such societies contrasting it with 'idle dissipation, or tumultuous conversation' which only corrupted the morals, weakened the intellectual powers, and destroyed the constitution.[33]

NOTTINGHAM SCIENTIFIC EDUCATION

Nottingham schoolmasters and mathematics teachers promoted and shaped local scientific culture, offering courses of lectures on philosophical and geographical subjects for moral and polite improvement, natural theology and practical education. Given the growth of the textile industry and the success enjoyed by local figures such as Richard Arkwright, James Hargreaves and Jedediah Strutt in applying novel technologies to manufacturing processes, there was a strong belief in the utility of scientific and mathematical knowledges.[34] Commercial competition encouraged Nottingham tutors and academies to offer practical sciences such as applied mathematics, geography and mechanics.[35] Some Nottinghamshire grammar schools also began to supplement the classics by offering scientific subjects, charging for these as optional extras where necessary in response to competition from private schools. The strength of old and new Dissent in Nottingham encouraged pedagogical experimentation including the foundation of educational institutions designed to help women and the labouring classes. The adult school established by William Singleton in a room belonging to the Methodist New Connexion and quickly taken over by the Quaker Samuel Fox in 1798 for example, has been regarded as one of the first such English institutions. Although a small establishment, it helped to inspire others including an artisans' library and mechanics institutes.[36]

One influential school where natural philosophy featured prominently in the curriculum was the Nottingham Academy conducted by Charles Wilkinson and then Rev. John Blanchard which was advertised as alluringly 'open to the fields and Sherwood Forest' to the north of the town. The academy taught a large range of scientific subjects, establishing a regional reputation.[37] There was considerable competition between schools each claiming to teach subjects using scientific instruments and other equipment to qualify 'gentlemen boarders' for university, business or the professions. Wilkinson extended the Nottingham Academy in 1778 claiming to have a full library and 'perhaps the completest philosophical and mechanical apparatus' in any English academy. The school had a classics department and also taught many optional subjects including French, writing, arithmetic, English, drawing and dancing.[38] One of its rivals was Rev. Richard Oldacres' Academy at Woodborough near Nottingham which taught many applied mathematical and scientific subjects including geometry, algebra, fluxions, surveying, navigation and astronomy using philosophical instruments. Oldacres had gone to Derby in 1741 where he took day and evening classes, probably with the tutor and natural philosopher John Arden, and 'made himself master of almost every part of mathematics and natural philosophy'. In 1779 Oldacres claimed to have outlaid a 'considerable expense in procuring an apparatus of philosophical instruments' for his school, 'for the purpose of forming the tender minds of his pupils' for the 'sublime study of the wonderful works of the omnipotent', closing the 'avenues of vice' and fostering a 'thirst for useful knowledge and consequently a love of virtue, strongly excited'.[39]

Some of the earliest teachers of the sciences in Nottinghamshire such as John Badder (*c*.1700–56), George Ingman, Thomas Sparrow (1700-*c*.60), Robert White (1694–1773), Oldacres and Francis Holliday (1717–87) were practical mathematicians and mechanics who taught astronomy, astrology, navigation and land-surveying. They contributed astronomical reports, mathematical problems and other pieces to newspapers, almanacs, the *Gentlemen's Magazine* and the *Ladies Diary*.[40] White the Bingham schoolteacher compiled almanacs for the Company of Stationers, prepared a celestial atlas in 1750 and was a proficient of the classics.[41] Thomas Peat (1707/8–80) a 'land-surveyor ... skilful astronomer, mathematician, and schoolmaster', gave public lectures on natural philosophy in Nottingham. Moving to the town from Ashley Hay near Wirksworth, Derbyshire around 1722, Peat was encouraged by Cornelius Wildbore, a master dyer and fellow worshipper at the High Pavement Presbyterian Chapel, Nottingham who leant him books. In 1740 with Badder of Trowell, Ingman and others he projected the *Gentlemen's Diary* and became anonymous editor of the *Poor Robin's Almanac*. Working with Badder, Peat surveyed Nottingham, publishing the map in 1744. He left Nottingham for some time but returned in 1778 to offer his services as schoolmaster, painter and land surveyor teaching a variety of subjects including drawing, geometry, mensuration, surveying, dialling and trigonometry. Badder and Peat were also stimulated by their association with their friend Deering who republished their survey in his *Nottinghamia*.[42]

Around 1740 Peat projected a course of 14 lectures on natural philosophy at Nottingham to be given at the convenience of subscribers in sessions of up to two hours each, the syllabus being published by Samuel Creswell at a later date. Peat emphasised the practical rather than polite aspects of the sciences. He contended that the many 'advantages ... accrued to the world from experimental philosophy', were evident in the 'frequent improvements' they had wrought to many aspects of society. There was 'not a single art or science' that could not be improved by the sciences which aimed to 'discover causes from ... effects' and to 'make art and nature conspire in subserving to the necessities, conveniences and ornaments of life'. Peat emphasised that natural science was 'better understood by *seeing* the experiments performed than by a long and tedious application to books'. Equipment utilised included 'a fine orrery, as improv'd by Messrs. Cole and Son' mathematical instrument makers to the King, a 'curious model of a fire engine', a friction machine, towers and a pyrometer.[43]

Peat's lectures were aimed at 'gentleman and ladies' and the experiments were 'rendered as plain and intelligible as possible' for beginners. The course moved from an exploration of the laws and attributes of matter and Newton's laws of motion to the practical mechanics of wedges, screws, planes, pulley, wheels and axles and the improvement of carriages. Hydrostatics was covered in two lectures which explored the construction of pumps, fire engines and determining 'the genuineness, etc. of liquors' and 'counterfeit coins'. Optics and light occupied three lectures and included the theories of different types of lenses, the structure of the eye and on 'the first day the sun shines clear, the experiment of the *camera obscura*'. Pneumatics took three lectures and featured an analysis of 'damps in mines', respiration, the motion of sound and 'the famous experiment with Otto von Guerick's hemispheres'. Highlights of the course were probably the final lecture on 'geography and astronomy explained by the use of globes' and the 'fine orrery' and a special lecture on electricity. Peat printed a separate handbill for the latter suggesting that he hoped it would attract a more popular audience unwilling to attend other lectures. At the Hand and Pen in Castlegate, he called for gentlemen and ladies who would be 'agreeably entertained with a variety of curious and surprising experiments' with the 'compleat electrical MACHINE'. Persons would be 'electris'd emitting fire from any part of their bodies by the near approach of any other person' and 'inflammable substances' would be 'fir'd by attraction and repulsion'. Also featured were 'the electrify'd *syphon*', 'very *curious experiments* with the *artificial magnet*' and the surprising *Leyden* experiment', probably a demonstration of Pieter van Musschenbroek's Leyden jar condenser for the storage of static charge, which dates the demonstration to 1746 or later when the discovery was made.[44]

Although Nottingham tutors and institutions were important in the town's intellectual culture, with their extensive apparatus and theatrical demonstrations involving globes and orreries, competing itinerant lecturers made a great impact and provided a sustained public platform for the sciences.[45] By the 1740s John Arden, William Griffis and Francis Midon were giving courses of lectures in the east midlands, Griffis apparently stimulating Peat to undertake the Nottingham lectures, possibly as a joint venture. Given Arden's residence in Derby between 1738 and 1752 and the size of the potential Nottingham audience, he probably lectured there in competition with Griffis.[46] Competition between lecturers increased as the expanding turnpike network facilitated travel in the region. Visiting lecturers strove to foster a good reputation and build up their audiences. Threats to their intellectual status and integrity were a serious matter and could potentially destroy their business in particular places.

An indication of the degree of competition at Nottingham by the 1770s is the extent to which some lecturers embellished or exaggerated their qualifications to enhance their reputation. In 1772 for instance Arden delivered a course of lectures on mechanics, astronomy, geography, use of the globes, hydrostatics, pneumatics and optics. However, he was forced to defend himself against serious charges of pedalling inaccurate information made in the newspapers by resident schoolmaster and rival scientific lecturer Charles Wilkinson who offered a course of lectures on the properties of matter, laws of motion, mechanics, astronomy, hydrostatics, pneumatics and electricity.[47] Wilkinson aggressively accused Arden of 'subverting the good order of reasoning' by 'buoying up his subscribers with fancies; feigning new systems, baffling and deluding their reason' and 'flying for relief to imaginary causes' by accounting for effects 'which never existed in nature'. These serious charges were rebutted in angry correspondence published in the newspapers by his son James.[48]

Wilkinson's competitive determination to maintain his reputation and the insecurity of private teaching are evident in further correspondence to the Nottingham newspapers in which he questioned the status and intellectual authority of another rival. In 1774 he complained that tutors at his academy who were retained 'upon large salaries' had 'too frequently ... been privately employed in their respective departments upon low terms to the great disadvantage of his establishment'.[49] Similarly, in 1775 Wilkinson maintained that because he had bought new equipment and given public lectures, 'it hath been invidiously propagated' that he intended to decline the academy and become an itinerant tutor.[50] His reassurance that the apparatus was directed for the use of pupils and the apparent misreading of his intentions underlines the competitiveness between tutors and the need to keep the school a priority. Assistant schoolmasters who left their original employer to set up rival schools were a particular threat and Wilkinson acted swiftly to question the philosophical credentials of one former assistant, Denis Delaruelle, who started a Nottingham school in 1774. When Delaruelle claimed in his advertisements to be a member of the Royal Academy at Paris, Wilkinson wrote to Samuel Creswell publisher of the *Nottingham Journal* claiming to be 'somewhat surprised to find him assume an honour in his public advertisements, which I had never heard him mention, and which I am apprehensive does not appertain to him'. He also wrote to Joseph de Lalande (1732–1807), distinguished professor of Astronomy at the Parisian Royal Academy of Sciences, who responded that he was surprised an 'unlettered person' should 'pretend to be a member of the academy of sciences' as 'we only have eight foreign members' amongst whom was Benjamin Franklin. He had 'never heard of any such person' as Delaruelle. Creswell printed a conciliatory paragraph suggesting that he was 'the innocent cause' of the quarrel because when Delaruelle employed him to print the advertisements he had advised 'that he published him a member of the academy'. Delaruelle only having meant that 'he had received his education in an academy of science in Paris' but he had misunderstood his 'broken English'.[51]

By the late eighteenth century, lecturers were making frequent visits to Nottingham demonstrating the size of the audience for the sciences. It became so common to attend lectures that parodies were popular such as Lowe's performance of George Savil Cary's 'Lecture on lectures', featuring 'surprising experiments and operations, and drawn swords, gold medals, lemons, boxes, looking glasses, etc'.[52] The competitive challenge was met through provision of more subjects, more elaborate demonstrations and a wider range of equipment. Pitt for example, for his lectures on natural and experimental philosophy in 1779, included a 'very extensive apparatus' which he claimed was 'the largest collection in the mechanical branch of any in the kingdom'. It included a grand orrery, planetarium, cometarium, globes, air pump, electrical machines, pile drivers, cranes, magnets, optical instruments, telescopes, microscopes, water and steam engines and models. Small-scale engines were used to demonstrate the application of steam power to commercial and industrial uses such as raising water and supplying water, pumping ships, and later, powering manufacturing machinery.[53] The emphasis upon industrial and manufacturing applications of natural philosophy and the economic benefits of these innovations calculated to appeal to the merchant and industrial interest in manufacturing towns is also evident in John Weaver's course of lectures of 1784. Weavor's list of apparatus was twice as long as Ardens and included wedges, pulleys, corn mills, steam engines, models of pumps, lenses, telescopes mirrors, five large electrical machines, orreries, planetariums and large globes. With the assistance of a mechanic in Sheffield he had also constructed 'a working model of a steam engine, on a construction (it is presumed) superior to former inventions of that kind' and which he promised would be a hundred pounds cheaper to construct and maintain with fuel.[54]

Competing lecturers tried to heighten the dramatic, comic, spatial and theatrical aspects of natural philosophy. For his lectures on astronomy using the 'astrotheatron, or large transparent orrery' of 15 square feet, 'music being a sister science to astronomy', recalling the Pythagorean music of the spheres, Long engaged the band of the eleventh regiment of light dragoons to perform 'favourite pieces, previous to, and in the interval of the lecture'.[55] A poem that appeared in the *Nottingham Journal*, ostensibly by a member of the audience, celebrated the 'sister fountains' that dispensed 'food to the mind and rapture to the sense':

> Whilst on our ears celestial measures roll,
> And solemn numbers harmonize the soul,
> Sublimely awful, to the wond'ring eyes
> The daring copies of creation rise.

Whilst praising the skill of Long's presentation, the anonymous author alluded to the natural theological benefits that the lectures conferred:

> Though great the learned artist's bold designs
> And nicely as in various parts combine'
> Sunk in the contrast to the mind's eye here
> His scientific labours disappear;-
> Enough that haply, with superior force,
> He marks the wheeling planet's destin'd course,
> Unfolds the volume of great nature's laws,
> And points (while truth the solemn semblance draws)
> The intellectual sight to the eternal cause.[56]

There was clearly a significant Georgian Nottingham audience for the sciences although it is difficult to determine the impact of specific aspects of natural philosophy upon individual behaviour without detailed knowledge of attendance at private schools and lectures. The impact of scientific ideas upon some of the most prominent Dissenting figures and reformers is evident and it is clear that some such as Peat, Wilkinson and Goodacre made a living from teaching such subjects. One example of the stimulus that intellectual culture could provide to industry is the inspiration it gave to the development of local bleaching and cotton spinning industries. Robert Hall (1756–1827), a cotton spinner from Basford near Nottingham was educated at the Nottingham Academy like William Strutt, another inventive east-midland textile manufacturer. Here, Hall was taught aspects of natural philosophy and European languages which enabled him to keep abreast of developments in chemistry applied to bleaching. Hall gained a detailed knowledge of the work of Antoine Lavoisier, Carl

Wilhelm Scheele, Claude Louis Bertholet, Thomas Henry and Joseph Black through his education, reading, correspondence and presumably attendance at philosophical lectures, which he utilised to become one amongst the first to introduce chlorine gas into the bleaching process in 1789. As we have seen, most visiting lecturers emphasised the utility of scientific knowledge for industry, John Warltire for example, included airs in his syllabus and had close connections with the Lunar Society. Hall corresponded with Thomas Henry at Manchester who told James Watt in 1790 that 'my method of beaching by vapour succeeds admirably in the hands of a bleacher (i.e. Hall) in the neighbourhood of Nottingham.' Further evidence about Hall's procedures came out when he spoke as a witness at a trial instigated by Charles Tennant and others in defence of his patented bleaching process. The process involved piping the chlorine gas into an enclosed receiver containing the goods to be bleached hanging on a frame. They were lowered into an agitated mixture of water and lime through which, after a 1791 modification designed to reduce dangerous emission, the gas rose.[57]

DISSENT AND GOVERNMENT

As we have noted with respect to county towns, English Dissenting and reforming groups pursued the sciences as one means of asserting their cultural and political identity and counter civil discrimination. Where local government and society was dominated by Tory and High Anglican Church oligarchy such activity provided an alternative public platform and the means through which socially marginal groups could assert themselves. Nottingham, however, was one of those towns where Dissenters exerted considerable influence in the governing oligarchy, although, as we shall see, they could still face vigorous local criticism for their religious and political opinions. The sympathetic John Aikin remarked that Nottingham was 'one of the few places in this kingdom' in which the 'principles of civil and religious liberty' regarded by many as 'fundamental to a free constitution and of the highest importance to human society' were 'allied to municipal power and magistracy'.[58]

Through his agents and symbolised by domination of Nottingham Castle in local topography the Duke of Newcastle, whose principal seat was at Clumber Park, enjoyed considerable powers of patronage and electoral influence in the county. The Duke and other Nottinghamshire aristocrats and gentry served as patrons and donors for charitable ventures such as the General Hospital and the Lunatic Asylum. However, represented by the rival hill upon which High Pavement Presbyterian Chapel and many merchant houses stood and the unusually large Market Place, Nottingham government was dominated by Dissenters to the extent that they held around a third of corporation positions despite the relatively small size of the High Pavement congregation. Multiple office holding was common and offices were exchanged between and within small family groups. Nottingham Dissenters did not always operate as a homogenous group and there was considerable debate within the corporation during the 1780s and 1790s, when enclosure controversies occurred and the election of Tory opponents as junior councilors was tolerated. At some 3,000 by 1800, the relatively large number of burgesses as a proportion of the urban population made the political system fairly democratic resulting in some turbulent and closely fought borough elections. Burgesses were entitled to elect members of the senior council when vacancies occurred, although these had to come from members of the clothing and livery. They also much more regularly voted in the usually contested elections to the six positions on the junior council.[59]

Most Nottingham Nonconformist political campaigning was led by members of the High Pavement Chapel, especially Rev. George Walker, minister from 1774 to 1798, and his friend the classicist and biblical scholar, Gilbert Wakefield, philosophers of national reputation and the most prominent Nottingham intellectuals. Wakefield was regarded as one of the greatest British classical authorities and communicated his ideas to many influential figures such as Charles James Fox. Walker was prominent in national Whig and Dissenting circles being 'intimately acquainted' with Joseph Priestley, highly regarded by Richard Price and friendly with Benjamin Franklin.[60] Subscribers to his first set of sermons came from across England and included book clubs and societies, ministers, clergy and other individuals in towns as far afield as Norwich, Manchester, London, Sheffield and Derby, reflecting the geographies of late-Georgian rational Dissent.[61] Walker and Wakefield were members of the book society at the White Lion tavern in Nottingham and organised an informal literary society composed 'of a few select individuals, accustomed to meet alternately at each other's house'. Members were 'generally of a description superior to what most provincial towns are capable of affording, men of cultivated understandings and of great moral worth'.[62]

Wakefield enjoyed the conviviality of literary circles commenting 'with a pensive pleasure, saddened by regret' on the 'delightful converse' that took place in the society of students and tutors connected with the Warrington Dissenting academy, where he had been employed before returning to Nottingham. Wakefield described the Nottingham society as being 'composed of a select number of friends, congenial in sentiment and disputations'.[63] The club included individuals with literary and scientific interests including Rev. Nathaniel Philips, Walker's colleague at High Pavement between 1778 and 1785, and Nicholas Clayton, minister between 1785 and 1794. Prominent Nonconformist professionals also probably attended such as the Presbyterian physician John Attenborough and the medical men Snowden White, John Storer and Thomas Hawksley. The Mansfield attorney Samuel Heywood, a close friend of Clayton, Walker and Wakefield, also joined the society. His early death in 1789 at the age of 34 was a profound shock to the members, Wakefield commenting that his loss was 'not confined to the narrow circle of our society' but one felt by the wider Nottingham community. It prompted Wakefield to compose a poetical epitaph celebrating Heywood's 'duty, gratitude, love and kindness' and his 'ease, urbanity and cheerfulness' at 'the request of our society'.[64]

Interconnections within the group and their shared interest in the sciences were stimulated by education and pedagogical activities. Philips was taught by David Jennings and others at the Hoxton Presbyterian academy and was 'a good astronomer and proficient in the physical sciences.' He subsequently took over the school that had been conducted by Anna Barbauld and her husband Rev. Rochemont Barbauld at Palgrave, Suffolk.[65] Similarly Clayton, who had been a student under Philip Doddridge at the Northampton Dissenting academy, was a proficient mathematician and natural philosopher and 'not merely a theoretical but a practical mechanic – an excellent workman with a lathe and in cabinet work'.[66] Wakefield regarded him as 'my very particular friend' and when Heywood died and Wakefield left Nottingham, Clayton told the latter of his 'deep regret for the absence of your society and … anxious solicitude for your welfare and happiness'.[67] Clayton's interest in mechanics was shared by Walker who composed an essay on the application of steam power to cotton spinning and was a partner and Nottingham agent for Major John Cartwright and Rev. Edmund Cartwright's pioneering woollen manufacturing mill at Retford, Nottinghamshire in addition to sharing the former's political platforms.[68]

The literary society helped to sustain Nottingham as an important Dissenting intellectual centre with links to the Manchester philosophers, Scottish universities, the Warrington Academy and leading metropolitan writers. Simpson, Walker and Clayton attended Glasgow University after their education in Nonconformist schools. Simpson was taught at Warrington where Priestley was tutor during the 1760s, and between 1772 and 1774 Walker was mathematical tutor at the academy. Wakefield was made an honorary member of the Manchester Literary and Philosophical Society after his 'Essay on the origin of alphabetical characters' was read and praised at two meetings.[69] In 1797 Walker was appointed professor of theology at Manchester College and in 1804 president of the Manchester Literary and Philosophical Society, labouring to keep the former going at great cost to his own health after the resignation of other tutors in the face of falling revenues and declining student discipline. Wakefield was classical tutor at Warrington between 1779 and 1783, whilst Clayton was the last principal tutor at the academy (1780–3), using his mechanical skill to produce 'some of the most accurate and highly finished' pieces of scientific apparatus for the laboratory. Walker, Wakefield and Clayton were all friends with Dr John Aikin, Wakefield composing the inscription for his monument at Cairo Street Chapel, whilst Aikin reciprocated by writing short biographies of Walker and Wakefield.[70]

REFORM CAMPAIGNS

Dissenting and Enlightenment progressive scientific ideas helped to inspire the Nottingham literary society coterie to adopt a strong reformist public platform during the 1780s and early 1790s supported by local business and intellectual networks. This was evident in their campaigns for urban improvement funded by enclosure, for non-denominational education, for political reforms and against religious intolerance. Whilst the reform campaigns enjoyed some success in peacetime, as the demands of war grew, the political reaction against reformers gathered pace. Middle-class sympathisers kept quiet and organised reform societies tended to go underground. The destruction of Priestley's laboratory by the Birmingham 'church and king' mob with the tacit connivance of local magistrates and Wakefield's imprisonment on the charge of seditious libel for publishing a tract against the British government's handling of the conflict, were two of the most notorious examples of the crackdown.[71]

Campaigns against the Test Acts and for parliamentary reform in Nottingham were led by Walker and Wakefield and supported by most of the corporation, whilst Walker was able to mobilise inter-denominational support from Dissenters in adjoining counties for a 'committee of the midland district'. He hoped that this 'chain of intercourse and communication' would advance 'in order through successive gradations to represent' the 'wisdom of the whole body', creating a 'permanent council of the Dissenters of the whole kingdom' as 'speculation alone is a feeble instrument, if it be not aided by action'.[72] The case for religious equality before the law was effectively argued in Walker's pamphlet, *The Dissenter's Plea* which Fox regarded as the best written on the subject of the Test Acts. The 'disgraceful' laws against Nonconformists had, 'for more than a century ... debased the character of Protestant Dissenters, by cutting them off from the common privileges of citizenship, and stigmatising them as outcasts of society'. On the basis of equality of rights Walker argued that citizenship entitled all to freedom of worship as an act of personal conscience without interference from civil authorities.[73]

Though social hierarchies were inevitable and had divine approval, all as citizens were entitled to democratic participation through manhood suffrage, whilst government needed to be freed from corruption and divisive factionalism. Walker spoke at a Nottinghamshire county meeting in 1782 at Mansfield in support of the extension of suffrage, shorter parliaments and against rotten boroughs and again in 1785 at another county meeting of freeholders in Newark in support of parliamentary reform when the platform included his friend Major Cartwright.[74] The reformers demanded that government power be reduced, rotten boroughs removed and seats redistributed so that the manufacturing interests of England could be better represented and the 'three civil orders' of the 'excellent' constitution be restored as a 'well-balanced depository of that supreme power which every government must somewhere repose'.[75] They sympathised with the early stages of the French Revolution and joined with the Derby philosophers in sending a delegation consisting of Dr William Brookes Johnson and Henry 'Redhead' York to the French assembly in 1792. Walker also composed an address from Nottingham and Derby Dissenting ministers in support of Priestley. A Nottingham petition in 1793 promoted by Walker proposed complete male suffrage and therefore went further than Fox and most of the Whigs nationally. This was signed by 2,500 citizens and presented to Parliament by Robert Smith one of the Nottingham MPs, but by 1794 the reform movement in Nottingham no longer commanded a public platform despite continuing hostility to the war. Reforming views, did, however, continue to be articulated by Dissenters and Whigs under the shelter of the relatively sympathetic corporation, although many also signed loyalist petitions.[76]

As president of the Manchester Literary and Philosophical Society, Walker held probably the most important office in Georgian English provincial science. Like Priestley he regarded rational reform of the constitution as part of broader social improvement inspired by Enlightenment natural philosophy and rational religion. Wakefield and Walker had detailed knowledge of the sciences as the latter's essays and surviving lecture notes on subjects including natural history, meteorology, hydrography, geology, mineralogy and zoology reveal.[77] The works of both are suffused with references to natural philosophy and analogies founded upon the sciences to support moral, theological and political arguments. This knowledge was used to attack materialist tendencies in Enlightenment philosophy in an analogous fashion to their opposition to extreme radicalism. Presbyterian emphasis upon the role of individual conscience and self-discovery entailed study of the natural world and rational biblical analysis. Campaigns for religious tolerance and representation fostered suspicion of overbearing civil and aristocratic authority and scepticism towards received systems of learning. This emphasis upon individual learning rather than received authority is manifest in the pedagogical philosophies of the Nottingham schools favoured by Dissenters such as those conducted by Charles Wilkinson, Rev. John Blanchard and Robert Goodacre, in particular, the latter's emphasis upon empiricism, experimental demonstration and discovery. It also partly explains the eagerness with which Nottingham Dissenters embraced non-denominational institutions and experimental educational institutions.[78]

Wakefield's celebration of the sciences was as strong as Walker. He asked 'what subject of human contemplation shall compare in grandeur with that which demonstrates the *trajectories*, the *periods*, the *distances*, the *dimensions*, the *velocities*, and *gravitation* of the planetary system?' Newtonian science showed that the diameter of the earth's orbit was 'but an evanescent point at the nearest fixed star to our system'. Likewise, 'the first beam of the sun's light, whose rapidity is inconceivable' might still be 'traversing the bosom of boundless space'. Language sank 'beneath contemplations so exalted', inspirational and revealing of the 'GREAT ARTIFICER of that WISDOM which could contrive this stupendous fabric' and the providence and power 'whose hand could launch into their orbits, bodies of a magnitude so prodigious!' Most of Wakefield's work, inspired by natural science and Newtonianism, concerned the application of Enlightenment ideas to classical, theological and political subjects. Wakefield tried to apply philosophical reasoning to classical studies through textual exegesis employing his profound knowledge of ancient languages and the classics to facilitate biblical analysis. Many of his works, notably the *Silva Critica*, aimed to unite theological, classical and philological learning and analyse biblical texts using continental scholarship and classical sources, utilising his knowledge of Hebrew, Syriac, Chaldean, Arabic, Persian and Coptic. Wakefield defended Christianity against the atheism and materialism promulgated by Thomas Paine in his *Age of Reason*, but supported a rational form of Christian theology shorn of public worship, doctrinal orthodoxies such as the trinity and incarnation, and national elements. His attacks upon Paine therefore won him few friends amongst church and king loyalists.[79] Wakefield's studies were similar to the textual and exegetical criticism promoted by Priestley in the *Theological Repository* to which the former contributed biblical translations, and reflected the quest for a rational theology in liberal Dissent.[80]

The search for a philosophic theology exploiting rational biblical and classical exegesis scholarship is evident in Wakefield's essay on alphabetical characters which was so popular with the members of the Manchester Literary and Philosophical Society that they heard it twice and elected him an honorary member. Wakefield celebrated the enlightened age when the 'human mind [had] acquired so much honour' by the introduction of 'astonishing improvements' in philosophy and science 'beyond the example of former ages'. However, just as Walker had argued that Enlightenment psychology and natural philosophy underlined the importance of the intervention of 'divine will', so Wakefield concluded that the elaboration of rational history and the classics demonstrated the need for divine intervention. The 'same ingenuity and strength of faculties' employed to 'investigate the sublime laws of the planetary system ... adjust the tides ... disentangle ... rays of light ... detect the electric fluid' and extend 'researches into the remotest region of mathematical science' could be applied to other endeavours such as enquiries into the origins of alphabetical characters.[81]

As we have seen, Walker was a skilled mathematician and mechanic whose essays embrace a range of subjects from classical literature to astronomy and psychology.[82] His knowledge of astronomy and psychology is evident in essays on the immateriality of the soul, on imitation and fashion and on the relationship between reasoning in natural and moral philosophy. Whilst he enthusiastically embraced much of the progressive Enlightenment philosophy espoused by writers such as Erasmus Darwin and Thomas Beddoes, Walker was hostile to the materialistic tendencies that he discerned in their work. Walker described Darwin as 'an author to whom all the secrets of nature appear to be revealed' and celebrating his 'manly intrepidity, and defiance of all antiquated notions'. In turn, Darwin accepted Wakefield's textual criticisms of the *Loves of the Plants*, condemned his imprisonment and urged him to call at Derby if he was 'ever released from the harpy claws of power'. In contrast to the developmental psychophysiology espoused in Darwin's *Zoonomia* and other works, Walker pointedly contended that the operations of Newtonianism manifest through the gravitational impulse and Hartleyan associational psychology supported, by analogy, the argument that the whole of the universe and the body were 'permeable by the soul'. Although equally as inspired as Darwin by the wonder and magnitude of the Newtonian universe as its boundaries were advanced by William Herschel's astronomical observations, Walker drew slightly different conclusions. The 'universal mind' willed 'the motion of the orbs of ten thousand thousand solar systems, and the endless succession of motions, which are constantly taking place in the vast extent of the universe'. Likewise, the human body was 'the theatre of the operations of the human soul', an argument more convincing and conferring 'more dignity on the subject, than any tale of vibrations and vibranticulae' or the operation of 'subtile' fluids such as the electricity in the manner advocated by philosophers such as Darwin.[83] Like Darwin and Priestley, Walker's psychology was founded upon associationism, especially the principle of imitation from which character and behaviour were derived. Walker contended that Darwin had been correct to argue that this was 'assuredly the source from which the whole character of animal creation' was acquired, but that he had been timid in 'not extending it to man also' as he proposed in his own 'simple theory of all animated nature'.[84]

OPPOSITION TO WHIG SCIENCE

Just as the linkage between progressive Enlightenment ideas, urban improvement and enclosure was seriously challenged even where Whigs and Dissenters were prominent in local government, so there were local challenges to the Lockean and Priestleyan philosophy articulated by Walker, Wakefield and their circle. The mathematician, philosopher and poet William Cockin (1736–1801), for example was probably the most distinguished tutor at Charles Wilkinson's Nottingham Academy and a scholar of national reputation who had taught in London academies and Lancaster Grammar School. A close friend of the artist George Romney, Cockin was the author of a mathematical textbook, treatises on language and natural philosophy and assisted Thomas West in the composition of his *Guide to the Lakes* (1778).[85] Cockin also compiled meteorological observations, publishing an account of one 'extraordinary appearance' in the *Transactions of the Royal Society* in 1780.[86] Given that Cockin was a critic of Walker's friend Priestley, of Locke and even Newton, his appointment would appear to indicate a deliberate attempt to differentiate the academy from local Dissenting intellectual circles and the corporation.

Cockin used a poem in praise of the Scottish philosopher James Beattie with lengthy notes to attack Priestley's writings, especially his assault upon the 'common sense' theories of Beattie, Thomas Reid and James Oswald, and deists, atheists, rational Christians or Socinians.[87] It was intended to 'caution the young adventurer in philosophy against the dangerous impositions of metaphysical subtlety and to impress on his mind a favourable idea of our most holy religion' against the 'quibbling infidel'. Cockin urged Beattie to

> Proceed, great sage; in this good work proceed,
> Expose more HUMES and hallowed be the deed!
> Hark! REID and OSWALD sounding with applause
> The inspiring call to this important cause;
> Then urge the charge till each apostate page
> Feel the keen vigour of they honest rage ...

He differentiated his educational philosophy from Whigs and Dissenters by condemning the metaphysical, scientific and psychological systems of Locke, Newton and Priestley which he regarded as misrepresenting natural complexity with overly simplistic mental and physical laws and properties. Their ideas were dominated by system and opinion unjustified by scripture and they ignored the fact that 'mankind are endued with certain innate practical principles', 'arbitrarily assuming' that matter is nothing but 'powers and properties'. Really 'admirable in nature' were the 'deeply hidden energies with which it is replete, the wisdom and contrivance shewn in the adjustment of its parts and operations, rather than the paucity of ... its first principles'. Cockin condemned Locke for the 'tendency and incompleteness of his theory' and for not 'introducing into the philosophy of human nature the clearness and satisfaction expected', just 'endless perplexity and doubt'. Although not long since 'almost every operation in the material world was accounted for by the *Newtonian attraction* and *repulsion*', Newtonianism was one of the 'erroneous systems of philosophy'. Subsequent experience had discovered that 'these favourite principles' were 'very defective' because of the degree of complexity exhibited even in 'the most narrow province of enquiry' which resisted the framing of simple universal laws. Priestley was also attacked for the 'indecency of language, which nothing surely of the gentlemen could adopt', for 'heaping no common abuse on some of the warmest friends of religion' and leading 'his pages with personal and illiberal terms', although Cockin admitted that time had 'greatly improved' his 'controversial style' and 'metaphysical acumen'.[88]

HENRY KIRK WHITE AND NOTTINGHAM SCIENCE

The existence of a philosophical society in Nottingham by the end of the 1790s, which was more formally constituted than Walker and Wakefield's literary club (though it may have been connected or inspired by it), is attested by the experiences of the poet Henry Kirk White who describes some of its meetings in detail. According to Robert Southey in his life of White based upon family records and discussions with those who knew him, by the age of 15, the poet was already a skilled mechanic who studied scientific subjects such as chemistry, astronomy and electricity. Aged 14 in 1799 he became 'ambitious of being admitted a member of a literary society then existing in Nottingham, but was objected to on account of his youth'. [89] White's election into the society and some of its activities are described in three letters sent to his brother in October and November 1799.[90] In the first, White describes the society as being 'called the scientific' to which he has been elected as professor of literature, though the youngest member. After he gave a two and three quarter hour lecture on the subject of genius he recorded that the meeting resolved that 'the most sincere thanks' be given to him. They praised his 'most instructive and entertaining lecture' and assuring him that they had 'never had the pleasure of a better lecture delivered from the chair'. He was then elected onto the committee although almost blackballed for his youth. Southey remarked that 'it may well appear strange that a society in so large a town as Nottingham' in order to acquire and diffuse knowledge and 'respectable enough to be provided with a good philosophical apparatus, should have chosen a boy, in the fifteenth year of his age, to deliver lectures to them upon general literature.'

According to White the society was very well equipped. It had 'every kind of philosophical apparatus' including an electrifying machine, air pumps, globes, a microscope, telescopes, a camera obscura, astronomical tables, chemical apparatus and various engines. In a later letter White states that the society 'is not in fact totally a literary or philosophical society' but is called 'a society for the promotion of useful knowledge'. It had a library and took in various periodicals including the *Reviewer*, the *Monthly Magazine* and the *Philosophical Magazine*. Lectures 'were now chiefly on the subjects of electricity and magnetism' and the society had 'an excellent electrifying machine' which provided 'both an amusement and instruction'. White provided a brief account of the history and basic ideas of 'the most astonishing science' of electricity as he was coming to understand it for his brother and drew a sketch of the society's electrical 'useful and elegant' cylinder belt-driven machine. The 'earth and animal creation in short everything which surrounds us' contained a quantity of 'the very subtle fluid called electricity' which was essential to life as 'without this we could not exist or live'. The fluid was 'brought into action' when it 'reacts certain bodies and expels others'. The society had also been discussing meteorological electricity. White remarked that clouds consisting of 'vapours' had been condensed and surcharged with this fluid so that on 'the first object it approaches it discharges the over plus which flies rapid by to it' which is what is described as lightning. Once the fluid had gone 'it produces a vacuum into which the air rushing causes the explosion called thunder', thus the 'wonderful appearance' was 'simply defined'.[91]

White's account suggests that the society was quite a formally constituted body. It had officers and special lecturers known as 'professors' for each subject area, a committee of governance, rules and regulations, a collection of scientific apparatus for demonstration and experimentation and a library taking major periodicals. The lack of publicity is perhaps understandable given the hostility towards Nonconformists, reformers and natural philosophy during the later 1790s exacerbated by the seditious meetings act of 1799. Other philosophical societies such as that founded by Richard Phillips at Leicester ceased to gather because of political pressure.[92]

CONCLUSION

This chapter has argued that natural philosophy was an important part of public culture and private activity in Georgian Nottingham. As a regional capital, historic county town and agricultural and manufacturing centre with a large and influential Dissenting population, mathematical and scientific education were valued for a variety of reasons which shaped the character of Nottingham science. The sciences were valued for their role in natural theology, as polite, unifying and fashionable knowledge for the middling sort and gentry, and for their apparent utility in agricultural, industrial and mercantile improvement. Nottingham's Whig reforming culture was nurtured and found a ready audience in the powerful and influential Dissenting groups such as the Presbyterians who dominated the town's government and encouraged pedagogical experimentation. Nottingham philosophical culture can therefore be compared with Manchester, Birmingham, Derby and other towns with rich Georgian scientific cultures, however like the latter, with Dissenters playing a major role in urban government. Nottingham philosophical, Dissenting and reforming circles fostered the careers of Wakefield and Walker, two local philosophers who utilised Whig and Dissenting networks to obtain a national platform for their ideas.

Despite the strength of this merchant Whig oligarchy, the economic character of Nottingham combined with the relatively large electorate produced a fluid and dynamic urban society determined to assert its provincial identity as well as follow the latest metropolitan fashions. It was possible to advance through trade and business encouraged by Nonconformist utilitarian precepts of individual and social improvement driven by local social influence and the continuing perception of national exclusion. Tutors such as Peat, Wilkinson and Goodacre stimulated and satisfied the varied demands for philosophical education, competing to offer an ever greater range of scientific subjects utilising larger and more sophisticated philosophical apparatus. Many itinerant and resident lecturers offered courses on mathematical and philosophical subjects including some of the most important itinerant lecturers such as Arden, Warltire and Booth. Considerable competition resulted in ever-more grandiose scientific theatre including the provision of music, more elaborate apparatus and dramatic demonstrations which helped to enlarge the audience for public science to include women and children. Local newspapers carried frequent letters, editorials and articles on scientific subjects especially meteorology, mathematics, chemistry and astronomy as well as advertising a large number of scientific books which catered for local business, agricultural and intellectual concerns. Local craftsmen sold scientific apparatus for use as ornaments and research tools whilst libraries circulated copies of scientific texts. Enlightenment scientific knowledge was utilised in the rhetoric of improvement, both urban and agricultural, as represented by the meetings and prizes of the local agricultural and plantation societies.

In the post-war period, the sciences in Nottingham gained a more sustained public platform through the activities of the Nottingham Subscription Library (1816) and scientific associations nurtured under its wings such as the Literary and Scientific Society and the Nottinghamshire Natural History Society (1836). Professionals such as the physicians Storer and White provided continuity with pre-war clubs and coteries. However, unlike eighteenth-century Nottingham philosophical clubs, the post-war philosophical associations publicised meetings in newspapers, published catalogues, advertised lectures, promoted public exhibitions and supported original scientific research such as George Green's electro-mathematical studies.[93]

8   The Sciences, Enclosure and Improvement

INTRODUCTION

The strong reciprocal relationship between the urban spaces of Enlightenment and the development of Enlightenment ideas has long been acknowledged, particularly as evidenced in attempts to foster rational, civic and governable urban spaces inspired by the sciences. The realities of Enlightenment urban living were, however, fragmented, contested and varied embodying contradictions inherent within intellectual endeavours as well as the operation of commercial forces. With respect to contested notions of improvement, for instance, more traditional 'Tory' views of the countryside sometimes competed with emphases upon modernity, utility and urbanity.[1]

This chapter explores two ways in which the sciences impacted upon the forms of Enlightenment townscapes: through the direct and indirect ways in which they shaped architecture and building and secondly through the inspiration they provided for campaigns of urban improvement.[2] It examines how the spatial character of towns shaped intellectual cultures and the reciprocal role of scientific ideas in urban development. The translation of an essentially rural notion of 'improvement' to urban contexts explains how it frequently encapsulated seemingly contradictory notions of tradition and modernity and provided rhetorical justification for various economic and social changes. Translations of rural to urban improvement were facilitated by commercial interests and aristocratic estate exploitation. Detailed analysis of local geographical, cultural, economic and political circumstances reveals how Georgian improvement sometimes married landed interests with those of progressive urban intelligent-sias.[3] Provincial *savants* such as those in Nottingham were inspired by notions of Enlightenment modernity and progress to promote urban improvement and enclosure whilst opponents were motivated by other dialectical aspects of Enlightenment thought, including perceptions of threatened individual and community rights. Enlightenment discourses supplemented and opposed economic arguments during local enclosure controversies stimulating passionate public debates, reconfiguring local loyalties and challenging established elites.[4]

ANCIENT OR MODERN?

Mental and physical reinvention and reinterpretation of spacial forms was at the heart of the reassessment of the antiquity in early-modern society. In a classic account, Plumb saw 'ever-expanding' and 'aggressive consumption' based on industrial production as the primary driving force of modernity challenging the medieval worldview and represented by phenomena as diverse as animal and plant breeding, the dissemination of scientific ideas and the growth of new forms of children's literature.[5] Urban change and development signified and encouraged social and cultural change precipitating 'further momentum for the growth of public sociability in the Augustan era' and fostering the 'physical context and space for the enactment of new forms of fashionable socialising'. The importance of Enlightenment scientific cultures in encouraging British urban improvement is evident in the development of Edinburgh, Manchester, Bath and London, where Ogborn associates the varied experiences of urban spaces from pleasure gardens to taxation offices with contradictory encounters with modernity.[6] Through emulation, pattern books and new editions of architectural works like Marcus Pollio Vitruvius' *De Architectura* and inspired by Enlightenment rationality, neo-classicism provided languages for improvement and modernity, a utilitarian aesthetic, a means of managing probability and a series of models for urban design and reconstruction. Enlightenment rationality fostered the development of urban spaces intended to embody neo-classical principles in individual structures and across townscapes. The parts of idealised classical towns were fragments that could be replicated in the ordered whole as the crescents, squares and circuses of Bath showed so well.

In the context of neo-classicism and a vastly enlarging worldview, the sciences helped to inspire and mould town improvement which was reflected aesthetically and practically in architecture, building and urban design. Various renaissance and Newtonian natural philosophies were invoked in support of Palladianism, whilst various philosophical ideas and practices of the middle and later eighteenth century were reflected and embodied in design, construction and townscape geographies. English Enlightenment urban development followed a different path from the continental experience because of piecemeal urban improvement by entrepreneurs, landowners and their agents and the complexities of local government. Urban government was managed by a variety of bodies including parishes, manors, corporations, guilds, charities and improvement commissions, most of which advanced their own interests, whilst some larger towns such as Manchester and Birmingham had no corporation and were governed by manorial courts. Individual developers and agents were responsible for the piecemeal transformation of Bath, for instance, yet shared aesthetics and expectations fostered an attractive illusion of unified coherence and purpose. The supposed rationality of improvement was, of course, celebrated by many Enlightenment philosophers. In his *New Atlantis*, Francis Bacon portrayed a highly urbanised society engaged in philosophical studies conducting experiments to facilitate agricultural and social improvements, signifying the union between rural and urban and translation of improvement from country to town. The reconstruction of London after the fire and the attempts to foster urban Enlightenment spaces in eighteenth-century Scotland provide two widely recognised examples. The role of Scottish philosophers in the intellectual, social and cultural life of the nation and urban improvement is well documented but the relationship between English scientific culture and urban change and development is less appreciated.[7]

Bacon was, of course, a pivotal figure in the association of between modernity and progress and the sciences, although he favoured phrases such as 'advancement', 'progression' of learning or 'innovation', rather than improvement. Bacon's work, of course, helped to inspire the Royal Society after the Restoration, the *Advancement of Learning* and *New Atlantis* suggesting how learned societies could drive progress. Bacon regarded geographical exploration beyond the Pillars of Hercules as a metaphor for the new intellectual spaces revealed by the sciences.[8] It 'would be a disgrace for mankind if the expanse of the material globe, the land, the seas, the stars, were opened up and brought to light', while, in contrast to this 'enormous expansion', the 'bounds of the intellectual globe should be restricted to what was known to the ancients'.[9] Although Bacon believed that rediscovering the wisdom of the ancients was valuable, he held a much less static image of the universe and human learning constantly employing metaphors and images of movement and exploration. His vision of the future was the great town in *New Atlantis* with its baths, hospitals, dispensaries, 'Conservator of Health', ordered streets, large brick buildings, towers, subterranean caverns, engine houses and firm government. The directors of the learned institution Saloman's House helped to govern the state, gathering knowledge and conducting experiments whilst sending representatives throughout the land to gather information and publicise discoveries supported by universities and large 'galleries' with their displays of inventions, collected objects and statues of inventors. Their purpose was to enquire into 'the knowledge of causes, and secret motions of things' to enlarge 'the bounds of human empire', and they became the utilitarian model for the Royal Society and prize-giving eighteenth-century learned societies.[10] In Bacon's vision, social and intellectual progress were furthered by scholarly experiments in large specialised 'houses', gathering and publicising information under the auspices of great learned institutions which facilitated government.

As befitted his political success, Bacon tended to distinguish between ordered and unchanging government and broader social improvements driven by the sciences, applied mathematics, mechanics and horticulture. Although he often tried rhetorically to distance himself from Aristotelian scholasticism and to distinguish this from enquiring natural science, much of Bacon's work was inspired by antiquity. The myth of New Atlantis partially reaffirmed belief in a past golden age of global trade and intellectual intercourse. This relationship with antiquity typified the early-modern world perspective as European disagreements concerning the alleged superiority of the ancients in art, architecture, mechanics and the sciences demonstrate. Old religions dominated, Freemasons and intellectuals sought to rediscover ancient wisdom, David Hume investigated whether ancient or modern populations were greater, Bacon, Gibbon and others regarded the age of the Antonines as probably the zenith of history whilst the superiority of modern politics, aesthetics, mechanics and even the sciences remained uncertain. Discussing the state of architecture William Gilpin lamented that 'the secret is lost ... the ancients had it ... if we could only discover their principles of proportion' and urged that the superior wisdom of antiquity be unlocked and rediscovered through philosophical studies of ancient mythology and language.[11] Despite the birth of original literary forms like the novel, eighteenth-century writers such as Joseph Addison, Samuel Johnson, Alexander Pope and the Augustins of course, remained steeped in classical models. Learned groups and societies frequently intermixed antiquarian, chorographical with scientific subjects as papers in the *Transactions* of the Royal Society and activities of gentlemen's societies at Spalding, Stamford, Peterborough and Lincoln demonstrate.

Of course ancient precedents and models provided a psychologically and rhetorically satisfying means of interpreting and softening the radicalism of modernity. Reconstructed ruins, contrived prospects highlighting antiquities, follies, classical temples and grottoes within the Georgian landscape garden simultaneously signified tradition, continuity, good taste and modernity. Richard Arkwright incorporated Palladian decorations in his Masson Mill at Cromford, Derbyshire whilst Josiah Wedgwood succeeded most effectively by producing established designs in the latest most technologically advanced manufactories, wedding traditional skills with novel technology for middling-sort consumers. One of the earliest English language discussions of the Linnaean system appeared in notes for an edition of Virgil's *Georgics* translated by John Martyn in 1741, including extracts from *Flora Lapponica*. Respecting the achievements of forbearers and ancestors rather than negating, destroying or insulting, the notion of improvement appeared unthreatening to tradition. As Johnson's definition of improvement as 'melioration; advancement of anything from good to better ... progress from good to better ... encrease' implied, this was consonant with Georgian respect for classical culture and older forms of government, monarchy and constitution.[12] Profound and radical changes such as new institutions and buildings and novel cultural forms were clothed in a classical language to disguise novelty and mollify opposition. Those like Thomas Paine who offended against such conventions were pilloried. Though usually contested, discourses of urban improvement often commanded powerful inclusive public platforms. Aided by the common inheritance of neoclassical aesthetics and urbanity promoted in public culture and scholarly endeavour and new wealth from trade and commerce these could, in the English provincial context, as Borsay has contended, challenge and bypass metropolitan cultural suzerainty.[13] As Gibbon – an MP and son of a London alderman – remarked of ancient Rome, the 'spirit of improvement' 'represented in the majestic edifices destined to the public use' had spread beyond Italy to enhance the grandeur of 'provincial towns', through the munificence of wealthy urbanites who 'esteemed it an honour ... to adorn the splendour of their age and country'.[14]

URBAN IMPROVEMENT

The sciences were promoted as part of broader polite culture by the urban revival and determined the nature of this transformation, helping to inspire and shape the character of spatial and institutional change. As Bacon had emphasised, 'science also must be known by works ... whence it follows that the improvement of man's mind and the improvement of his lot are one and the same thing.' The concept of improvement originating, as we have seen, in the country, underscores the degree to which towns remained intimately associated – and often indistinguishable from – rural surrounds. Although its meaning was contested, improvement came to designate a large number of innovations, the perception being as Samuel Johnson said, that the age was 'running mad with innovation'. For Hume, 'the spirit of the age' impacted upon 'all the arts', the 'minds of men, being once roused from their lethargy, and put into a fermentation, turn themselves on all sides, and carry improvements into every art and science'.[15] In the urban context, 'improvement' included the development of 'superior' paved streets, squares, walks, gardens, lighting, water supply, canals and the construction of new civic buildings. Often driven by aristocratic landowners and their agents whose estates included urban areas, the transformations achieved by urban improvement are encapsulated in Humphry Repton's suggested changes to Knutsford, Cheshire (Figure 31). The idea also came to embrace innovations in poor law administration.[16]

Fig. 35. View of the Town of Knutsford, from the background of which the road proceeds to Tatton Park

[Fig. 36. View of the Town of Knutsford, as proposed to be improved and shewing, in the background, a new entrance-gate and lodges to Tatton Park.]

31 H. Repton, view of the town of Knutsford before and after improvement
from J. C. Loudon ed., *The Landscape Gardening and Landscape Architecture
of the Late Humphry Repton Esq.* (1840)

Most accounts of post-Restoration improvement have, following Plumb, tended to emphasise the importance of political stability from the late seventeenth century, or demographic and economic factors, such as urban population growth, release of new capital, and metropolitan emulation. Improvement has typically been associated with a growing and increasingly economically and culturally confident middling sort of traders, professionals and urban gentry, adopting the values and cultures of the landed aristocracy to remodel urban environments, fostering towns as rising centres of polite culture and society.[17] This was not a uniform process and there was considerable variation in the rate and character of British urban development. Much of the earliest remodelling of English towns and establishment of local improvement commissions occurred in the south and east before midland and northern manufacturing towns after 1750, a process continuing into the nineteenth century. Jones and Falkus hold that improvement was largely as a result of greater efficiency in urban government in 'handling the growth of inland trade, the profits made, and the attractiveness of investing in public works with the return secured on the rates'. They emphasise London's importance as a role model and how the process of urban remodelling after 1670 began, often using the opportunity of fire devastation, as a 'slightly lagged (and regionally scattered) effect of the activity in agriculture and commerce which was released by the Restoration settlement'.[18] Court and aristocratic patronage fostered many urban developments but the growing importance of middling-sort consumption, competition, emulation and aspirations of the middling sorts manifest through cultural expression, and aping of the aristocracy provided considerable impetus. On the other hand, this could also foster opposition, expecially where wealth, rights and privileges appeared to be threatened.[19] In order to appreciate the forms of town improvement it is necessary to look beyond economic factors and consider the relationship between Enlightenment and urban change.

Metropolitan reconstruction, of course, provided an influential stimulus and yardstick for provincial urban improvement. Christopher Wren (1632–1723), chief surveyor of London after the Great Fire, was as much Baconian natural philosopher and mathematician as architect. Wren's mechanical skills were evident when young through his design of a rain gauge and pneumatic engine. He subsequently served as Professor of Astronomy at Gresham College from 1657 and Savilian Professor of Astronomy at Oxford between 1661 and 1673 and helped to found the Royal Society, serving as third president. Wren made important contributions to mathematics and geometry that were praised by Newton and conducted experimental research into the laws of impact, on pendulums and in meteorology.

He was one of the earliest philosophers to utilise microsopes and telescopes to study and draw insects and the moon, inspiring Robert
Hooke (1635–1703) to produce his *Micrographia*. Wren's original geometric grid plan for the redesign of London after the fire was rejected,
but with Hooke as his assistant, he utilized applied mathematics to
redesign and rebuild St Paul's Cathedral and reshape the metropolis,
treating some buildings as giant scientific instruments.[20] Under the
commissioner's plan for metropolitan reconstruction, widths of
streets, classes of houses and methods of construction were standardised, with only brick and stone being permitted as materials. The revitalised city, with its neo-classical public buildings, terraced houses,
gridiron squares, parks and gardens, was the most influential model
for the transformation of provincial towns (Figures 32, 33, 34 and
35).[21]

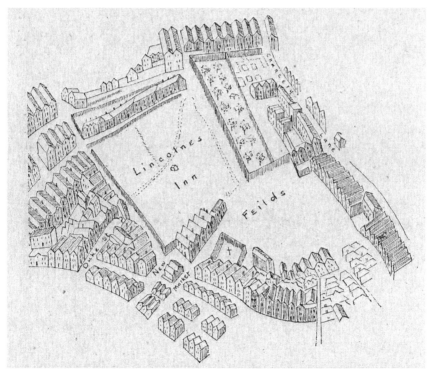

32 Plan of Lincoln's Inn Fields in 1658 from Y. J. Sexby, *Municipal Parks,
Gardens and Open Spaces* (London, 1898)

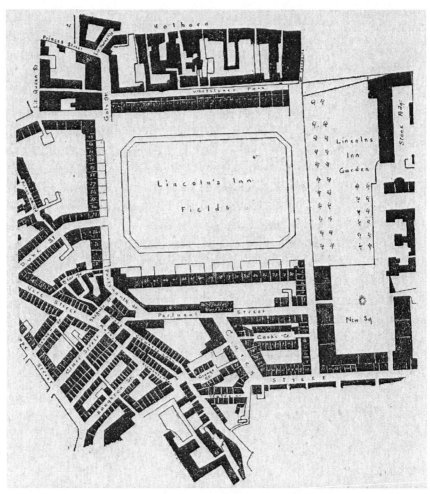

33 Lincoln's Inn Fields improved in 1780 from Y. J. Sexby, *Municipal Parks, Gardens and Open Spaces* (London, 1898)

Ogborn regards metropolitan urban improvement as a powerful manifestation of modernity. Like Habermas, Porter and others he sees London as unique because of its size, character and unprecedented growth, as the centre of a growing global merchant trade and empire presaging the birth of the modern world. Exploring spatial manifestations of change in various sites including hospitals, streets, pleasure gardens and excise and register offices using theories of modernity such as the work of Giddens, Ogborn emphasises the operation of various processes demarcating modernity including individualisation, public sphere formation, commodification, bureaucratic rationalisation, state formation, altered spatial and temporal conceptions and communication innovation. He concludes that the metropolitan experience demonstrates that modernity was a fractured and complex phenomenon rather than an overarching linear process with many countervailing trends and inconsistencies requiring geographical contextualisation.[22]

34 Leicester Square in 1700 from Y. J. Sexby, *Municipal Parks,
Gardens and Open Spaces* (London, 1898)

The impact of the sciences upon urban change and improvement is evident in Enlightenment Scotland. Emerson has argued that 'virtually every civic improvement' in Edinburgh owed something to the town's philosophical community, including the Royal Infirmary, Dispensary, observatory, Botanic Garden, Exchange and the harbor. Inspiration from the Enlightenment sciences is also evident in the improvement of Glasgow, Aberdeen and other towns.[23] The intense interest in the relationship between commercial success, polite sociability, scientific ideas and urban redevelopment evinced by Scottish intellectuals was articulated by Hume who helped to draft improvement proposals for Edinburgh in 1752 and whose essay 'Of Refinement in the Arts' makes the case for such changes. [24] The 1752 proposals, which instance the success of some London developments, demonstrate the interconnections between economic, social and intellectual endeavours in furthering the 'spirit of industry and improvement' throughout the British Isles. Removing old narrow, crowded and dirty lanes and providing magnificent new public buildings, improved capital cities would 'naturally become the centre of trade and commerce, of learning and the arts, of politeness, and of refinement of every kind' which would 'diffuse themselves through the nation, and universally promote the spirit of industry and improvement'.[25] A general Edinburgh improvement act will enlarge and beautify the city by opening new streets to the north and south, removing markets and shambles, creating a canal with walks and terraces, the expenses being defrayed by a national rate. Gentry families would be enticed from country seats into the city by excluding the poor, and in effect, appropriating features from country estates into the town to make them feel at home with such as grand neo-classical architecture, creating illusions of space and the pastoral through the provision of squares, gardens, walks and open spaces and tree planting.[26]

This vision was partly realised through various developments and James Craig's elitist design for the 'New Town', which, although quite conservative and rigidly symmetrical was innovative in its sensitivity to the site and influenced other British urban improvements.[27] There was no overall town planning but rather construction around different centres such as the cathedral, castle and markets; the new town was geometric, ordered and geometrical on a grid plan system, open to carriages and sedan chairs rather than mere pedestrian rambling. There were fewer irregularities, hidden courts or alleys, the relatively spacious streets were open to the sky with much greater exposure to light and air than in the old town, including squares, gardens and walks and there was greater uniformity in building styles, alignments and materials. According to Philo 'the resulting geometric ordering of the New Town can perhaps be described as a paramount geography of reason, a starkly 'reasoned geography'. Similarly for Daiches, the New Town was 'planned to achieve' architecturally and spatially 'the ideals of order, elegance, rationality, progress and proper social relationships represented by the Scottish Enlightenment', though of course, implementation and interpretation varied according to social and commercial circumstances.[28]

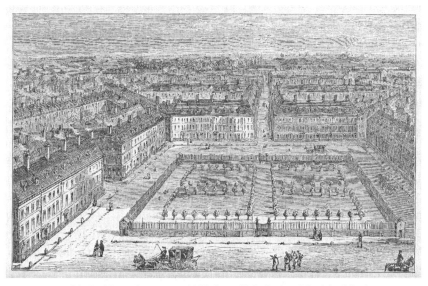

35 Red Lion Square in 1800 from Y. J. Sexby, *Municipal Parks, Gardens and Open Spaces* (London, 1898)

Reconstruction on the scale of post-earthquake Lisbon did not exist in Enlightenment England but the sciences did impact upon provincial urban improvement in various ways including road and canal projects and experiments with new forms of construction and lighting, such as those undertaken by various Lunar Society members with industrial interests. Lightning conductors, more efficient oil burning lamps, architectural and structural innovations including the use of iron and other new building materials, chimneys and heating systems, the application of modern mechanics to domestic economy, devices to reduce smoke, rationalisation of the town watch through reorganisation and the watchman's clock all illustrate how the sciences could facilitate urban change. A good example is the improvement of public buildings such as prisons, hospitals, lunatic asylums and schools. Geological, topographical and meteorological conditions, psychological and medical theories and other factors impacted upon location, building design, functions and governance. Although Jeremy Bentham's Panopticon was never constructed, it reflected and encouraged attention to such factors in the construction of semi-public and public buildings in terms of layout, design, heating, ventilation, building materials, degrees of isolation and regulation.[29]

Application of ideas and practices inspired by the sciences was fostered by medical men, many of whom were Scottish or had been educated in Scotland. At Bath, they were amongst the most vociferous voices supporters of urban improvements, encouraged by their major stake in attracting the sick as lifeblood of the local economy. Here the urban economy depended upon the clinically validated efficacy and constant recommendation of the mineral waters by medical professionals. At Norwich, Derby, Nottingham and other towns, improvement proposals during the 1780s were partly justified with reference to the medical benefits of widening and paving streets and removing signs that impeded the 'free circulation of air' and in the case of the former, the removal of old church yards to the edge of the city where decaying bodies could not cause 'pestiferous disorders'. Whilst this was not necessarily explicit in proposals, greater residential and public differentiation between rich and poor provided another attraction such as the tendency for the rich, as in Nottingham and Lincoln, to inhabit the higher areas of the town. At Norwich it was suggested by William Chase that the old church yards could be 'formed into handsome grass-plats, both for use and ornament', squares and public gardens created, whilst the islands in the river that tended to 'collect the filth of that neighbourhood' and create a nuisance could be removed, helping to attract the genteel back into the city, increasing commerce and trade. Furthermore, for the health benefits of the middling sort, in 'so large and opulent a city as this' 'a neat and commodious suite of baths rooms' ought to be erected, with 'bathing and exercise' being 'equally conducive to health'.[30]

Just as the Enlightenment sciences encouraged imposition of order to the natural world through the ideas, instruments, processes and institutions associated with the Newtonians and Linnaeans, so urban improvement often represented rationalisation and sometimes mas-culinisation of the geographies of human experience. In this way urban redevelopment can be perceived as being part of a broader masculinisation of philosophical endeavour as 'external' 'nature' be-came increasingly differentiated from urbanity, the town as mental space became differentiated from rural life, from the seasons and rhythms of medieval society.[31] Like 'nature' and applications of ex-perimental and instrumental procedures, Newtonian laws and Lin-naean taxa, the urban environment was now to be tamed and shaped by masculine rationality and commercialism, ordered and manipu-lated according to dictates of instrumental rationality, controlled space, driving out unplanned streets, wavy lines and asymmetries of the wooden vernacular, making way for the strong firm lines and columns of stone neo-classical rationality, just as natural knowledge increasingly elaborated gender attributes and excluded women from the masculine public spheres.[32] Radicals of all political and religious affiliations embraced shared and contested neo-classical discourses which could harmonise with the new rationality of planning and de-sign inspired by Enlightenment sciences.

The Emphasis upon the public civic value of architecture, design and engineering also demonstrates the impact of neo-classicism and Enlightenment purposeful rationality. The Baconian rhetoric of the public utility of natural knowledge and mechanics dominated meet-ings and public pronouncements of many literary and philosophical associations reinforced by a congruent Enlightenment utilitarian aes-thetic. This was encouraged by the notion of the 'economy' of nature promoted by the sciences and natural theology, which emphasised empirically validated 'laws' of nature and was congruent with com-mercialistic and trading interests that justified urban improvements on grounds of economy. However, it was also justified by a rational aesthetic equating beauty with utility and by the psychophysiology of associationism which accounted for beauty and sublimity in terms of individual psychology rather than merely abstract ideas. Just as the post-Renaissance symmetry of the formal garden surrendered to picturesque landscape gardens, so the Platonic mathematical ration-ality of Palladianism was succeeded by the second generation of neo-classical, picturesque, exotic and romantic forms encouraged and justified by aspects of associational psychology and philosophy. Successful architecture and building became determined by calculations of fitness and utilitarian success informed by philosophy and the sciences. Hume, whose *Treatise* strove to 'introduce the experimental method of reasoning into moral subjects', remarked there that 'most of the works of art are esteem'd beautiful, in proportion to their fitness for the use of man, and even many of the productions of nature derive their beauty from that source', equating the utility of art and architecture with the highest form of beauty. Hence, 'the convenience of a house, the fertility of a field, the strength of a horse, the capacity, security, and swift-sailing of a vessel, form the principal beauty of these several objects.' Likewise, in his *Enquiry Concerning the principles of Morals*, Hume exclaimed that 'a ship appears more beautiful to an artist, or one moderately skilled in navigation, where its prow is wide and swelling beyond its poop, than if it were framed with a precise geometrical regularity, in contradiction to all the laws of mechanics.' Similarly buildings, 'whose doors and windows were exact squares, would hurt the eye by that very proportion; as ill adapted to the figure of a human creature, for whose service the fabric was intended.'[33]

In his *Elements of Criticism*, Hume's friend Henry Home, Lord Kames considered that both gardens and architecture raised the emotions of beauty through regularity, order, proportion, symmetry, simplicity, colour and utility, arguing that it was 'the perfection of every work of art that it fulfils the purpose for which it is intended; and every other beauty, in opposition, is improper'.[34] Both William Hogarth and Hume used the example of a ship and how the dimensions, regularity, uniformity and symmetry were used as general guides during construction, the most 'beautiful' being that which sailed the best, whilst in his *Analysis of Beauty*, following Socratic models, Hogarth argued that the characteristics that produced beauty were 'fitness, variety, uniformity, simplicity, intricacy, and quantity'. Fitness 'of the parts to the design for which every individual thing is form'd, either by art or nature' he considered first as, 'of the greatest consequence to the beauty of the whole, hence 'dimensions of pillars, arches, etc. for the support of great weight' and all the orders in architecture' were regulated by this attention to propriety.[35] Similarly, in his *Aesthetics*, Adam Smith emphasised the importance of regularity, symmetry, balance, conformity and uniformity in the beautiful, finding the shape and form of gothic architecture 'not agreeable', which in architectural terms translated as 'the exact resemblance of the corresponding parts of the same objects ... as in the opposite wings of the same building'.[36]

However, all these writers emphasised that if uniformity, regularity and symmetry were carried to excess, they created tedium through repetition rather than architectural beauty. Burke went further, ridiculing the equation of beauty with utility and contending that a combination of effects including variety and novelty was required, much depending upon subjective factors.[37] In his *Elements of Criticism*, Home considered that architecture was 'not far advanced beyond its infant state' arguing that to reach maturity required much 'greater variety of parts and ornaments' than available at present. The standard classical language was good at expressing beauty and grandeur but few other emotions. Empirical work was needed to bring architecture 'to perfection' and 'ascertain the precise impression' made by individual features and their role in determining emotions as had already been partially achieved in gardening where the 'several emotions' excited by different features 'either singly or in combination' were dissembled with 'some degree of precision'.[38]

The psychophysiological basis of beauty was emphasised by Archibald Alison who argued that 'abstract or ideal standards destroyed the functions of works of art'. It was through the 'trains of thought ... produced by objects of taste' and the spontaneous stimulus to the imagination that works became beautiful, 'the sublimity or beauty of forms' arising from 'the associations we connect with them, or the qualities of which they are expressive to us'. Richard Payne Knight went further and declared that proportion depended 'entirely upon the association of ideas' and not upon 'either abstract reason or organic sensation'. Otherwise, 'like harmony in sound or colour, it would result equally from the same comparative relations in all objects; which is so far from being the case, that the same relative dimensions, which make one animal beautiful, make another absolutely ugly.'[39] Refusing to follow Knight in rejecting objective aesthetic qualities in full, Humphry Repton strove to justify his schemes by invoking associational psychology and aspects of Newtonian natural philosophy. Whilst concentrating on the improvement of country seats, Repton paid close attention to the interrelated modernisation of town and country houses and urban gardens. In his *Observations on Landscape Gardening* (1803) he contended that associations through habit conditioned attitudes towards the colour and quality of construction materials for houses, so small red and lime-whited houses offended the eye through displeasure from 'common red bricks' and the 'meanness of a lath and plaster building' respectively. In support he explicated Isaac Milner's 'Theory of colours and shadows' in order to demonstrate that it is 'the choice of colours which so often distinguishes good from bad taste' in manufactures, furniture and dress (Figures 36, 37, and 38). The introduction of colour artificially 'was 'not the effect of chance, or fancy, but guided by certain general laws of nature'.[40]

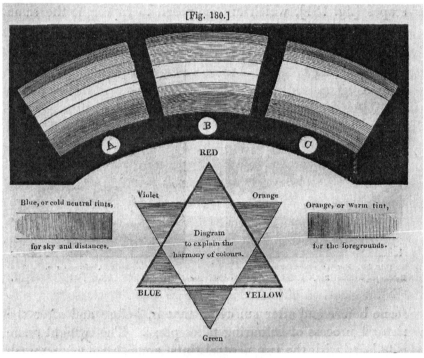

36 Diagram to explain the harmony of colours from J. C. Loudon,
*The Landscape Gardening and Landscape Architecture of the Late
Humphry Repton Esq.* (1840)

37 H. Repton, the symbols of ornamental gardening from J. C. Loudon, *The Landscape Gardening and Landscape Architecture of the Late Humphry Repton Esq.* (1840)

[Fig. 181. Morning, after the sun is risen.   Morning twilight, before sun-rise.]

Relative proportions: Red, 45; orange, 27; yellow, 48; green, 60; blue, 60; indigo, 40; violet, 80—360 parts, according to Sir Isaac Newton's scale of quantity.

38 Relative proportions of colour at different times of the morning from J. C. Loudon ed., *The Landscape Gardening and Landscape Architecture of the Late Humphry Repton Esq.* (1840)

After London, Bath was probably the most influential English townscape in the second half of the eighteenth century. On the face of it, Bath improvement owed little to the sciences, and the wealth generated through the spa and resort supported by the corporation has usually been regarded as primarily determining the architectural and spatial character of change. The relative uniformity of Bath squares, circuses, terraces and public buildings and physical and aesthetic relationships provide an impression of unified design often attributed to John Wood the elder and his son John Wood the younger. Although Wood senior's architectural vision set an important precedent followed by later entrepreneurs and builders, the schemes associated with his name were assembled piecemeal over many years. In reality Bath's 'new landscape' resulted from 'a host of decisions made by thousands of independent developers and craftsmen working throughout the century', only possible because 'a powerful architectural tradition emerged' which 'informally conditioned' the work of individual builders.[41] This tradition had special strength because of Bath's status as pre-eminent English spa and resort town sanctioned by antiquity and tradition. With its confident inspiration from the

238

sciences, mathematics and antiquarianism, Wood's work exemplifies the admixture of tradition and modernity embodied by Georgian notions of improvement exemplified by the activities of associations such as the Royal Society, the Society of Arts, Society of Antiquaries and gentlemen's societies. It is also evident in the improvement of York which was also promoted as an ancient centre. The Bath 'Circus', for instance, demonstrates inspiration from reinterpretations of antiquity as the emergence of a modern townscape. Just as Wren and Hook's post-fire London structures were originally designed partly to function as scientific instruments, so the Circus and the Crescent were inspired by ancient observatories. Wood's antiquarian and scientific concerns also shaped his influential designs for other buildings including the Bristol Exchange.[42]

ENCLOSURE AND URBAN IMPROVEMENT

Even where broad agreement existed within elites, translations of Enlightenment idealisations of improvement into urban spatial realizations were highly contested due to religious, economy and political tensions, economic difficulties and other factors. Where urban elites were divided and common lands under threat from enclosure as in Nottingham and Derby during the 1780s and 1790s, quarrels over improvement were even greater, reflecting tensions, contradictions and complexities within Enlightenment ideas invoked by opponents and supporters alike. The enclosure of common fields surrounding towns was often controversial and frequently exploited by rival groups prepared to employ the rhetoric of ancient rights and privileges or progress and improvement. Considerable care was taken by some landowners to serve the interests of tennants, but tales of imperious landlords uprooting local communities in the name of estate improvement portrayed in works such as Oliver Goldsmith's *Deserted Village* (1770) caused concern. Some common lands had originally belonged to religious houses and were donated or appropriated to towns as a result of monastic dissolution. Common land enclosures gathered pace after 1750 in response to economic, aesthetic and social demands. Improvements often, but not invariably on the metropolitan model, were thought to enhance the fashionable status and gentility of provincial towns though usually at the cost of greater social segregation. Enclosure was sometimes achieved by general consensus; however, where opposition arose, promoters used the expensive means of campaigning for a local parliamentary act, which necessitated support from local members of parliament. Support of local landowners, influential urban gentry, professionals, middling sort and local government bodies was usually necessary, although commissions could become rival power centres. Once established, improvement commissions were sometimes innovative in urban government, fostering distinctive middling-sort civic political identities.[43]

At Derby, for example, most members of the Philosophical Society supported enclosure of common lands to pay for an improvement commission, although they were vociferously opposed by an influential Tory-led group with wider political support. Erasmus Darwin president of the Derby Philosophical Society published a broadside in support of the enclosure campaign and used his authority as physician to articulate the case for improvement made by many provincial Enlightenment progressives. He and William Strutt also encouraged their friend Richard Lovell Edgeworth to contribute a supporting broadside. With its endorsement of the social, economic, cultural and governmental benefits of improvement Darwin's arguments are similar to those proposed by John Gwynn in his *London and Westminster Improved* (1766) and the plan for the improvement of Edinburgh, reflecting the familiarity of improvement discourses in British society as well as probably Darwin's residence in the latter during the 1750s as a medical student with James Keir. These emphasised the many benefits to be derived from more rational street development and the importance of an improved town for attracting the gentry and becoming 'the centre of trade and commerce, of learning and the arts, of politeness, and of refinement of every kind' which would 'diffuse themselves through the nation, and universally promote the spirit of industry and improvement'.[44] Besides taking part in the paper campaign, Darwin attended early meetings of the improvement commission with his son, Erasmus, writing to Josah Wedwood in 1791 to enquire about the right kinds of bricks to be used, costs and how they might be obtained. This correspondence demonstrates clearly the extent to which philosophical supporters of improvement strove to apply Enlightenment rationality to solve problems of paving, lighting and construction.[45]

Urban growth encouraged improvement campaigns in towns such as Nottingham which benefited from the opening up of road, river and canal communications and commerce from the River Trent (See figures 29 and 30).[46] By the mid eighteenth century, textile industries dominated in the town whilst copper and iron working, china manufactory and mineral and spa working were also important, encouraged by proximity to coal and lead mining districts.[47] Driven by immigration and a high birth rate, the Nottingham rose by about 60 per cent between 1779 and 1801 to 28,861.[48] Localised improvement was fostered by parishes, the corporation and subscription ventures including a new water supply system, the former retaining responsible for paving, lighting, watching and aspects of fire fighting.[49] The need for general urban improvement for the benefit of trade and commerce was accepted by most religious and political groups, although the precise nature of measures, responsibly for implementation and sources of funding were highly contested, particularly where enclosure of common lands was projected. Some measures clearly benefited wealthier inhabitants, as only the main streets tended to be paved, cleaned and illuminated, whilst piped water supplies were limited to subscribers, others remaining dependent on rivers and wells. The poorer sorts tended to be concentrated in the courts and alleys of lower-lying districts towards the Trent. Ancient rights accorded to common and pasturage on surrounding common lands were more important to poorer burgesses and inhabitants who had more to lose from enclosure and sale than the middling sort. Usually on the periphery of urban areas spatially and culturally, common lands frequently acted as tolerated sites for illicit social and cultural activities (Figure 39). Enclosure was therefore an act of elite moral policing and control which excited common resistance.[50]

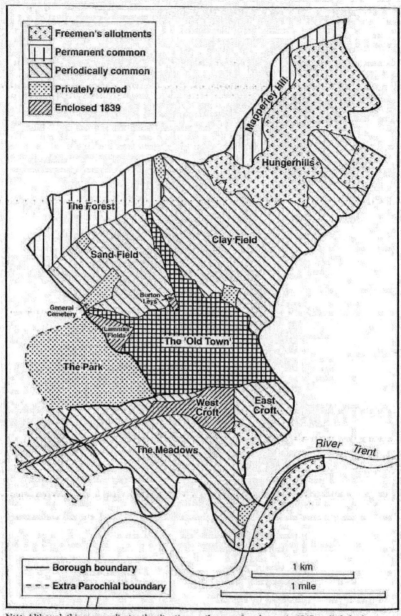

Legend:
- Freemen's allotments
- Permanent common
- Periodically common
- Privately owned
- Enclosed 1839

Mapperley Hill

Hungerhills

The Forest

Clay Field

Sand Field

General Cemetery

Burton Leys

Lammas Fields

The 'Old Town'

The Park

West Croft

East Croft

The Meadows

River Trent

— Borough boundary

- - - Extra Parochial boundary

1 km

1 mile

Note: Although this map replicates the situation on the eve of enclosure in 1845, so little had changed that it is almost exactly as depicted on the 1609 Crown Survey Map

39 Common lands surrounding Nottingham from J. V. Beckett *et al.* eds., *A Centenary History of Nottingham* (Manchester, 1997)

A major campaign to enclose the common lands to pay for improvement was undertaken at Nottingham during the 1780s, involving public meetings, referendums of the inhabitants, counter-petitions and a paper war of broadsides, handbills and letters to the newspapers. Additional proposals to raise revenue through various forms of taxation and funds obtained by the development of a Nottingham canal also circulated, although the disagreements split the corporation. The rejection of Samuel Heywood, a pro-enclosure candidate, for election to the corporation was greeted with the ringing of church bells. During the election the hostility of burgesses was demonstrated as an effigy was carried around 'with a plan of the enclosure to the polling booth and their destroyed'. A gardener was then 'chaired through the Market Place and several streets, with cabbages and carrots fixed on poles, attended by numbers in the interest of the favoured candidate'. Then a flag was produced ornamented with a blue border on which was written: 'NO STOOPS AND RAILS – NO ENCLOSURE – NO PAVING THE STREETS WITH THE BURGESSES PROPERTY – NO BRIBERY NOR CORRUPTION – COLISHAW AND LIBERTY FOREVER.'[51]

Although support for enclosure cut across religious and political affiliations and social classes, reformers and Dissenters tended to be the most enthusiastic. The campaign was led by some senior urban gentry, corporation figures, clergy, professionals, textile manufacturers, wealthy merchants and intellectuals inspired by Enlightenment progressivism as well as economic interest. Notwithstanding the common hall vote against enclosure, a number of prominent corporation figures supported enclosure notably wealthy urban gentry such as John Sherwin, John Fellows and Thomas Smith. Support also came from hosiers, merchants and professionals including Samuel Turner, an attorney and reformer and close associate of Walker, Cartwright and Heywood, who publicly voiced details of the plan at the town meeting in 1786.[52] As at Norwich and Derby, Enlightenment medical theories provided important justifications for urban improvements on grounds of public health. Nottingham medical men who favoured enclosure included John Storer, Charles Pennington, Thomas Hawksley and the Presbyterian physician Snowden White. As we have seen, the physicians Storer and White were prominent leaders of local scientific culture. Pennington, another physician, practised in Nottingham for over forty years from the 1780s, serving as first physician to the General Lunatic Asylum (1811). Hawksley a chemist and uncle of the engineer Thomas, was 'highly esteemed for his benevolence and charitable efforts' according to his memorial tablet in St Nicholas's Church.[53]

The significance of the sciences and scientific culture in the Nottingham campaign is evident from the activities of Walker and Wakefield and their relationship to Samuel Heywood. Although young, Samuel Heywood was a leading member of the High Pavement congregation who rented the house that had been occupied by Walker from the chapel trustees from 1781 and attended vestry meetings from 1779, serving as chapel warden. An intimate friend of Walker and Wakefield, he was equally inspired by Enlightenment progressivism and rational Nonconformity, associating urban improvement with social and political campaigns for repeal of the Test and Corporation acts, abolition of slavery and franchise extension.[54] On behalf of the Presbyterian congregation Walker and his associates supported the establishment of an interdenominational Dissenting academy that would educate ministers 'in the cause of truth and liberal sentiment' but considered the choice of a metropolitan location as ill advised. They accepted that London was 'the most important' place in the kingdom, but argued that 'some central spot' in the country was to be preferred for the school where the 'folly, extravagance, dissipation and vice are not so triumphant as in London'.[55] Heywood chaired and sat on various committees including those designed to concert measures against the Corporation and Test Acts and the slave trade in 1787.[56] He attended the literary and philosophical club formed by Walker and Wakefield and his premature death at the age of 34 was profoundly felt by the members. Heywood supported extension of suffrage and shorter parliaments and other national political reforms, speaking at Nottinghamshire reform meetings with Cartwright and Walker.[57] Walker and his friends simultaneously asserted Enlightenment universalism with provincial identity, regarding the social and cultural vitality of the provinces as a vital counterweight to metropolitan domination and immorality. Hence their desire for reform of ancient political institutions and systems was partially manifest in a critique of the nature of Nottingham itself as a culturally founded spatial entity by promoting a radical reappraisal of land usage and boundaries.

CONCLUSION

English urban renaissance townscapes have often been regarded as providing the context for Enlightenment cultures, for instance through the provision of public and semi-public arenas and institutions such as taverns, theatres, coffee houses and market places. The emphasis is usually upon relationships between economic, social and spatial changes rather than the impact of competing ideas. In this chapter, however, it has been argued that there was also an important sense in which Enlightenment ideas shaped the character, experiences of – and behavious within – urban places and in English as much as Scottish towns. The inspiration of the sciences is most evident in the case of urban improvement, which, as we have seen, often appealed to elites of varied political and religious affiliations. Gentry and professional groups, including lawyers, medical men and clergy came to believe in the utility of natural knowledge for urban as they had for agricultural improvement. Medical men led campaigns for enclosure and improvement in order to expand and reinvigorate their scientific knowledge, reinforce professional status and differentiate themselves from quacks. They also used their position as scientific 'experts' to agitate for urban improvement and attempt to shape the forms of development, for instance by supporting the founding of institutions such as infirmaries and improvement commissions, and public health measures.

The sciences impacted upon both the form and nature of urban improvements and improvement campaigns, for instance through enthusiasm for the application of mathematical rationality to construction and urban design. At Nottingham, although economic arguments were employed by both sides, enclosure and improvement were encouraged by the progressive Enlightenment philosophy articulated by local intellectuals such as Wakefield, Walker, Darwin and friends such as Edgeworth. Fostered by intellectual conviviality, they supported enclosure on the basis that urban improvement facilitated economic, social and intellectual progress. Opponents were also motivated by major Enlightenment concerns in contending that enclosure would destroy precious ancient rights and able to make emotional appeals for the protection of these from local oligarchies. These dialectical intellectual counter-currents help to explain, in a much more satisfactory way than economic selfishness alone, why opinion during the enclosure controversies transcended political and religious affiliations. The enclosure controversies at Nottingham and Derby conducted in various contexts including town halls, taverns and printed material combined with the referendums and large number of individuals mobilised to vote demonstrate, like borough elections, the sophistication of local urban political discourse.

9 Placing Electricity and Meteorology: Abraham Bennet (1750–1799)[1]

INTRODUCTION

Announcing the discovery of the pile in 1800 which became known as the 'Voltaic pile' and subsequently the electric battery, Alessandro Volta (1745–1827) paid tribute to three British electricians who had stimulated his work; these were William Nicholson (1753–1815), Tiberius Cavallo (1749–1809) and Abraham Bennet. Bennet remains by far the least known of the three despite recognition from Humphry Davy, Erasmus Darwin and other contemporaries. At one level as a natural philosopher or 'electrician' Bennet was a rather isolated provincial figure residing in the small town of Wirksworth in the inaccessible upland Derbyshire Peak ministering, as a lowly parish curate in place of an absentee vicar, to the needs of local sheep farmers, lead miners and gentry. On the other hand, we will see how Bennet moved within much wider regional, national and international Enlightenment philosophical networks; he also enjoyed major patronage from Whig aristocratic families, encouraging him to design scientific instruments and undertake an influential series of electrical and magnetic experiments. He counted European natural philosophers such as Darwin, Joseph Priestley and Volta as friends and used a mechanical revolving version of his 'doubler', a device that augmented electric charge, to devise a concept of 'adhesive electricity', which encouraged the latter in the formulation of his contact theory of electromotivity. This helped to ensure that descriptions of Bennet's discoveries and publications were disseminated to British and European readers. These factors lift him from the provincial and local into national scientific status in the world of electricity, as the rapid international use of Bennet's gold-leaf electroscope demonstrates. This chapter will contend that it was studies of meteorology in the Derbyshire and Staffordshire Peak that stimulated Bennet's work in electricity and magnetism, revealing much about the relationships between national and provincial cultures and the geographies of English Enlightenment sciences.[2]

The chapter takes inspiration from three recent studies of eighteenth-century electricity and meteorology with important historical-geographical dimensions: Pancaldi's *Volta: Science and Culture in the Age of Enlightenment* (2003), Jankovic's *Reading the Skies: A Cultural History of English Weather, 1650–1820* (2000) and Golinski's *British Weather and the Climate of Enlightenment* (2007) which each reveal much about the social and cultural contexts of Enlightenment sciences. In their incisive analyses of the sciences and cultures of meteorology, climate and electricity, in which Bennet was closely involved, each reveals the importance of places and localities, instrumentation and national and international philosophical communities in the interpretation, dissemination and utilisation of scientific knowledge and practices. In addition to examining the path to the discovery of the pile, Pancaldi's study of Volta underscores the vital stimulation provided by international philosophical communities with interests in electricity. These networks provided Volta with international philosophical recognition during a period of great political instability. These three books also confirm the importance of the Enlightenment quest for order manifest in attempts to understand, control and harness natural powers for social and economic improvement, including collecting everyday local knowledge, beliefs and practices of weather conditions. As we have seen, Golinski also contends that whilst Enlightenment has often been interpreted as overwhelmingly a doomed struggle to control nature, the experience of meteorology brought home to many natural philosophers and their audiences the limitations of reason, inviting celebration and wonder as much as utilitarian rational responses. Masses of meteorological observations were collected in notebooks and private diaries inspired by the hope that empirical analysis and experiment would facilitate weather predictions and possibly even allow human interventions for the benefit of agriculture, medicine and industry. Electricity and meteorology were closely interrelated, many regarding the former as responsible for changes in atmospheric conditions, rainfall and other meteorological phenomenon. As the careers of Bennet and Volta demonstrate, one means in which electricity and meteorology were studied, interpreted and rationalised was through the development of scientific instruments, especially those capable of detecting small quantities of electricity.[3]

ENLIGHTENMENT METEOROLOGY

Voluminous surviving correspondence concerning the weather, weather diaries, numerous descriptions in local newspapers, contemporary pamphlets, tracts, and other printed sources all attest to Georgian fascination with the weather.[4] Enlightenment weather studies took the form of increasingly regular observations, keeping of systematic records as well as more continuous and consistent use of instruments, including thermometers and barometers. Often assumed to be a forceful power in nature, electricity was widely regarded as a form of fire and later subtle fluid or effluvia, and therefore instruments for detecting and measuring the fluid were also employed. Arguably, the Aristotelian vision of meteorology as the study of meteors or all spectacular and striking phenomena believed to take place in the atmosphere, such as shooting stars, continued to dominate this aspect of European natural philosophy until the Enlightenment. The extra-terrestrial origin of meteors was not always appreciated and Aristotelian meteorology was combined with local, folk and popular beliefs and practices. In contrast with the nineteenth century, there was greater emphasis upon locally based studies of unusual atmospheric phenomenon.

Georgian meteorology was closely intertwined with antiquarian, chorographical and natural historical studies, and information concerning all these frequently appeared in county histories and topographies. As Jankovic emphasises, these included Charles Leigh's *Natural History of Lancashire, Cheshire, and the Peak in Derbyshire* (1700),

Robert Plot's natural histories of Oxfordshire (1677) and Staffordshire (1686), William Borlase's *Natural History of Cornwall* (1758) and Gilbert White's *Natural History of Selborne* (1790). In Jankovic's interpretation, there was a tension in English meteorology between cultures of provincial landed society and metropolitan institutions led by the metropolitan Royal Society. English weather observations were intended to facilitate agricultural and horticultural productivity and economic improvement, as the records kept by gardeners and exhortations in gardening literature suggest. This reinforced the providential and patriotic view of the climate as fundamentally moderate, benign, temperate, productive and prosperous, like the organically evolved and divinely sanctioned constitution. For Golinski, the importance of countryside, local tradition and individual experience informed local and national natural philosophy encouraged by notions of improvement, interfacing between tradition and totalising Enlightenment rationality.

On this interpretation, meteorology facilitated assertions of local and landed identities, ensuring recognition of their 'specialness', which encouraged attention to the marvelous rather than mundane and the continuation of Aristotelian meteorology. As we shall see, Bennet's Derbyshire meteorological studies, which helped to inspire his electrical and instrumental work, were motivated by a desire to understand the role of electricity in changing weather conditions and need to detect and experimentally investigate these processes. As such they derived much from this tradition exemplified by observations from many country clergymen and gentry partly inspired by natural theology and partly the desire to celebrate their locality whilst eschewing the provincial and parochial. There were always therefore, interconnections between the spaces of the Peak countryside where Bennet sought to detect, observe and measure electricity and his instrumental and experimental work at home and in various places including Birmingham, Derby and London. As Jankovic has demonstrated, philosophers such as the Cornishman William Borlase, also situated far from the metropolis in a mining county, 'bridged a gap between the meteoric tradition' recording 'public meteors of the parish and county', the 'instrumental weather record' and the 'physical parameters of the scientific atmosphere'.[5] Similarly, Bennet used a variety of instruments to conduct observations and experiments

amongst the Peak hills and in his laboratory using everyday household and workshop and local objects, including by-products of the
local lead-mining and iron-working industries, securing a national
and international audience for his publications and transcending his
provinciality. Whilst his work was shaped by the Peak, Bennet argued
that his discoveries had universal relevance and application. In terms
of national agricultural improvement, studies of local agriculture,
natural history and meteorology acknowledged and celebrated local-
ness and difference whilst claiming to contribute towards universal
Enlightenment progress.[6] This is embodied, as we shall see, in the
agricultural improvement concerns of Bennet's most important patrons, including the Dukes of Devonshire and Bedford and Sir Joseph
Banks, who paid close attention to the development of his estates at
Overton near Ashover in the Peak.[7]

CAREER AND PATRONAGE

Born in 1749, the son of Abraham Bennet, schoolmaster of Whaley
Lane, Cheshire, and Ann Fallowes of Cheadle, Abraham Bennet was
baptised at Taxal, Derbyshire, on 20 December. He was probably the
eldest son, but had at least one brother, named William. Abraham
married and his widow, Jane, was to survive him for 27 years, dying
at Mappleton in July 1826. They had six daughters and two sons. In
September 1775, he was ordained in London and appointed to cur-
acies at Tideswell and a year later at Wirksworth, on a double stipend
of £60 per annum, also taking charge of the local grammar school,
whilst in 1796 he was appointed Rector of Fenny Bentley, Derbyshire
(Figure 41). A memorial tablet in Wirksworth church records that he
was additionally 'Domestic Chaplain to his grace the Duke of Devonshire; Perpetual Curate of Woburn and Librarian to his grace the
Duke of Bedford'.[8] Bennet's clerical office, of course, required that he
continuously seek patronage and his scientific work was equally dependent upon such support. His most important patrons were Rever-
end Richard Kaye, FRS, Dean of Lincoln and absentee Vicar of Wirksworth (1787–90), Sir Joseph Banks, President of the Royal Society; the
dukes of Devonshire and Bedford, George Adams, instrument maker

251

to the King, and members of the Gell family, local Wirksworth gentry of Hopton Hall. Bennet also managed to obtain the support of other Derbyshire natural philosophers, members of the Lunar Society and of the Manchester Literary and Philosophical Society. The Royal Society served as the usual official forum for the announcement of his discoveries and sanctioned their authenticity in the scientific community. His second paper in the *Transactions*, for instance, was communicated to the Royal Society by the Kaye to whom the *New Experiments* were also dedicated. Bennet was also able to use his skills as an instrument maker and experimentalist to cultivate contacts in national and international philosophy from the 1780s, as his experimental work in London demonstrates.

40 Chatsworth House, Derbyshire, residence of the Duke
and Duchess of Devonshire (*c*.1800)

Bennet's clerical and scientific career and patrons attest to his Whig sympathies. William Cavendish, fifth Duke of Devonshire and Francis Russell, fifth Duke of Bedford were the two most powerful and influential English Whig magnates. Closely associated with the Prince of Wales, both were firm supporters and patrons of Charles James Fox, whilst Georgiana Cavendish, the Duchess of Devonshire was, of course, notorious for her open campaigning for the party. Whilst the Cavendish family led fashionable Whig society from their metropolitan base at Devonshire House, the Duke of Bedford led the Whigs in the House of Lords, earning Burke's ire for his unquestioned loyalty to Fox, even when the latter continued to welcome the French Revolution. As major landowners and supporters of urban, agricultural and horticultural improvement, the Russell and Cavendish families were also enthusiastic developers of their estates, providing important stimulation for Bennet, their patronage enabling him to undertake electrical and meteorological studies with apparent agricultural applications. The fifth Duke of Devonshire's second cousin was Henry Cavendish (1731–1810), perhaps the most distinguished, rigorous, sophisticated and diffident Georgian natural philosopher, who conducted numerous electrical experiments, including many upon the conducting power of various substances, some of which were published in the *Philosophical Transactions*. Although a famously private and painfully shy individual, it seems likely that he would have encouraged Bennet's electrical experiments. The Duke and Duchess of Devonshire strove to develop Buxton, Derbyshire as the leading northern English spa town, constructing a crescent to rival Bath and gigantic stables, whilst, as we have noted, Georgiana took a keen interest in natural philosophy and electricity, forming a mineralogical collection. The duke and duchess took lessons from Bakewell geologist White Watson who sold them mineral cabinets and topographic inlaid strata using local samples, designing a tufa grotto lined with minerals and fossils for the Chatsworth gardens (Figure 40).[9]

The Duke of Bedford also spent large sums on major building projects on his London estates and the development of Woburn, Bedfordshire, and was probably the leading aristocratic promoter of agricultural improvement on the extensive family estates, being regarded as a relatively benevolent paternalistic landlord. An original member of the Board of Agriculture set up in 1793 and first president of the Smithfield Club in 1798, he established a large model farm at Woburn experimenting with cattle and sheep breeding and crop growing, offering prizes as an inducement for further improvements. The Duke employed various innovators to promote improvement which help to explain his interest in Bennet's work as electrician, inventor and meteorologist. These included geologist John Farey, gardener and arboriculturist William Pontey and mechanic Robert Salmon who designed a model farm and invented machines for ploughing, sowing, reaping, pruning, haymaking, canal equipment, surgical instruments and even a 'humane man trap'. As Librarian and Perpetual Curate at Woburn and private chaplain to the Cavendish family, Bennet therefore had support from – and close access to – the two wealthiest and most influential progressive Whig families.[10]

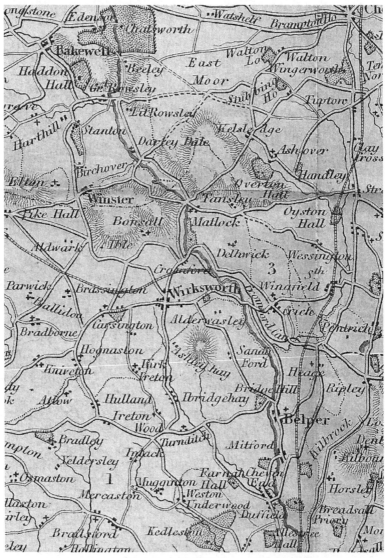

41 Detail of the Wirksworth area from map of Derbyshire,
W. Adam, *Gem of the Peak* (1840)

Aristocratic patronage was balanced with the friendship and assist-ance of English midland and Manchester philosophical coteries, which included Erasmus Darwin, White Watson and the members of the Lunar and Derby Philosophical societies. The Lunar members helped Bennet to publish his research and supported his nomination as FRS in 1789, although the relative harmony of the philosophical community represented by the Royal Society was disturbed by the political turbulence of the revolutionary era whilst declining health also impeded Bennet's research.[11]

WIRKSWORTH AND THE PEAK

Bennet's scientific career was shaped by his residence at Wirksworth, situated on the southern edge of the Derbyshire Peak region. The weather of the region is strikingly different from that of adjacent low-lands primarily because of high precipitation levels induced by eleva-tion and lower temperatures, and the diversified topography, which causes much local climatic variation. With lower levels of direct sun-light than the midlands to the south, higher rainfall, harsher winters and strong winds, all of which have impacted upon regional geology and soils, weather observations were at once problematic but also strikingly varied (Figure 42).[12] Wirksworth was the capital of the local lead mining industry, although a new important industrial develop-ment was the construction of the Haarlem cotton mill on the banks of the River Ecclesbourne by Richard Arkwright around 1778, which employed about 200 and was probably the first-ever successful such steam-powered venture. According to John Aikin the town lay 'in a bottom, eternally enveloped in smoke from the neighbouring lead and calamine work', which had continued since at least the Anglo-Saxon period. Besides the large church, town hall, free school, Dis-senting meeting house and alms houses, the importance of lead mining is evident from the fact that the town contained the Barmote Court for the wapentake in the Barmote Hall presided over by the Barmaster. The court met biannually to resolve disputes and verify staked claims with juries of twelve experienced miners. Within the hall was the fourteen pint heavy bronze oblong measuring vessel which signified the special laws and customs of the lead mining industry, including a requirement that landowners provide enough land and water to allow exploitation of newly discovered veins by their finders. The industry was at its height between the seventeenth and early nineteenth centuries, with Derbyshire producing about 5 to 6 thousand tons per year by the end of the eighteenth century. The lead miners were regarded as a highly distinct and at times truculent group, with their own customs, clothing, language and ashen grey skin colour, whilst buying, selling and transportation was in the hands of local merchants, some of whom lived in Wirksworth.[13]

42 Engraving of Peak Cavern, Derbyshire by J. Smith
from a drawing by E. Dayes (1803)

Many of Bennet's congregation were miners or connected to the industry in other ways and the exploitation of lead provided numerous metal and mineral by-products of the extraction and smelting processes which he utilised in experiments. Obtained from veins and joints within vertical fissures in the Carboniferous Limestone, Derbyshire lead tended to have its own individuating characteristics, mostly occurring in the form of galena or lead sulphide with little silver. As veins closer to the surface were exploited shafts got deeper, which stimulated the development of mine engineering, particularly the provision of pumps for breathing and drainage, whilst the cutting of soughs from the seventeenth century was another major financial and engineering achievement. The Meerbrook Sough, for instance, begun in 1772, stretched for miles right under Wirksworth, and took decades to cut. Important changes in mining and smelting techniques occurred in the period with reverberatory furnaces or cupolas replacing more traditional ore hearth smelting furnaces. Introduced from the 1730s, the former kept the ore apart from the coal used to fuel the temperature change necessary to melt it, much of which was brought over from nearby Derbyshire coal mines. Lead mining had a major impact upon the Peak landscape with the miles of soughs, worker's cottages, engine houses, workshops and smelting works with their cupolas, lengthy spoil banks full of discarded minerals associated with lead and cast aside from veins including sparkling galena, fluorspar, zinc blende, barytes and calcite crystals. The industry also stimulated local geological investigation, for instance, through attention to the contexts in which veins and joints occurred. Local and visiting natural philosophers and mineral traders including John Whitehurst, White Watson and Erasmus Darwin made close observations of the impact of mining utilising the knowledge and experience of miners and others connected to the industry, collecting specimens and exploring caves and fissures.[14]

Whilst from a metropolitan perspective Bennet had disappeared to a remote and distant part of England, as the artist Joseph Wright of Derby demonstrated in his views of Arkwright's mill at Cromford and rural iron working, the Peak was one of the most industrially innovative regions of the country offering major stimulation to his experimental work. Without the mining and cotton mills, Wirksworth would undoubtedly have been a quiet and fairly inaccessible small provincial market town on the edge of the Peak. The presence of these industries however, provided Bennet with a large range of minerals, metals and equipment with local mechanical, engineering and geological expertise, all of which stimulated and shaped his electrical and meteorological work.

ELECTRICITY AND METEOROLOGY

Various phenomena had always betrayed the existence of electric charge produced in different ways, such as rubbing wool on glass. Threads and pieces of leaf brass had been used in the early eighteenth century because they would diverge if electrified. The first real electrometer was invented by John Canton (1718–72) and used a pair of pith balls hung on fine linen threads. On the air being electrified in a room, the balls would diverge. The invention of the Leyden jar showed that electricity could be 'stored' and perhaps the strength of the charge estimated.[15] Electrical theories followed the suggestions in the queries of Isaac Newton's *Optics* (1723), which had speculated about the properties of the mysterious ether apparently inhabiting the universe and seeming to obey the laws of attraction and repulsion. The electrical theory of Benjamin Franklin (1706–90) which became widely known after 1750 – primarily in connection with experiments culminating in the invention of the lightning conductor – was no exception. According to Franklin's theory, largely accepted by Bennet, there were three states of electric charge, positive, negative and the state of equilibrium. In the positive state, charged bodies held an abundance of the electrical 'fluid' and in the negative state, there was a privation or lack of fluid. The state of equilibrium was the natural state of charge of a body. The Franklinist system suffered from one obvious problem, how to explain the fact that negatively charged bodies repulsed each other. One answer was a rival theory which held that there were actually two electrical fluids, the vitreous (positive) and the resinous (negative) fluid, thus, the charging of a body by rubbing would represent the rushing of one kind of 'fluid' into the place previously occupied by the other. The two theories were incommensurable, as no experiments had been devised which could conclusively prove either.[16]

A series of experiments and observations of fluids in motion in the home and surrounding countryside, including water globules on powdered garden leaves and ponds and exploding inflammable air bubbles laced with 'nitred paper' and gunpowder, appeared to confirm Bennet's belief that the principles of attraction and repulsion determined the behaviour of all fluids. He accounted for the repulsion of wet and dry corks in water by the 'attraction of intervening fluids', suggesting that the 'elasticity of the air, electricity and all other elastic fluids' might be 'explained in the same manner'. Quoting Priestley's observation that even seemingly improbable theories could suggest 'useful experiments' if they suited all the facts, he argued that the phenomenon of attraction and repulsion was explained by a 'mixture of a system of fluids of which some are capable of permeating glass and other solids' such as light, heat and magnetism. Hence, if the 'electrical fluid' was forced upon one surface of glass then a 'finer fluid or system of fluids' which constituted its elasticity were 'pressed out' and rarified on the other un-insulated surface.[17]

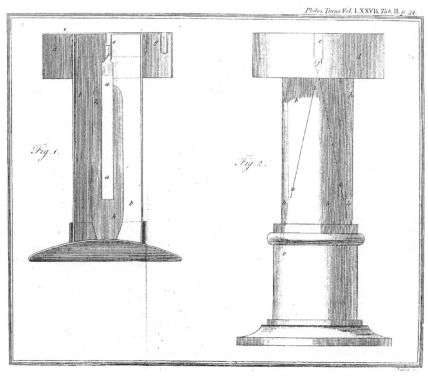

43 A. Bennet, engraving of gold-leaf electroscope from
Philosophical Transactions, 77 (1787)

Time of obſervation.	Weather.	Winds.	Baro-meter.	Thermo-meter.	Hygro-meter.	Electri-city.	Inſtruments to collect.
1787.			In.	°	In.		
Jan. 23. 11 o'clock morning	Cloudy	W. moderate	29.3	44	$0\frac{1}{8}$ M	Poſitive	Torch and bottle
24. ½ paſt 10 morning	Thin clouds	N. moderate	29.44	42	$0\frac{2}{8}$ M	Poſitive	D°
25. 11 morning	Thin clouds	E. gentle	29.51	37	$0\frac{1}{8}$ D	Poſitive	D°
25. 12 at night	Hard froſt and aurora borealis	E. gentle	29.50	38	$0\frac{6}{8}$ D	Poſitive	D°
26. 11 morning	Clear froſt	N. gentle	29.5	38	$0\frac{2}{8}$ D	Poſitive	D°
26. ½ paſt 8 at night	Beginning to ſnow	N.E. briſk	29.5	37	$0\frac{1}{8}$ D	Poſitive	Umbrella
27. 11 morning	Cloudy and likely to ſnow	N.E. gentle	29.2	35	$0\frac{2}{8}$ D	Poſitive	Torch and bottle
28. 10 morning	Snowing	S.E. ſtrong	28.8	28	$0\frac{5}{8}$ D	Negative	Umbrella and torch
28. 40 min. paſt 12 noon	Snowing very little	S.E briſk	28.5	28	$0\frac{1}{8}$ D	Poſitive	Torch and bottle
28. 11 at night.	Fair, but over-caſt	S. gentle	29.06	37	0	Poſitive	D°
29. 10 morning	Small rain and miſt on the hills	S. briſk	29.06	38	$3\frac{1}{8}$ M	Negative	D°
29. ½ paſt 10 morning	Rain ceaſed	D°	29.06	38	$3\frac{6}{8}$ M	Poſitive	D°
29. 3 afternoon.	Few drops of rain	S. ſtrong	29.06	41	$5\frac{3}{8}$ M	Poſitive	D°
30. 45 min. paſt 10 morning	Fair, but over-caſt	S. ſtrong	29.45	44	$6\frac{4}{8}$ M	Poſitive	D°
30. 11 morning	D°	D°	29.44	46	$7\frac{1}{8}$ M	Poſitive	D°
31. 10 morning	Fair and clear.	S.W. gentle	29 55	44	$3\frac{1}{8}$ M	Poſitive	D°
Feb. 1. 9 morning	Thick miſt with fine drops of rain	Calm	29.53	46	$2\frac{6}{8}$ M	Poſitive	Umbrella and torch
1. 11 morning	Thick miſt	S.W. gentle	D°	D°	D°	Poſitive	Kite
2. 11 morning	Thick miſt	S.W. gentle	29.4	44	$1\frac{3}{8}$ M	Poſitive	Torch and bottle
3. 10 morning	Heavy clouds	S.W. briſk	29.2	46	$3\frac{2}{8}$ M	Poſitive	D°
3. ½ paſt 11 morning	Small rain	S.W. ſtrong	D°	D°	D°	Negative	Umbrella
4. 8 morning	Clear and froſty	Calm	29.3	42	$0\frac{1}{8}$ D	Poſitive	Torch and bottle
5. ½ paſt 10 morning	Thick miſt on the hills	S.E. gentle	29.26	38	$0\frac{1}{8}$ M	Poſitive	D°
6. 10 morning	Thick miſt and few drops of rain	S. gentle	29.12	40	1 M	Poſitive	D°
7. ½ paſt 10 morning	Cloudy	S.W. briſk	28.7	46	$5\frac{6}{8}$ M	Poſitive	D°
7. ½ paſt 2 afternoon	Small rain	S.W. briſk	28.69	47	$4\frac{7}{8}$ M	Negative	Umbrella
8. ½ paſt nine morning	Clear and froſty	Calm	29.28	40	$0\frac{1}{8}$ M	Poſitive	Torch and bottle
9. 12 noon	Cloudy after rain	S. ſtrong	28.75	46	4 M	Poſitive	D°
9. ½ paſt 1 afternoon	Rain	S. ſtrong	28.68	46	$4\frac{1}{8}$ M	Negative	Umbrella
10. 9 morning	Rain	S. gentle	28.64	44	$1\frac{1}{8}$ M	Negative	Umbrella
11. 12 noon	Beginning to rain	S. briſk	28.34	42	2 M	Negative	Torch and bottle
12. 12 noon	Fair, with heavy clouds	S. ſtrong	27.8	44	$2\frac{1}{8}$ M	Poſitive	D°
13. ½ paſt 9 morning	Very ſmall rain, thin clouds	N.W. briſk	28.1	44	$1\frac{4}{8}$ M	Negative	D°
14. and 15. abſent.							
16. 10 morn.	Small rain	S.W. gentle	28.83	47	$3\frac{1}{8}$ M	Negative	D°
17. 10 morning.	Very few thin clouds	W. gentle	29.42	44	0	Poſitive	D°
18. 1 afternoon	Few white clouds	W. briſk	29 42	48	1 M	Poſitive	D°
19. 11 morning	Few diſtant clouds	N.W. gentle	29.53	47	$1\frac{4}{8}$ M	Poſitive	D°
20. 11 morning	Over-caſt	N. gentle	29.47	46	$1\frac{4}{8}$ M	Poſitive	D°
21. 3 afternoon	Heavy clouds	N. gentle	29.42	47	$1\frac{1}{8}$ M	Poſitive	D°
22. 11 morning	Few white clouds.	N.E gentle	29.4	40	$0\frac{6}{8}$ D	Poſitive	D°
23. ½ paſt 10 morning	Over-caſt	E. briſk	29.38	40	$1\frac{4}{8}$ D	Poſitive	D°
24. 10 morning	Clear and froſty	S.E. gentle	29.32	42	$0\frac{4}{8}$ D	Poſitive	D°
25. ½ paſt 12 noon	Few clouds	N.W. ſtrong	29 23	46	1 M	Poſitive	Lantern

44 A. Bennet, section from Derbyshire electro-weather diary, *Philosophical Transactions*, 77 (1787)

Bennet's electroscope was based on another instrument made by his friend Tiberius Cavallo. Instead of threads or pith balls, Cavallo used silver wire terminated by pieces of cork contained in a glass bottle and held in place by a glass tube. A wire ran from the tube to the large brass cap and strips of tinfoil (earthed) allowed the electricity to be 'conveyed off' when the corks touched (Figure 43). Many of these innovations were adopted by Bennet, but his electroscope was larger with a 5-inch tall glass case. Two inches in diameter, it rested on a wood or metal base. Two slips of leaf-gold were suspended in the glass, and the peg and tube holding them touched the outer cap. Two pieces of tinfoil were fastened on opposite sides of the internal surface of the glass.[18] Bennet's electroscope was important because the glass case allowed atmospheric electricity to be easily detected without interference from air currents. The instrument was more sensitive than other kinds because it was larger and because the gold leaves

were finer and lighter than other materials. Bennet put the electroscope to use in a series of experiments detecting the presence of
charge, such as when different types of powder were blown at the
electrometer and found to register electricity. Stimulated by the highly varied and localised weather of the Peak, including the high
rainfall and severe frosts which seemed to offer considerable advantages for exploring the relationship between electricity and meteorology, he compiled a diary of charges present in different conditions
(Figure 44). To carry out some of these tests Bennet utilised an extraordinary apparatus consisting of a candle (found to increase the detection sensitivity) held in a small lantern fixed to a tinned iron funnel and fastened to the end of a 10-foot deal rod. To the lower end of
the funnel a brass wire was fastened which could be used to communicate the electricity inside to the cap of the electroscope. With the
exploding gunpowder bubbles, this must have been an interesting
spectacle for the locals as their curate carried it about.[19]

DETECTING ATMOSPHERIC CHARGE: THE ELECTRICAL DOUBLER

For the detection of even smaller charges, Bennet found that he could
use Volta's condenser. In 1778, Volta had announced the invention
of the electrophorus which was a 'perpetual' electrical creator consisting of a dish of metal containing the dielectric cake, a wooden shield
covered with tinfoil and an insulating handle. The dielectric cake was
a mixture of turpentine, resin and wax. Operation was simple. The
plate was charged from a machine and then discharged by touching
the shield and plate together or alternately. When the shield was
removed its negative charge could be given to the hook of a Leyden
jar, then replaced, touched, brought back to the hook, etc., until the
condenser was sufficiently charged. Volta's condenser was in fact an
electrophore that had a layer of varnish as cake. Bennet used both the
larger and smaller of Volta's condensers remarking on their 'amazing
power'. The idea of using both actually originated from Cavallo, but,
even then, Bennet found that some charges still could not be detected. Solving this problem led to his second important invention, the
doubler of electricity.[20]

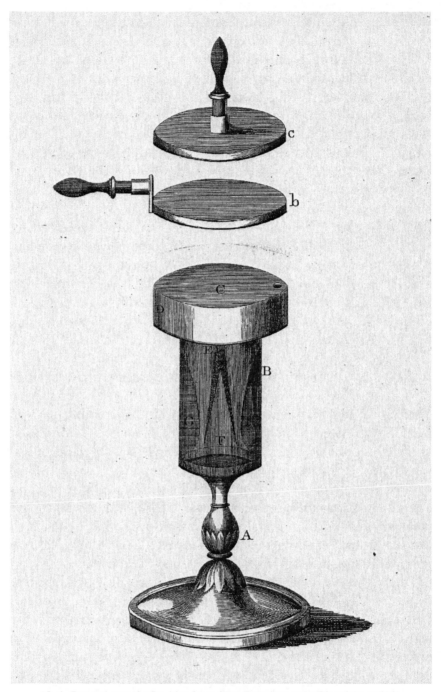

45 A. Bennet, simple doubler from *New Experiments on Electricity* (1789),
courtesy of Derby Local Studies Library

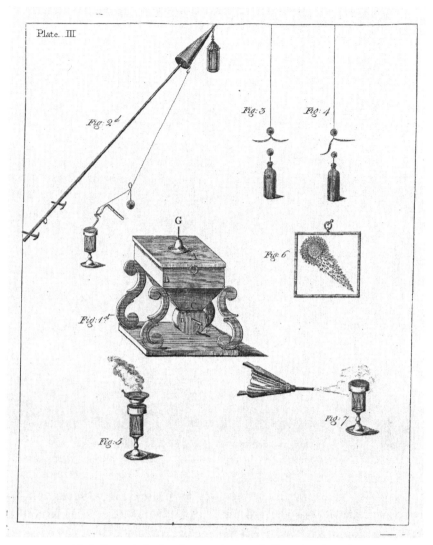

46 A. Bennet, deal rod apparatus aand other experiments,
from *Philosophical Transactions*, 77 (1787)

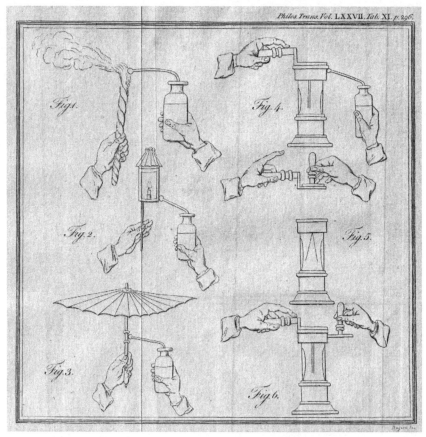

47 Experiments with the simple doubler from A. Bennet
New Experiments on Electricity (1789), courtesy of Derby Local Studies Library

The discovery of the doubler was announced in a paper sent to the Royal Society by Richard Kaye in 1787 (Figures 45, 46 and 47). This instrument, the 'simple doubler', consisted of two polished brass plates (b and c) with insulating handles one in the middle and the other on the side (Figure 45). The plates were varnished on the under-side, with the handles insulated by glass covered with sealing wax. Collected electricity in the Leyden jar was applied to the cap of the gold-leaf electroscope upon which was placed the plate b, touching it with the forefinger stretched over the insulating nut. Thus the electricity 'spreads upon the cap', which served as a condensing plate and electrified the plate b contrarily (because it was earthed) with the varnish interposed as a charged electric.[21] The jar was then removed and the forefinger lifted up. The plate b was separated from the cap and the plate c placed on its upper side and touched by stretching a finger over the nut of its insulating handle. Then the last plate was electrified contrary to b, and the finger was removed with the plate c separated from b. It would then be evident 'to electricians that the electricity of the cap and that of the plate c will be of the same kind and nearly of equal quantity'.[22] Thus the original charge was effectively doubled. Next the edge of the plate c was applied to the side of the cap, and touching and placing b as before, the electricity of c and that of the cap both acted on the plate b. The result was that in Bennet's terms, the 'intensity' of its contrary electricity became equal to both. When c was removed in an un-electrified state, and the forefingers were taken off from b, then b was lifted up. Once c had been placed upon it, the process could proceed as before, repeated until the gold leaves diverged. Bennet took the fact that the gold leaves diverged about twice the distance on each operation as a rough indication that the electricity was being doubled.

There were problems. Cavallo announced that though the merit of Bennet's invention was considerable, the use of it was 'far from precise and certain' and it was 'not an instrument to be depended upon'.[23] The primary purpose of the doubler was manifesting small charges by augmentation, but it was found that the doubler produced electricity even when no charge was given to it in the first place. Suspecting friction to be a cause, Cavallo designed a collector of electricity which needed no touching during doubling. Bennet too had noticed the problem of adherent or 'spontaneous' charge and suspecting friction, he tried to devise a doubler with sliding or revolving parts. But before he had finished, William Nicholson, a London teacher and instrument maker, sent him an ingenious revolving doubler that he had constructed (Figure 48). This consisted of two insulated and immovable plates about 2 inches in diameter and a moveable plate, also insulated, which revolved in a vertical plane parallel to the two other plates passing alternately. A ball 'I' was made heavier on one side than the other and placed on the axis opposite the handle to counterbalance plate B so it could be stopped at any part of its revolution. The plate A was constantly insulated and received the communicated electricity. Plate B revolved and when opposite to plate A, the connecting wires at the end of the crosspiece D touched the pins of A and C at E and F. A wire proceeding from the plate B touched the middle piece G, which was supported by a brass conducting pillar in connection with the earth. In this position, if electricity was given to A, then B would acquire a contrary state and revolving further – the wires also moving with it by means of the same insulating axis – the plates were again insulated till the plate B was opposite to C. Then the wire at H touched the pin on C, earthing it and giving the same kind of electricity as that of A.[24] By moving the handle still further, B was again brought opposite to A with the connecting wires joining A and C. These both acted on B, which was earthed as before hence nearly doubling 'its intensity'. Simultaneously, the electricity of C was absorbed into A because of the increased capacity of A while opposed to B. This was capable of acquiring a contrary state because it was earthed 'sufficient to balance the influential atmospheres of both plates'. By continuing to revolve the plate B the process was performed 'in a very expeditious and accurate manner'.[25]

Nicholson's doubler was found by Bennet to still retain its spontaneous charge, which he thought was due to 'the increased capacity' of approximating parallel plates that 'might attract and retain their charge tho' neither of them were insulated'.[26] The idea that substances always contained a residual charge either positive or negative regardless of whether they were insulated or not proved to be a very fruitful line of research. Bennet tried various methods to deprive the doubler of all spontaneous charge, such as earthing and rapidly rotating it before any experiment. In one experiment a copper plate was applied to plate A while A and B were parallel (so that B was earthed). After only five revolutions of B the gold leaf diverged negatively by a quarter of an inch. But where was this electricity coming from? Bennet concluded that different substances had different 'adhesive affinities' to the electrical fluid. They could be either positive or negative, the charge being attracted by the position of different plates in parallel.[27] After finding that different types of flowers on the plates could change the charge produced, he drew the following pregnant conclusion 'it easily occurred, that if the spontaneous electricity in the beginning of the process was sufficiently weak, the mere contact of metals or other substances having a different adhesive affinity with the electrical fluid might also change it.'[28] This was a momentous discovery in the history of electricity, and Bennet confirmed his supposition in a series of experiments with different substances. First, with the plate B parallel to A but insulated, A was touched with a steel blade and B touched with softened iron wire. After 16 revolutions the gold leaf diverged positively. On reversal of the experiment, with the knife applied to B and the soft wire to A, negative charge was registered. Similarly, other metals were applied to the plates and the charge doubled. Reversal of the metals changed the state of the final charge; therefore Bennet proved to his satisfaction that the metals were causing the charge.

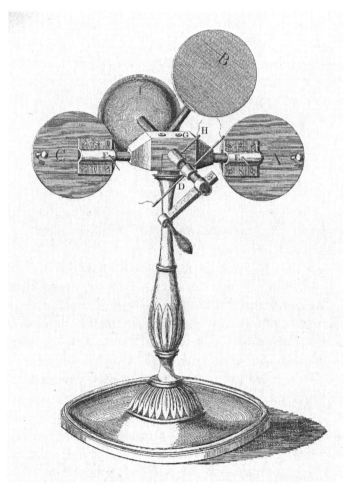

48 William Nicholson's revolving doubler from A. Bennet, *New Experiments on Electricity* (1789), courtesy of Derby Local Studies Library

Various single metals were then tested on different plates. In one experiment the metal was applied to plates A and C and the cross-piece, then to plate B with B standing in the lower part of its plane. The results of these single contact experiments were twofold and of crucial importance. Bennet had identified a method of determining the nature of the 'adhesive affinity' of electricity to different metals, in other words, the natural electrical state of metals. Second, as the number of revolutions approximated inversely to the strength of the 'adhesive affinity' then metals could theoretically be graded according to the strength of affinity. When lead ore was applied to plates A, C and the crosspiece, positive electricity was registered. Zinc produced negative charge, so the 'adhesive affinity' was positive for lead ore and negative for zinc. Gold, silver, copper and brass were also found to be positive, while tin and zinc were negative. Bennet widened his theory to include non-metal substances with pure antimony, bismuth, tutenag, and different woods and types of stone producing positive charge. He proved that the shape of the substance affected the strength of the charge registered, with thin plates of zinc being stronger than a large lump.[29]

BENNET AND VOLTA

Bennet's experiments helped Volta in the formulation of his contact theory, though Volta's work had already impacted upon British electricians. The publication and dissemination of the *New Experiments on Electricity* by subscription, which summarised his electrical work, and knowledge of Bennet's papers in the *Transactions*, impacted upon European natural philosophy. The work described the gold-leaf electroscope and Bennet's general electrical theory, giving an account of experiments with a Lichtenberg electrophorus with reproductions of chalk figures and patterns that he produced. The phenomenon of electricity caused by evaporation – for example from heated metals plunged into water – was illustrated, followed by a description of the doubler. The book concluded with a theory of atmospheric electricity and how different electrical states were associated with different weather conditions, but the most important sections described the mechanisation of the doubling process and the experiments on 'adhesive electricity'. Bennet, Cavallo and Nicholson had met Volta in London in 1782 when he demonstrated the work of his condenser during a European tour. Volta described his experimental demonstrations and the reaction of London philosophers in his paper published in the *Philosophical Transactions* where he noted that 'the experiment on the evaporation of water' which had not worked very well in Paris 'succeeded much better in London'. Volta threw water upon lighted coals kept in an insulating chasing-dish emphasising that the electricity of the evaporation never failed to 'electrify the chasing-dish negatively and strongly enough for detection with a simple electrometer', sometimes causing a spark with a condenser. He noted that the 'first experiment of this sort was made at Mr Bennet's who is a great lover of electricity in presence of Mr Bennet, Mr Cavallo and Mr [Richard] Kirwan, members of the Royal Society, and of Mr [Adam] Walker, lecturer of experimental philosophy.'[30]

In the *Commentarius* of 1791, Luigi Galvani (1737–98) demonstrated how frogs' legs jerked under certain electrical conditions, such as when touched by metals. This was taken as proof that animal electricity existed which was secreted by the brain and distributed through the nerves causing motion. Volta disagreed, arguing that the mere contact between metals generated a charge, with animal parts being unnecessary, and proceeded to rank metals according to their electromotive power; he wondered, like Bennet, if metals were mere passive agents. Indeed the frogs' legs served for Volta as an analogous position to that of the revolving doubler in Bennet's experiments. Volta found that two metals in contact with the legs produced no convulsions, unlike one metal. After various experiments, including the application of metals to the tongue (producing an acid taste), Volta made this explicit statement: 'metals are thus not only perfect conductors, but motors of electricity'. The earliest announcement of this theory was June 1792 three years after Bennet's *New Experiments* had been published.[31] From 1796, Volta dispensed with the frogs' legs for good, using Nicholson's doubler instead to detect small contact charge. In one conclusive experiment, metal strips of zinc and silver were held together and allowed to touch a condenser plate, with negative charge being registered. He then used Nicholson's doubler to test the effects of two metals in contact. These experiments led to the invention of the pile, which was thought by Volta to be decisive proof that Galvani was wrong. Thus the letter to Banks was entitled 'On electricity generated by the contact of conducting substances' and Volta paid little attention to the chemical effects of the pile.[32]

In 1788, it was emphasised in the *Analytical Review* that the 'property of the doubler to exhibit electric signs without previous communication of electricity' had been originally observed by Bennet who 'conceived that it might be removed by repeating the process … beginning from another plate'.[33] Full descriptions of Bennet's and Nicholson's work with the doubler were provided in standard contemporary works such as George Adam's *Essay on Electricity*.[34] George Singer emphasised the importance of Bennet's work on detecting small atmospheric charges including his use of the fishing rod apparatus, funnel and wire as a 'sensible electrometer' using a torch or lamp to facilitate detection. The doubler was a 'contrivance 'of considerable ingenuity, by which the powers of the simple condensers are far exceeded' although this 'increased sensibility' at the cost of a 'tendency to produce the electrical states spontaneously', thereby producing 'equivocal results'.[35] Nicholson, the inventor of the revolving doubler wrote that: 'With regard to the principle of the electric-motors of Signor Volta, I must observe that Bennet made many direct experiments by the application of different metals, by the single contact and double touch, to the plates of the doubler, followed by the production of electricity, which were published in his *New Experiments*.' He did not know the date of Volta's experiments but believed them 'to be much later than those of the same kind by Bennet. This last philosopher, as well as Cavallo appears to think that different bodies have different attractions or capacities for electricity.'[36] Similarly, in addition to the gold-leaf electroscope, it was the experiments on the doubler and 'adhesive electricity' that other contemporary and subsequent electricians considered to be Bennet's greatest achievement.

Cavallo's experiments were made after Bennet, so that as Nicholson indicated, it was Bennet who anticipated Volta's concept of electromotive potential. Volta's experiments with the doubler, notably those using different metals in contact, were strikingly similar and other contemporaries recognised this, including Davy and Darwin.[37] Bennet's concept of adhesive affinity was similar to Volta's electromotive potential of substances. When he spoke of the fact that 'the mere contact of metals or other substances having a different adhesive affinity with the electrical fluid might also change it' and the 'method of single contact' which 'appeared to cause a positive charge'[38] this equated to Volta's electromotivity. Only the terminology was different. Certainly Volta was a much greater experimenter and his formulations were more precise; but when he investigated the electromotive potential of metals and other substances he was following a lead already taken by Bennet, who had extended his concept of 'adhesive affinity' to non-metals such as wood, and substances such as bismuth, antinomy and tutenag. These he had then graded according to the strength of their adhesive affinity. Doubtless Volta's primary motivation for investigating contact properties came from the rivalry with the Galvanists. In 1792 Volta stated that 'metals are thus not only perfect conductors, but motors of electricity' which was a 'new virtue of metals, which no one has yet suspected, and which I have been led to discover'.[39] This was three years after Bennet's book, which if Volta had closely read would have prevented him from making this statement. Therefore the relationship between Volta and Bennet is not straightforward. Volta subscribed to the *New Experiments*, but may have not read it until 1795, or perhaps he read it quickly and was only reminded of it after Cavallo's work. Whatever the precise details, the closeness of Volta's experiments with the doubler to those undertaken by Bennet confirm the influence, as does Volta's own tribute to Bennet in the letter to Banks announcing the discovery of the pile.[40]

ELECTRICAL NETWORKS

Analysis of the *New Experiments* subscription list reveals the geographies of Bennet's scientific contacts. The most important relationships were with the Lunar Society, Manchester philosophers and various Derbyshire philosophers, notably Erasmus Darwin, many of whom subscribed. He had close connections with the nearby Derby Philosophical Society the members of which subscribed individually and collectively to the *New Experiments*.[41] Priestley received the dedi-

271

cation of Bennet's 1786 paper on the gold-leaf electroscope, which already had 'the honour' of his 'approbation'. Bennet had carried his electrometer 'from Birmingham to London, [and] another from Wirksworth to Etruria in a portmanteau on horseback, yet without injury'.[42] This was a tour of the Lunar map, taking in Derby, Birmingham and Josiah Wedgwood's Etruria. Bennet used an electrophorus to create Lichtenburg figures – beautiful patterns made on the resinous electrophorus by drawing over it the knob of a charged glass – which were shown by projecting fine powdered resin over the plate. One such figure was represented in the frontispiece of the *New Experiments*. Wedgwood had the idea of commercially producing the figures by using fine powdered enamel instead of resin and then baking the plate or vessel, hence Bennet's visit to Etruria.[43] In 1785, John Southern, one of the 'Soho Group' of Birmingham inventors and industrialists, published *A Treatise upon Aerostatic Machines*, one of the earliest English books on balloon construction. He was a 'special subscriber' to Bennet's *New Experiments* and included in his treatise a description of a process for manufacturing inflammable air and a method of pasting sheets together for making experimental paper balloons given to him by Bennet. In Bennet's notebook there is evidence that he worked on the problems of lighting, likewise a Lunar interest. He recorded a design for 'a convenient fountain lamp' using a glass vial placed in a socket of tin with the neck downwards and a looking glass to concentrate the light which was illustrated with five drawings. From about 1784 to 1786, lamp designs often featured in Lunar correspondence. Aimé Argand (1755–1803) had created a new oil lamp incorporating a tubular wick and glass chimney. An upward draught of air reduced smoke and smell while increasing the light. Matthew Boulton and James Watt became involved in the manufacture and Wedgwood corresponded with Darwin on the subject of creating marketable lamps.[44] William Nicholson, the creator of the revolving doubler, was also closely involved with the Lunar Society and the design of lamps, having worked for Wedgwood's pottery company and been a member of John Hyacinth Magellan's London Philosophical Society in the 1780s with Wedgwood, John Whitehurst and some other Lunar members.[45]

Darwin was the most important of Bennet's philosophical friends and encouraged him to become interested in electricity and meteorology.[46] Darwin's letters and books reveal a close collaboration between himself and Bennet during the 1780s. Darwin had invented a mechanical doubler, which he passed to Bennet for development, and he continually praised Bennet's work in his books, probably encouraging Joseph Johnson the publisher to sell the *New Experiments* in London. Darwin had known Benjamin Franklin since the 1750s and in his last letter to him of 1787, he described Bennet's doubler.[47] In the *Zoonomia*, Darwin singled out Bennet's 'ingenious' experiments with metals and the doubler, calling it 'the greatest discovery made in that science since the Coated Jar and the eduction of lightning from the Skies'.[48] In the *Temple of Nature*, Darwin again described the doubler and associated the metal contact experiments with galvanism and Volta's pile. Darwin's first Royal Society paper had concerned electricity, being an attempt to refute a theory that cloud formation was only the result of the electric fire forcing vapour particles into the air and that rain was the result of the loss of electricity – hence lightning.[49] Later he argued that cloud formation was due to adiabatic expansion, the expansion of a gas (in this case water vapour) from a region of high pressure to a region of low pressure; in conditions of low pressure heat was removed from the water, which then condensed to form precipitation. By contrast, in the *New Experiments*, Bennet tended to stress the electrical origins of various natural phenomena, such as weather conditions, the aurora borealis and meteors. He confirmed that ascending water vapour was electrified positively and so felt justified in interpreting lightning as the release of the charge from clouds and therefore the cause of rain.[50] Darwin dealt extensively with electrical phenomena in the *Economy of Vegetation*, ranging from the electroscope to electrical fish and lightning, and he encouraged Bennet to investigate meteorological electricity in order to promote agricultural and social improvement. The electroscope and doubler were contrived 'for the purpose of more easily making an electro-meteorological diary' so that they could throw some light upon 'the causes of the sudden changes of aerial currents, a circumstance of so much importance to the early growth and maturity of vegetation'.[51]

Analogies and shared concepts and terminology abound in the works of both Bennet and Darwin, such as the analogy of a corks attracting or repelling each other in water and smeared with oil as an illustration of electricity flowing from points, which appears in Bennet's notebook, the 'Letter on attraction and repulsion' sent to Thomas Percival and published in the *Manchester Memoirs*, Darwin's *Economy of Vegetation* and the *Temple of Nature*, demonstrating how closely the two worked together on electro-meteorology.[52] But Bennet was an individual philosopher and not merely an actor reading a Darwinian monologue. Their magnetic theories reveal an interchange of ideas and a theoretical divergence between the two. In his final Royal Society paper, Bennet announced the discovery of a magnetometer for the detection of minute forces (Figure 49). Previous magnetometers had used materials that twisted out of shape, such as cotton; but Bennet's utilised spider's thread,[53] which had remarkable tenuity and once took 18,500 revolutions before the thread broke, with the line never deviating from the meridian. Darwin borrowed Cavallo's book on magnetism from the Derby Philosophical Society library and asked Bennet to investigate Cavallo's claim that inflammable air caused magnetism using the magnetometer. Later 'at the request of Dr Darwin' Bennet repeated another experiment of Cavallo's which claimed that iron filings increased their magnetic attractions by effervescence with diluted hydrochloric acid. Cavallo, Bennet and Darwin were testing the degree to which chemical changes induced by heat would generate or release electric or magnetic fluids.[54]

One of the most important experiments that Bennet undertook with the magnetometer tried to take advantage of its sensibility to detect the momentum of light. This was predictable if it were an aetherial fluid and Bennet seems to have accepted that the vibration theory of light was confirmed by the behaviour of electricity. Although he did not provide a very detailed description of his experiments, he suggested that light and heat might not consist of emitted particles but actually be 'vibrations made in the universally diffused caloric or matter of heat or fluid of light'. Cantor has argued that this came to be regarded as a decisive experiment for wave theory. For instance in the mid-nineteenth century, Humphry Lloyd noted that it was 'now universally conceded that no sensible effect of the impulse of light has ever been perceived' and the 'experiments of Mr Bennet seem to be decisive on this point'.[55]

During the 1790s, Darwin and Strutt moved away from Bennet's Franklinist unitarian position towards a dualist electrical theory. This situation was mirrored by magnetism. In the *Temple of Nature* Darwin advocated two magnetic fluids, an 'arctic' and an 'antarctic', partly on the basis of a kind of symmetry with his electrical dualism. Another difference between Bennet and Darwin concerned the importance of electricity in earthquakes. Bennet's electrical experiments and observations in the meteorological tradition encouraged him to emphasise the strength of the aetherial fluid in nature whereas Darwin's geological studies and observations fostered scepticism. After an earthquake in November 1795 that appeared to centre on Derbyshire and Nottinghamshire, Bennet sent an account to the Royal Society which was printed in a paper by Edward Whittaker Gray. Bennet suggested that the 'circumstances seem to favour the supposition of earthquakes being caused by electricity, but it is only from a collection of numerous facts, that any rational theory can be formed upon the subject'.[56] The theory that earthquakes were caused by electricity had been held by the antiquarian William Stukeley and repeated by Priestley. Evidence for the electric origins of earthquakes was said to include the appearance of fireballs – one of which had reportedly been seen at Derby – wind direction, the fact that vegetables grew more quickly, the sight of a bright aurora borealis and even medical complaints. Bennet provided descriptions of the earthquake to his Derbyshire friends including Rev. Peach of Edensor, John Chatterton of Derby and White Watson of Bakewell. The Chattertons, John I (1742–1800) and John II (1771–1857), were plumbers and glaziers living in Derby. John II, presumably Bennet's correspondent, was a chemist, inventor, friend of Darwin and Philosophical Society member.[57] As we have seen, Watson produced stratigraphical sections of Derbyshire inlaid with actual rock and mineral samples and was a corresponding member of the London Mineralogical Society and author of *A Delineation of the Strata of Derbyshire* (1811).[58]

In the *Economy of Vegetation*, Darwin had described the earth as a 'large mass of burning lava' in 'basaltic caves imprisn'd deep' with 'vaulted roofs of adamantine rock'.[59] He argued that the evidence for the existence of the 'billowy lavas' came from the heat found in mines and his own observations on warm springs such as St Anne's Well at Buxton, which he contributed to Pilkington's *View of Derbyshire*. Whitehurst held that volcanic activity deep in the Earth's crust caused many geological phenomena. Strata were thrust up into mountains and the size and depth of oceans, rivers and valleys were the result of pressure from these forces.[60] Darwin accepted this position and following Whitehurst saw evidence in the geology of Derbyshire. It was these 'central fires' of fluid lava that caused earthquakes, like a stroke on liquid in a bladder which would be felt on the other side. Thus Bennet and Darwin had different views of the causes of earthquakes, with Bennet suggesting electricity to be involved while Darwin saw heat from fluid lava to be important. Related to this, the two had different theories about the cause of the Earth's magnetism, Darwin holding that molten iron in the Earth's core caused the field, while Bennet thought that a magnetic atmosphere existed over the Earth, being rarified at one pole and condensed at the other.[61] An electrical theory for earthquake origin was less ideologically challenging than a gradualistic developmental geological theory based on central volcanic fires which challenged the Mosaic account.

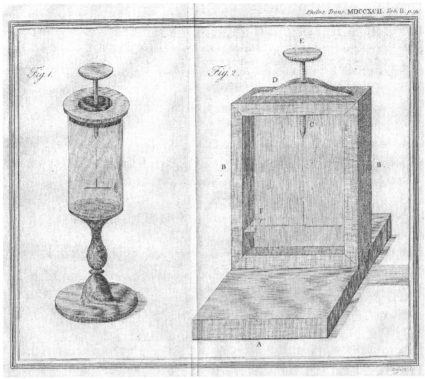

Philos. Trans. MDCCXCII. *Tab.* II. *p.9b*

Fig. 1.

Fig. 2.

49 A. Bennet, spider's thread magnometer, *Philosophical Transactions*, 82 (1792)

LOCALITY AND EXPERIMENT

Contemporaries recognised the impact of Bennet's Peak location upon his philosophical work, although they did not believe that it discredited it. The electro-chemist George John Singer, for example, noted that dry red lead placed upon the cap of Bennet's gold-leaf electrometer had caused the leaves to open with positive electricity. Although this was completely against his expectations, Singer did not question the authority of Bennet's findings in Wirksworth but considered it a 'curious fact that all the specimens of red lead I have hitherto tried [to] produce negative electricity when projected on the cap of the electrometer, though they are attracted by the negatively electrified surface in Lichtenberg's experiment'. Singer was only able to

277

explain this 'anomaly' by supposing that 'the electricity of the red lead is different when it is projected with another powder'. He was puzzled because the separation of the mixed powder of red lead and sulphur, or red lead and resin, 'has always taken place when I have projected them on a surface charged with both states of electricity'. However, red lead, sulphur or resin, separately sifted on the electrometer, had 'invariably occasioned it to diverge negatively'. The 'general accuracy' of Bennet's experiments and the 'coincidence of the greater number of them' with his own experience led Singer to 'believe that the red lead he employed really produced the described effect. He argued that the difference 'in that article' probably resulted from 'various methods of manufacture'. This was because in Derbyshire 'where Mr Bennet resided, red lead is manufactured by the direct oxidation of the metal' whereas a 'considerable portion of that sold in London is said to be made from litharge, and is considered as less pure'. This variation seemed to Singer to account for the different results obtained by himself and Bennet; however, it 'by no means' explained the 'singular phenomenon of a negatively electrified powder being determined to a negative surface when at the same distance from one that is positive'. Singer placed so much confidence in Bennet as an experimentalist that differences of local conditions had to be acknowledged and cited to explain apparent experimental differences. This also underscores Singer's recognition of the implications of local differences upon experimental results, especially between London and the industrial regions where different geological and industrial conditions and processes prevailed.[62]

CONCLUSION

It has been argued that inspiration for Bennet's work came from a number of sources. Most important were observations and experiments conducted in the Derbyshire Peak, which utilised local knowledges, materials and conditions. However, Bennet transcended his localism and provinciality through a number of means. He used close contacts with local and regional philosophical friends and networks, especially Darwin and the members of the Derby, Manchester and

Birmingham philosophical circles as a means of gaining credibility and recognition in national and international Enlightenment scientific circles. It was through Priestley, for instance, that he gained entry recognition from – and entry to – the Royal Society whilst Darwin praised and publicised his work in his books, conducted experiments jointly with him and utilised his apparatus. Bennet and Darwin also shared knowledge and ideas of practices with each other and experimented together, although they came to share different views concerning problems such as the origin of earthquakes and global magnetism. As we have seen, Bennet not only cultivated scientific contacts but also, as a curate, relied greatly upon patronage in the Anglican Church and from the gentry and aristocracy. As the subscription list to Bennet's *New Experiments* makes clear, whilst at one level understanding of Georgian electricity and meteorology requires internalistic studies of ideas and practices, they cannot be fully understood without careful attention to place. Of national and international impact and importance, Bennet's electrical work was, in many important ways, shaped by local geological, meteorological, topographical, economic and industrial factors.

Final Conclusions

The case studies at the heart of this book have enabled us to explore the role of new spaces for scientific culture, physically, intellectually and metaphorically, and the importance of these in shaping modern public spheres. As we have seen, central to this were relationships between encounters with nature, Enlightenment epistemologies, natural theology and the production and consumption of scientific ideas, practices and processes in different spaces. Simultaneously, the characteristics of these places shaped the nature, perceptions, interpretations and experiences of the sciences. We have investigated the impact of discursive boundaries – and processes of knowledge authentication – upon the complex of spaces within associations such as the relationship between botanical taxonomies and the changing configuration of museum and garden displays and presentation of public experimental demonstrations. We have taken a tour through the Georgian home, town, garden, countryside and a variety of scientific associations, developing and interrogating a series of themes along the way, especially concerning the contested relationship between Enlightenment sciences and notions of improvement. Driven by its unique political, economic and demographic importance, London remained, of course the main place for the production, marketing and experience of the sciences, however, metropolitan scientific culture did not dominate in the way that it came to do during the nineteenth century. As we have seen towns such as Norwich, York and Nottingham and their counties participated in a virtual republic of science mediated through local, regional topographical, industrial, economic and social characteristics, such as the strength of commercial or Dissenting interests. Whilst perceptions of international scholarly participation remained vital, divergence, contradiction, contingency and context were equally vital determinants in shaping varied Georgian scientific endeavours.

At the regional level the commercial and industrial interests of northern and midland area helped to determine the utilitarian character of scientific culture fostered by provincial *savant* circles such as the members of the Manchester and Derby philosophical societies and the Lunar Society. The hothouses of Darwin and his Lunar and Derby philosophical friends for example, took full advantage of the latest industrial processes attendant upon transport improvements to provide efficient heating systems and glass. Overall, mirroring the geographies of wider urban renaissance culture, the pre-eminence of the metropolitan Enlightenment was contested in the fluid spaces of provincial Georgian science, identity, relative isolation and the continued importance of ascribed aristocratic, gentlemanly and professional status nurturing distinctive public cultures. These tensions and interactions between metropolitan and provincial natural philosophers and their audiences are well demonstrated by electrical and meteorological endeavours. Abraham Bennet's electro-meteorological work was driven and shaped by his Peak location, especially the distinctive geological, climatic, topographical and industrial character of the region. On the other hand, philosophers such as Darwin, Priestley, Bennet and their Lunar friends transcended localism and provinciality with the aid of patronage and support from Midland and Manchester philosophical networks.

It is worth briefly reflecting on some of the differences between the geographies of eighteenth- and nineteenth-century scientific culture, emphasising the impact of familiar processes including industrialisation, political reform and revolution, class differentiation, metropolitan domination, professionalisation, institutionalisation and specialisation within the sciences. Whilst there was much continuity between Enlightenment and nineteenth-century scientific cultures, a comparison serves to illuminate individuating characteristics and highlight the differences. It is useful to re-emphasise that by cultures of science we are here referring to the production, experience and dissemination of scientific ideas and practices and changing conceptual spaces. Of the changes that had an impact on the places where scientific knowledge was produced, disseminated, changed and challenged, industrialisation was probably the most profound because of its multiple ramifications. Technological developments included the transport revolutions represented by canals, roads, civil engineering and ultimately the railways and the impact of radically improved methods of industrial production. The former transformed the possibilities for collecting and communicating scientific ideas, equipment and objects around Britain and the world, the latter facilitating standardisation, mass production and larger markets for scientific equipment, books and other objects which dramatically increased potential audiences for the sciences and the rapidity and immediacy of production, differential interpretation, contestation and exchange of scientific ideas and practices. The Victorians were, of course, fully aware of these changes. According to the philosopher Herbert Spencer in 1857, besides the many social and economic developments induced by locomotive engines its progressive effects included the improvement of science and engineering, the 'prompt transmission of letters and of news' and the 'dissemination of cheap literature through railway book stalls' which made the 'pulse of the nation faster'.[1]

Changes more directly within the sciences also had a major impact. Older designations of natural philosophy and natural history tended to break down during the course of the nineteenth century as other subjects defined as distinctive disciplines by the end of the previous century, such as chemistry, came to carry different meanings and associations. Changing experiences of place hastened this process as Victorian chemistry, for example, became more defined as an institutionally based laboratory discipline with its own practices. Professionalisation in the sciences was marked by the formalisation of qualifications, expansion of university and college education, demarcation of laboratories as places for doing science, creation of professional and scientific societies and increasing exclusion of amateurs, although they remained important. This impacted upon – and was hastened by – institutionalisation. Arguably, few new kinds of scientific institution were created during the Victorian period, however, older forms familiar from the Enlightenment such as museums, scientific societies, botanical gardens and universities underwent considerable changes in terms of organisation, knowledge production, dissemination and popularisation. Many of the changes in the spatial character and experiences of science associated with these institutions were hastened by changes in the public character and audiences for science. Some of the most characteristic Victorian scientific institutions were the great metropolitan and provincial museums and libraries of cities such as Manchester, Liverpool, Birmingham, Leeds and Nottingham which offered opportunities for a much wider social participation in the sciences, including greater opportunities for the working classes and women and helped to shape organisations and reclassifications of exhibitions.

Many of the changes impacting upon the geographies of English scientific cultures after 1820 can be illustrated by comparing Enlightenment with nineteenth-century botany.[2] Crudely, between the 1820s and 1850s, the aristocratic Anglican establishment domination of the sciences represented by the unreformed Royal Society and Oxford and Cambridge universities, which placed pre-eminent emphasis upon patriotic utility and natural theology, was challenged by the new wealth of the middle classes and urban gentry. One manifestation of the attendant transformation of places where botanical knowledge was produced and disseminated was the growing popularity of field collecting in the countryside, rivers, the seashore and other places and the acquisition of living specimens, which began to supplant the dried herbariums of Enlightenment taxonomy. Another manifestation was the challenge to the domination of Oxford and Cambridge by new universities providing scientific education in London, Durham, Nottingham and other cities by the second half of the nineteenth century. The Oxford and Cambridge botanical gardens also developed into more formally constituted institutions no longer regarded as peripheral to university life as botany established itself beyond medical education. In both places, instead of occasional – and poorly supported – botanical lectures and individual initiatives such as James Donn's catalogues, courses were provided on a regular basis, gardens being more fully integrated into the plant taxonomic and physiological studies of students.[3] At Oxford the appointment of Charles Daubeny as Sherardian Professor in 1834 had a major impact on the design and management of the botanical garden which was more fully integrated into professionalising national science. As a founder member of the British Association for the Advancement of Science in 1831 and President of the British Association from 1836,

Daubeny strove to place natural history and the Oxford Botanical garden in the mainstream of science, for instance, by providing systematic displays intended to demonstrate Darwinian evolutionary theory.[4] At Cambridge University around the same time, transformation of the gardens was also facilitated – and symbolised by – the appointment of a new Professor, Stephen Henslow, who also began to offer regular botanical lectures to students utilising the botanical garden in ways that had never been possible for his predecessor Thomas Martyn.[5]

The redesign of university botanical gardens, however, in some respects, encouraged by the professionalisation of science and formalisation of scientific education, resulted in a greater exclusion of the amateur public as they became places more explicitly devoted to research and scholarship, although carefully regulated public access was usually maintained. At the same time botany and the less exclusive endeavour of natural history became distinguished from the emerging discipline of biology as metropolitan institutions such as the Royal Institution, London universities, British Museum, reformed Royal Society and national scientific societies strove to assert their domination. On the other hand, as university botanical gardens lost some of their informal semi-public status the foundation of botanical societies at Liverpool, Hull and other towns fostered new audiences for botany and the growing Victorian fascination with natural history was reflected in the proliferation of popular works on the subject and the activities of field clubs (Figure 23). New botanical gardens were created in Sheffield, Manchester, Plymouth, Birmingham and other places from the 1820s induced by rational recreational, utilitarian and natural theological concerns which also saw the creation of museums, libraries, mechanics' institutes and other civic scientific institutions. The idea of forming systematic botanical collections in private and gardens was also promoted by landscape gardeners such as John Claudius Loudon in his *Encyclopaedia of Gardening, Gardener's Magazine* and other publications, partly for the education of gardeners. A further development resulting from changing perceptions towards natural history and the importation of colonial plants was the creation of specialist forms of botanical gardens such as arboretums and pinetums which provided systematic, and usually labelled, representative collections of trees and shrubs. Helping to satisfy the impetus for collecting, arboretums were developed in the context of country estates, institutional botanical gardens, commercial nurseries and even burial grounds, whilst the desire for botanical diversity also impacted upon the planting and arrangement of collections in public parks.[6]

Changes in the perceptions and practices of botany and natural history are evident at university, institutional, commercial, semi-public and public botanical gardens and arboretums. Developments in taxonomy facilitated by the formation and management of systematic collections, notably the promotion of various forms of numerical and natural systems, continued to have a major impact and facilitated the replacement of Linnaean schemes as did greater attention to plant physiology and anatomy. The supplanting of Linnaean collections with those arranged according to natural systems, which required more specialist botanical knowledge to understand and interpret, served to further masculinise and differentiate the science from Georgian amateur botanical cultures. It would be wrong however, to exaggerate the domination of natural systems and for the first half of the century at least, encouraged by the demands of patrons and visitors, Linnaean arrangements remained popular sometimes adjoining natural schemes, and it remained unclear which criteria for taxonomic distinctions would come to dominate botany. In some places there was considerable resistance to these changes and at Glasnevin, Dublin for instance, Samuel Litton the Professor of Botany opposed Ninian Niven the new head gardener's attempts to replace Linnaean with natural arrangements.[7] At Oxford, plants and features that seemed to impede botanical experiment and observation such as a double yew hedge which proceeded through the centre of the garden were removed whilst the gated squares in which physic plants were kept were also reorganised. Daubeny set about reducing the domination of medicine significantly changing the title from 'Physic' to 'Botanical' and setting aside a section as an experimental garden for discovering 'the effects of soils, or chemical agents upon vegetation' and for 'other researches of a similar description'. The Daubeny laboratory was constructed within the gardens in 1848 by Magdalen College whilst plots were laid out to facilitate agricultural, horticultural and chemical experiments.[8]

At Cambridge after much agitation a new larger site for the botanical garden was found on the edge of the town and in 1831 the University obtained an Act of Parliament which allowed it to exchange with Trinity Hall and obtain almost 38 acres for a new botanical garden situated beside London Road on the outskirts of the town, although planting did not begin until 1846 because of legal problems. The small old garden had been diminished by the erection of the lecture room for the botanical and Jacksonian professors and further additions in 1834 and it was widely regarded as inadequate to house even a tiny fraction of the many global plants being imported into Britain. The new garden was designed by Edward Lapidge and although only partially executed, his plan included large glasshouse, systematic beds, and arboretum and specialist collections, Henslow pressing for as large a number of species as possible to take advantage of the extensive new spaces, facilitate taxonomic and physiological research and firmly establish botany as a serious professional discipline. The formal design was complemented by picturesque planting including a lake with an island of American plants and trees, sinuous paths and an arboretum around the periphery.[9] At Oxford and Cambridge evolutionary ideas had some impact upon the botanical gardens, indeed both Daubeny and Henslow provided considerable encouragement to Charles Darwin. Daubeny was also a strong proponent of evolution and used the Oxford garden explicitly to promote his views giving a party there in 1860 to celebrate what he regarded as the victory of the Darwinians after the famous debate between Huxley and Wilberforce at the Oxford British Association meeting that year which he had introduced as President of the Botany and Zoology sections. The impact of evolutionary ideas upon the spatial character of botanical gardens should not, however, be exaggerated, as Victorian taxonomy remained wedded to post-Enlightenment progressivism and natural arrangements in practice varied little from Darwinian ones, the differences concerning more the origins of taxa than their current manifestation in the natural world.

Changes in the aesthetics of landscape gardening impacted upon the design and management of botanical gardens. The picturesque made an increasingly large mark upon public and semi-public botanical gardens and arboretums, although from the mid-nineteenth century there was also a return to Italianate and formal styles which would have been recognisable to the promoters of the seventeenth-century Oxford garden. At Cambridge the centre of the new and much enlarged botanical garden was dominated by a grand avenue of exotic trees including Wellingtonia and other large evergreens providing a promenade for visitors following contemporary landscape-gardening practice, which came to dominate views of the gardens. The demands of patronage and public access also had a major impact, including the necessity for facilities such as lodges, toilets, gates, fences and provision for popular entertainment like refreshments and bands. Whilst encouraging the formation of systematic collections and the introduction of exotic plants, Loudon tried to reconcile botanical research, education and rational recreation with the picturesque, an effort exemplified by his creation of the 'gardenesque', which intended to demonstrate individual specimens to best effect by isolated planting upon mounds. At the Derby Arboretum (1840) for instance, encouraged and financed by Joseph Strutt, a local cotton baron, Loudon provided a labelled collection of over a thousand hardy plants, trees and shrubs within a mere 15 acres, whilst picturesque interest was provided using a series of mounds, sinuous paths, lodges, benches, vases and other features. Other landscape gardeners made similar efforts. Joseph Paxton for instance, planted a pinetum and arboretum at Chatsworth for the Duke of Devonshire in which specimen trees and shrubs were intermixed with picturesque landscape-gardening features such as waterfalls, streams and giant rock formations. At the Sheffield Botanic Garden founded in 1834, the Scottish landscape gardener Robert Marnock placed isolated trees and shrubs on mounds amidst other features including large hot-houses, rockwork, terraces and urns filled with colourful flowers, which he subsequently incorporated into the Regent's Park Botanical garden as curator. At Kew the old botanical garden was eventually replaced by a much larger new garden landscaped in a picturesque style which became the national botanical garden at the centre of an imperial network of gardens to a much greater degree than under Banks. At the same time public access became assured as a requirement of government funding and like the hugely successful mechanics' institute exhibitions of the 1830s and 1840s and 'Great Exhibition' of 1851, demonstrated the breadth of Victorian audiences for science which included all social classes, both sexes and numerous foreign and colonial visitors.[10]

There was also, of course, much continuity between the geographies of Georgian and Victorian scientific cultures. If we take the case of botany and botanical gardens again, many of the competing theories that shaped these endeavours originated during the Enlightenment and reflected continued progressivism and confidence. Arguably the most influential work in English natural history remained Gilbert's White's *Natural History of Selborne* which fostered collecting and field observation throughout the countryside. Similarly through publications such as the *Gardener's Magazine* Loudon succeeded in popularising and changing the subject in England between the 1820s and mid-1840s, inspired by his post-Enlightenment progressivist belief that social and geographical expansion of knowledge producers and audiences would facilitate improvement. Nineteenth-century botanical gardens strove to satisfy a vision devised, articulated and implemented by Erasmus Darwin and other botanists in the previous century which aimed to combine science with aesthetics and pleasure. The Linnaean and natural systems which, as we have seen, shaped the co-ordination of systematic collections were products of an eighteenth-century worldview whilst nineteenth-century evolutionism remained wedded to Enlightenment developmentalism as the progressivism of Charles Darwin and Herbert Spencer demonstrates. With their public municipal character and audience, English botanical societies such as those at Liverpool and Hull and, to a lesser degree, the Cambridge Botanical Garden, were products of Enlightenment which pre-dated the mechanic's institute movement of the 1820s.

It has only been possible here to hint at some of the main changes in the geographies of nineteenth-century English scientific cultures using the example of botany and botanical gardens. A similar exercise could have been undertaken for other sciences such as astronomy, physics and chemistry which the aforementioned processes also transformed. If we take the discipline and practices later designated as physics for instance, we can see how practical and institutional concerns and instrumentalism fostered a discipline increasingly requiring formal university education, the resources of specially designed laboratories and the practices of educational, national metropolitan and larger scientific institutions. This resulted in a greater concentration and formalisation of the places where physics could be done compared to the domestic and smaller associational spaces of the Georgian era. Opportunities for domestic-centred experiment in some of the physical sciences were reduced just as domestic life came to be dominated by leisure and family rather than work and manufacture. Where once astronomers such as William Herschel or electricians such as Bennet had been able to undertake major work from their homes, it was now much more difficult. Some aspects of astronomy such as observations of nebulae required powerful telescopes which were harder for amateurs to procure, whilst with respect to electricity and galvanism, the difficulty and expense of constructing Voltaic piles meant that only institutions were able to obtain ones large enough to conduct some types of experiment. With the professionalisation and specialisation of the sciences and ideological reinforcement of the separate spheres from the early nineteenth century, domestic scientific activity began to be associated with amateurs and dilettantes, women and the young, provinces rather than metropolis in distinction to serious science associated with universities, museums and scientific societies. Despite these perceptions, domestically centred science continued to provide a fundamentally important dimension to scientific culture and science, not least in meteorology, astronomy, natural history and botany, and to challenge institutional elite assumptions.

Notes

INTRODUCTION

1   *The Spectator*, 8 vols. (London, 1712–15), VI, no. 420, 131–6.
2   B. M. Stafford, *Body Criticism: Imaging the Unseen in Enlightenment Art and Medicine* (Cambridge, MA, 1991), 131, quoted in M. Ogborn, *Spaces of Modernity: London's Geographies, 1680–1780* (London, 1998), 27.
3   C. Withers, *Placing the Enlightenment: Thinking Geographically about the Age of Reason* (Chicago, 2007), 16.
4   Studies include: A. Wolf, *A History of Science, Technology and Philosophy in the 18th Century*, 2 vols., revised edition (New York, 1961); A. E. Musson and E. Robinson, *Science and Technology in the Industrial Revolution* (Manchester, 1969); T. L. Hankins, *Science and the Enlightenment* (Cambridge, 1985); J. Golinski, *Science as Public Culture: Chemistry and Enlightenment in Britain, 1760–1820* (Cambridge, 1992); W. Clark, J. Golinski and S. Schaffer eds., *The Sciences in Enlightened Europe* (Chicago, 1999); R. Porter, *Enlightenment: Britain and the Creation of the Modern World* (London, 2000); R. Porter ed., *The Cambridge History of Science, vol. 4: Eighteenth-Century Science* (Cambridge, 2003); R. Watts, *Women in Science: A Social and Cultural History* (Abingdon, 2007).
5   Ogborn, *Spaces of Modernity*.
6   R. Mayhew, 'Mapping science's imagined community: geography as a Republic of Letters, 1600–1800', *British Journal for the History of Science*, 38 (2005), 73–92.
7   'Astronomy', anonymous coloured print *c*.1760s, Science Museum collection, London, Science and Society Picture Library, ref. 10198603; P. Fara, *Sympathetic Attractions: Magnetic Practices, Beliefs and Symbolism in Eighteenth-Century* (Princeton, 1996), 56–8; J. Golinski, *Science as Public Culture: Chemistry and Enlightenment in Britain 1760–1820*, 201–2.
8   J. Golinski, *Making Natural Knowledge: Constructivism and the History of Science* (Cambridge, 1998); Ogborn, *Spaces of Modernity*; V. Jankovic, *Reading the Skies; A Cultural History of English Weather, 1650–1820* (Chicago, 2000); C. W. J. Withers, *Geography, Science and National Identity: Scotland Since 1520* (Cambridge, 2002); D. N. Livingstone, *Putting Science in its Place* (Chicago, 2003); see also the essays on the historical geographies of science and S. Naylor, 'Introduction: historical geographies of science – places, contexts, cartographies', in the special issue of *British Journal for the History of Science*, 38 (2005); Withers, *Placing the Enlightenment*; J. Golinski, *British Weather and the Climate of Enlightenment* (Chicago, 2007).
9   D. Gregory, 'Interventions in the historical geography of modernity: social theory, spatiality and the politics of representation', *Geografiska Annaler*, 73 (1991), 17–44; P. Howell, 'Public space and the public sphere: political theory and the historical

geography of modernity', *Environment and Planning D: Society and Space*, 11 (1993), 303–22; Ogborn, *Spaces of Modernity*; R. J. Morris and S. Gunn eds., *Identities in Space: Contested Terrains in the Western City Since 1850* (Aldershot, 2001).

10   S. Shapin, 'Placing the view from nowhere: Historical and sociological problems in the location of science, *Transactions of the Institute of British Geographers*, 23 (1998), 5–12, p. 6; A. Ophir and S. Shapin, 'The place of knowledge: the spatial setting and its relation to the production of knowledge, *Science in Context*, 4 (1991), special issue; N. Thrift, F. Driver and D. Livingstone, 'The geography of truth', and D. N. Livingstone, 'The spaces of knowledge: contributions towards a historical geography of science', *Environment and Planning D: Society and Space*, 13 (1995), 1–3, 5–34; D. Livingstone, *Putting Science in its Place: Geographies of Scientific Knowledge* (Chicago, 2003); Naylor, 'Introduction: historical geographies of science'.

11   B. Latour and S. Woolgar, *Laboratory Life: The Construction of Scientific Facts* (Princeton, 1981); K. Knorr-Cetina, *The Manufacture of Knowledge: An Essay on the Constructivist and Contextual Nature of Science* (New York, 1981); S. Shapin, *A Social History of Truth: Science and Civility in Seventeenth-Century England* (Chicago, 1995); J. Agar and C. Smith eds., *Making Space for Science: Territorial Themes in the Shaping of Knowledge* (London, 1998).

12   E. Hooper-Greenhill, *Museums and the Shaping of Knowledge* (London, 1992); C. Yanni, *Nature's Museums: Victorian Science and the Architecture of Display* (London, 1999); S. Alberti, 'Placing nature: natural history collections and their owners in nineteenth-century provincial England', *British Journal for the History of Science*, 35 (2002), 291–311.

13   Golinski, *Making Natural Knowledge*, 32–40.

14   H. Torrens, keynote address, 'Was county-wide natural history pioneered in the Midlands', Society for the History of Natural History, Spring conference (26 April 2002); S. Naylor, 'The field, the museum, and the lecture hall: the places of natural history in Victorian Cornwall', *Transactions of the Institute of British Geographers*, 27 (2002), 494–513.

15   R. Porter and M. Teich eds., *The Enlightenment in National Context* (Cambridge, 1991) and D. N. Livingstone, C. W. J. Withers eds., *Geography and Enlightenment* (Chicago, 1999); C. W. J. Withers, *Geography, Science and National Identity*; H. T. Buckle, *History of Civilisation in England*, 3 vols. (London, 1869), III, 281–482, especially 391–7, where he gives the example of the differences between the methods employed in Scottish, German and English geology and emphasises the differences resulting from localised fieldwork in each case. Herbert Butterfield associated the scientific revolution with global exploration, travel books, liberal Protestantism, stock projecting and other phenomena from the seventeenth century redolent of a 'growing modernity' (H. Butterfield, *The Origins of Modern Science: 1300–1800* (London, 1951), 159–74).

16   M. Billinge, 'Hegemony, class and power in late-Georgian and early-Victorian England', in, R. H. Baker and D. Gregory eds., *Explorations in Historical Geography* (Cambridge, 1984), 28–67.

17   I. Inkster, 'Introduction', I. Inkster and J. B. Morrell eds., *Metropolis and Province: Science in British Culture* (London, 1983), 11–54.

18   Jankovic, *Reading the Skies*, 3–9.

19   V. Jankovic, 'The place of nature and the nature of place: the chorographic challenge to the history of British provincial science', *History of Science*, 38 (2000), 80–113; Jankovic, *Reading the Skies*; Golinski, *British Weather*, xii–xiv.

20   P. R. Gross, N. Levitt and M. Lewis eds., *The Flight from Science and Reason* (New York, 1996), especially 257–98; P. R. Gross, *Higher Superstition: The Academic Left*

and its Quarrels with Science (Baltimore, 1998); A. D. Sokal and J. Bricmont, *Fashionable Nonsense: Post-modern Intellectuals' Abuse of Science* (London, 1999).

21 J. Money, 'The Masonic moment; or, ritual, replica, and credit: John Wilkes, the Macaroni parson, and the making of the middle-class mind', *Journal of British Studies*, 32 (1993), 359–95.

CHAPTER 1

1 S. Shapin and A. Ophir, 'The place of knowledge: a methodological survey', *Science in Context*, 4 (1991), 3–21; D. E. Sopher, 'The landscape of home: myth, experience, social meaning', in D. W. Meinig ed., *The Interpretation of Ordinary Landscapes* (Oxford, 1979), 129–49, pp. 134–5; E. Relph, *Place and Placelessness* (London, 1976); Yi-Fu Tuan, *Topophilia: A Study of Environmental Perception, Attitudes, and Values* (Englewood Cliffs, NJ, 1974); Yi-Fu Tuan, 'Place: an experiential perspective', *Geographical Review*, 65 (1975), 151–65; D. Geoffrey Hayward, 'Home as an environmental; and psychological concept', *Landscape*, 20 (1975), 2–9; J. Douglas Porteous, 'Home: the territorial core', *Geographical Review*, 66 (1976), 383–90; R. Porter, 'The terraqueous globe', in G. S. Rousseau and R. Porter eds., *The Ferment of Knowledge* (Cambridge, 1980), 285–324, p. 300, quotation from Jankovic below; V. Jankovic, 'The place of nature and the nature of place: the chorographic challenge to the history of British provincial science', *History of Science*, 38 (2000), 82, 102–3; V. Jankovic, *Reading the Skies: A Cultural History of English Weather, 1650–1820* (Chicago, 2000); J. S. Duncan and D. Lambert, 'Landscapes of home' in J. S. Duncan, N. C. Johnson and R. H. Schein eds., *A Companion to Cultural Geography* (Oxford, 2003), 382–403; D. N. Livingstone, *Putting Science in its Place: Geographies of Scientific Knowledge* (Chicago, 2003), 21–9; S. Mallett, 'Understanding home: a critical review of the literature', *The Sociological Review*, 52 (2004), 62–89; A. Blunt, 'Cultural geography: cultural geographies of home', *Progress in Human Geography*, 29 (2005), 505–15; A. Blunt and R. Dowling, *Home* (London, 2006); J. Golinski, *British Weather and the Climate of Enlightenment* (Chicago, 2007).

2 L. Weatherill, *Consumer Behaviour and Material Culture in Britain, 1660–1760*, second edition (London, 1996); A. N. Walters, 'Science and politeness in eighteenth-century England', *History of Science*, 35 (1997), 121–54; P. Langford, *A Polite and Commercial People: England, 1727–1783*, second edition (Oxford, 1998); M. Berg, *Luxury and Pleasure in Eighteenth-Century Britain*, second edition (Oxford, 2007); J. Stobart, *Spaces of Consumption: Shopping and Leisure in the English Town, c.1680–1830* (London, 2007).

3 H. C. Millburn and J. R. King, *Geared to the Stars: The Evolution of Planetariums, Orreries and Astronomical Clocks* (Toronto, 1978); One example is the grand orrery modified by Thomas Wright in 1733, King George III Collection, Science Museum, London, inventory number: 1927–1659.

4 R. Steele, *The Englishman*, no. 11 (29 October 1713), 70–3; S. Daniels, *Joseph Wright* (London, 1998), 34–9; P. Elliott, *The Derby Philosophers: Science and Culture in British Urban Society, 1700–1850* (Manchester, 2009), 54–8.

5 Inventories at Norfolk County Record Office, Norwich, information kindly supplied by Amy Barnett of Northampton University who cites a letter from M. Pocock to Rev. and Mrs Reading (11 October 1771) which describes Edward Nairne (1726–1806) of Corn Hill, London as the best instrument maker and lists specifications and prices of 2 globes and various telescopes from £5 to £105 (Norfolk County Record Office, BOL 2/42 (i) c); A. Barnett, 'In with the new: novel goods in domestic provincial England, c.1700–1790', in B. Blondé, N. Coquery, J. Stobart and I. Van Damme eds., *Fashioning Old and New: Changing Consumer Patterns in*

Western Europe (1650–1900) (Turnhout, 2009), 81–94; T. Fawcett, 'Popular science in eighteenth-century Norwich', *History Today*, 22 (1972), 590–95; T. Fawcett, 'Measuring the provincial Enlightenment: the case of Norwich', *Eighteenth-Century Life*, 8 (1982), 13–27; Weatherill, *Consumer Behaviour and Material Culture*.

6   Will of George Wegg, Essex County Record Office (Acc. C47 CPL 280), reference from J. Cooper and C. R. Elrington eds., *Victoria County History of Essex*, 10 vols. (London, 1903–2001) IX, 169–75; Information concerning Greenwood and a digital image of the inventory kindly supplied by Lesley Hoskins, Department of Geography, Queen Mary College, University of London (L. Hoskins, 'Household Goods and Domestic Cultures in England & Wales 1840–1880', forthcoming thesis, Queen Mary, University of London). It is possible that the instruments may be related to the Hackney Dissenting college in existence in the area during the 1780s and 1790.

7   Sale inventory, *Derby Mercury* (26 September 1837); Elliott, *Derby Philosophers*, 146, 271, 273.

8   B. Latour and S. Woolgar, *Laboratory Life: The construction of scientific facts* (Princeton, 1981); K. Knorr-Cetina, *The Manufacture of Knowledge: An essay on the constructivist and contextual nature of science* (New York, 1981); S. Shapin, 'The house of experiment in seventeenth-century England', *Isis*, 79 (1988), 373–404; S. Shapin, *A Social History of Truth: Science and Civility in Seventeenth-Century England* (Chicago, 1995), 383–410; J. Agar and C. Smith eds., *Making Space for Science: Territorial Themes in the Shaping of Knowledge* (London, 1998).

9   Walters, 'Science and politeness', 121–54; Langford, *Polite and Commercial People*, 60–121.

10   Pair of terrestrial and celestial globes by George Adams dated 1766, Science Museum collection, London, Science and Society Picture Library, ref. 10190768.

11   Portrait of a gentleman in his study with scientific instruments, oil painting on canvas *c.*1750, Science Museum collection, London, Science and Society Picture Library, ref. 10240879.

12   S. Mason, *The Hardware Man's Daughter: Matthew Boutlon and his 'Dear Girl'* (Chichester, 2005), 105–12; P. M. Jones, *Industrial Enlightenment: Science, Technology and Culture in Birmingham and the West Midlands 1760–1820* (Manchester, 2008), 57–60; S. Mason, *Matthew Boulton, Selling what all the World Desires* (New Haven, 2009), 12.

13   J. Secord, 'Newton in the nursery: Tom Telescope and the philosophy of tops and balls', *History of Science*, 23 (1985), 127–51.

14   J. Priestley, *A Description of a New Chart of History* (London, 1786); J. Priestley, *A Description of a Chart of Biography*, fourth edition (London, 1770).

15   Secord, 'Newton in the nursery'.

16   J. Oldfield, *An Essay Towards the Improvement of Reason* (London, 1707), 18–25, 365; J. W. A. Smith, *The Birth of Modern Education* (London, 1954), 121–6.

17   I. Watts, *The Improvement of the Mind* (London, 1795); I. Parker *Dissenting Academies in England* (Cambridge, 1914), 62, 138; H. McLachlan, *English Education under the Test Acts* (Manchester, 1931), 11, 49–52; Smith, *Modern Education*, 41–5, 87–90; M. Bryant, *The London Experience of Secondary Education* (London, 1986), 101.

18   I. Watts: *The Knowledge of the Heavens and Earth Made Easy: Or, the first principles of Astronomy and Geography explained by the use of globes and maps* ([1725] fourth edition, London, 1745), xvi–xx.

19   D. Jennings, *Introduction to the Use of the Globes and the Orrery: with the application of astronomy to chronology* (London, 1747); Watts, *Improvement of the Mind*, 156–62.

20   Watts, *Improvement of the Mind*, 17–18, 26.

21   Watts, *Improvement of the Mind*, 20–23.

22 Watts, *Improvement of the Mind*, xi.
23 A. Vickery, *The Gentlemen's Daughter: Women's Lives in Georgian England* (New Haven, 1998).
24 G. Summerson, *Georgian London* (London, 1962); D. M. George, *London Life in the Eighteenth Century* (London, 1965), 73–115; E. Burton, *The Georgians at Home* (London, 1973); E. Mercer, *English Vernacular Houses* (London, 1975), 26–32; M. Laithwaite, 'Totnes houses 1500–1800', in P. Clark, ed., *The Transformation of English Provincial Towns, 1600–1800* (London, 1985), 62–98; J. Chambers, *The English House* (London, 1985), 125–93; J. Summerson, *Architecture in Britain, 1530–1830*, eighth edition (London, 1991), 354–65; P. Borsay, *The English Urban Renaissance, 1660–1770* (Oxford, 1991), 41–59; P. Earle, *The Making of the English Middle Class* (London, 1991), 205–39.
25 Bryant, *London Experience*, 311–58; S. Skedd, 'The Education of Women in Hanoverian Britain, *c.*1760–1820', DPhil thesis, University of Oxford (1997); S. Skedd, 'Women teachers and the experience of girl's schooling in England', in H. Barker and E. Chalus eds., *Gender in Eighteenth-Century England* (Harlow, 1997; J. Goodman, 'Troubling histories and theories: gender and the history of education' and R. Watts, 'Science and women in the history of education: expanding the archive', *History of Education*, 32 (2003), 157–74, 189–200; M. Cohen, 'Gender and method in eighteenth-century English education', *History of Education*, 33 (2004): 585–95; '"To think, to compare, to combine, to methodize": notes towards rethinking girls' education in the eighteenth century', in S. Knott and B. Taylor eds., *Women, Gender and Education* (London, 2005); R. Watts, *Women in Science: A Social and Cultural History* (London, 2007), 55–98.
26 L. Schiebinger, *The Mind Has no Sex?: Women in the Origins of Modern Science* (Cambridge MA, 1989), 214–41.
27 M. Bryant, *A Compendious System of Astronomy* (London, 1797); N. Hans, *New Trends in Education in the Eighteenth Century* (London, 1951), 203–4.
28 J. Herschel, *Memoir and Correspondence of Caroline Herschel* (London, 1876); C. A. Lubbock ed., *The Herschel Chronicle* (London, 1933); L. Schiebinger, *The Mind Has no Sex?* , 260–4; M. Hoskin, 'Caroline Lucretia Herschel (1750–1848)', DNB.
29 J. Britton and E. W. Brayley, *The Beauties of England and Wales*, 19 vols. (London, 1801–15), III, 378; J. Pearson, *Stags and Serpents: The story of the house of Cavendish* (London, 1983), 86–115; T. Brighton, *The Discovery of the Peak District* (Chichester, 2004), 74–5; M. P. Cooper, 'The Devonshire Mineral Collection of Chatsworth: an 18th century survivor and its restoration', *Mineralogical Record* (April, 2005); D. King-Hele ed., *The Collected Letters of Erasmus Darwin* (Cambridge, 2007), 556–60.
30 S. Jones, *Rudiments of Reason: or the Young Experimental Philosopher*, 3 vols. (London, 1793), vol. 1, iv–v.
31 R. L. and M. Edgeworth, *Practical Education*, 2 vols. (London, 1798), I, 28.
32 Edgeworths, *Practical Education*, I, 23.
33 J. E. Smith, *An Introduction to Physiological and Systematical Botany*, fifth edition (London, 1825), xvii; F. A. Rowden, *A Poetical Introduction to the Study of Botany*, third edition (London, 1818), vii–x, quoted in A. B. Shteir, *Cultivating Women, Cultivating Science: Flora's Daughters and Botany in England, 1760–1860* (Baltimore, 1996), 62–4; Schiebinger, *The Mind Has no Sex?*, 241–4.
34 Jankovic, *Reading the Skies*; Jankovic, 'The place of nature'; for a different emphasis, Golinski, *British Weather*.
35 M. Craven, *John Whitehurst* (Ashbourne, 1997); Jankovic, 'The place of nature', 86–8; Jankovic, *Reading the Skies*, 105–13; R. Hamblyn, *The Invention of Clouds* (London, 2001), 58; Golinski, *British Weather*.

36 M. Edgeworth, *Letters for Literary Ladies* (1795), 66, quoted in J. Uglow, 'But what about the women? The Lunar Society's attitude to women and science, and to the education of girls' in C. U. M. Smith and R. Arnott eds., *The Genius of Erasmus Darwin* (Aldershot, 2005), 179–94, p. 177; Schiebinger, *The Mind Has no Sex?*, 240–1.
37 Edgeworths, *Practical Education*, I, 25.
38 Edgeworths, *Practical Education*, I, 25–6.
39 W. D. Hackmann, *Electricity from Glass: The History of the Frictional Electrical Machine 1600–1850* (Netherlands, 1978).
40 A. and N. Clow, *The Chemical Revolution: A Contribution to Social Technology* (London, 1952), 270.
41 P. F. Mottelay, *Bibliographical History of Electricity and Magnetism* (London, 1922), 153–5, 218–21; J. L. Heilbron, *Electricity in the 17th and 18th Centuries: A Study in Early Modern Physics*, revised edition (New York, 1999), 234–6, 242–9, 430–3.
42 J. Priestley, *History and Present State of Electricity*, first edition (London, 1767), preface, 10.
43 J. Priestley, *History and Present State of Electricity*, third edition (London, 1775), II, 134–7, quoted in S. Schaffer, 'Natural philosophy and public spectacle in the eighteenth century', *History of Science*, 21 (1983), 1–43, p. 8.
44 J. Wesley, *The Desideratum, or Electricity Made Plain and Useful* (London, 1760); P. Bertucci, 'Revealing sparks: John Wesley and the religious utility of electrical healing', *British Journal for the History of Science*, 39 (2006), 341–62; P. Elliott, '"More subtle than the electric aura": Georgian medical electricity, the spirit and animation and the development of Erasmus Darwin's psychophysiology', *Medical History*, 52 (2008), 195–220.
45 K. Whitaker, 'The culture of curiosity', in N. Jardine, J. A. Secord and E. C. Spary eds., *Cultures of Natural History* (Cambridge, 1996), 75–90; L. Daston and K. Park, *Wonders and the Order of Nature* (New York, 1998), 255–301; Livingstone, *Putting Science in its Place*, 29–32.
46 R. Greene, *Descriptive Catalogue of the Rarities in the Mr Greene's Museum at Lichfield* (Lichfield, 1773), *A Particular and Descriptive Catalogue of the natural, and artificial Rarities in the Lichfield Museum collected (in the space of forty years) by Richard Greene* (Lichfield, 1782), *A Particular and Descriptive Catalogue of the Curiosities, natural and artificial, in the Lichfield Museum. Collected (in the space of forty-six years) by Richard Greene* (Lichfield, 1786); J. Boswell, *The Life of Samuel Johnson LLD*, 4 vols., edited by G. B. Hill and revised by L. F. Powell (Oxford, 1934), III, 465–6; A. MacGregor ed., *Sir Hans Sloane: Collector, Scientist, Antiquary* (London, 1994); H. S. Torrens, 'Natural history in eighteenth-century museums in Britain', in R. G. A. Anderson *et al.* eds., *Enlightening the British: Knowledge, Discovery and the Museum* (Oxford, 2003), 81–91.
47 J. Nichols, *Literary Anecdotes of the Eighteenth Century*, 9 vols. (London, 1812), IX, 351–68; R. Surtees, *The History and Antiquities of the County Palatine of Durham*, 4 vols. (Durham, 1816–40), III, 370–3; G. Townshend Fox, *Synopsis of the Newcastle Museum late the Allan Formerly the Tunstall, or Wycliffe Museum* (Newcastle, 1827), preface v–xxi, 21–37, 176–8; W. H. D. Longstaffe, *The History and Antiquities of the Parish of Darlington* (Darlington, 1854), division vi, pp iii–xxxii.
48 J. E. McClellan III, *Science Reorganised: Scientific Societies in the Eighteenth Century* (New York, 1985).
49 D. King-Hele, *The Collected Letters of Erasmus Darwin* (Cambridge, 2007), 394.
50 Walters, 'Science and politeness'; Secord, 'Newton in the nursery', 134–5.
51 Daniels, *Joseph Wright*, 34–9; Elliott, *Derby Philosophers*, 54–5.
52 J. Ferguson, *Analysis of a Course of Lectures on Mechanics, Pneumatics, Hydrostatics and Astronomy* (London, 1761).

53 *Derby Mercury* (25 December 1778).
54 *Derby Mercury* (29 November 1739, 22 January 1740, 22 September 1743).
55 J. Uglow, *The Lunar Men: the Friends who Made the Future* (London, 2002), iv, 124–5; P. Jones, *Industrial Enlightenment: Science and Technology in Birmingham and the West Midlands 1760–1820* (Manchester 2008), 82–94.
56 King-Hele, *Darwin*, 188–9; King-Hele ed., *Collected Letters*, 448–50.
57 A. E. Musson and E. Robinson, *Science and Technology in the Industrial Revolution* (Manchester, 1969), 88–96.
58 A. Thackray, 'Natural knowledge in cultural context: the Manchester model', *American Historical Review*, 79 (1974), 672–709.
59 Elliott, *Derby Philosophers*, 69–85, 138.
60 M. Benjamin ed., *Science and Sensibility: Gender and Scientific Enquiry, 1780–1945*; D. F. Noble, *A World without Women: The Christian Clerical Culture of Western Science* (New York, 1992); L. and S. Sheets-Pyenson, *Servants of Nature: A History of Scientific Institutions, Enterprises and Sensibilities* (London, 1999), 335–49; Watts, *Women in Science*, 99–133.

CHAPTER 2

1 Some of this material was presented at a Derbyshire branch meeting of the British Association for the Advancement of Science at the University of Derby on 28 April 2009, and I am grateful to Ian Turner and the members of the audience for critical comments.
2 H. E. Gruber, *Darwin on Man: A Psychological Study of Scientific Creativity* (London, 1974), 148–9.
3 E. Darwin, *The Temple of Nature; or the Origins of Society* (London, 1803), 29.
4 D. King-Hele, *Erasmus Darwin: A Life of Unequalled Achievement* (London, 1999); R. Schofield, *Lunar Society of Birmingham* (Oxford, 1963); A. E. Musson and E. Robinson, *Science and Technology in the Industrial Revolution* (Manchester, 1969); P. Sturges, 'The membership of the Derby Philosophical Society, 1783–1802', *Midland History*, 4 (1978), 212–29; P. Elliott, *The Derby Philosophers: Science and Culture in English Urban Society, 1700–1850* (Manchester, 2009); J. Uglow, *The Lunar Men: the Friends who Made the Future* (London, 2003); C. U. M. Smith and R. Arnott eds., *The Genius of Erasmus Darwin* (Aldershot, 2005); D. King-Hele ed., *The Collected Letters of Erasmus Darwin* (Cambridge, 2007), 279.
5 A. Seward, *Memoirs of the Life of Dr Darwin* (London, 1804), 125–6.
6 Seward, *Life of Dr Darwin*, 126–7.
7 C. Darwin, *Life of Erasmus Darwin*, ed. by D. King-Hele (Cambridge, 2003), 31.
8 W. Shenstone, 'Unconnected thoughts on gardening', 'A description of the Leasowes, the seat of the late William Shenstone, Esquire' and 'verses to Mr Shenstone in W. Shenstone, *Works in Verse and Prose*, 2 vols. (London, 1764), II, 125–47, 333–72; C. Thacker, *The Genius of Gardening* (London, 1994), 197–201; J. Dixon Hunt, *The Picturesque Garden in Europe* (London, 2002), 49–50.
9 Darwin, *Temple of Nature*, 100–1; B. C. Gelpi, 'Significant exposure: the turn of the century breast', *Nineteenth-Century Contexts*, 20 (1997), 125–45.
10 King-Hele ed., *Collected Letters*, 315–16, 320.
11 Darwin, letter to Thomas Cadell (5 April 1781), letter to Joseph Banks (13 September 1781) in King-Hele ed., *Collected Letters*, 183, 186–8.
12 King-Hele ed., *Collected Letters*, 211–18; M. Craven and M. Stanley, *The Derbyshire Country House* revised edition (Ashbourne, 2004), 51–3.
13 E. Darwin, manuscript notebook, Cambridge University Library, DAR 227.2.11; Uglow, *Lunar Men*, 383; King-Hele ed., *Collected Letters*, 279; P. Elliott, 'Erasmus Darwin's Trees' in L. Auricchio, E. H. Cook, and G. Pacini eds., *Arboreal Values in*

Europe and North America, 1660–1830 (forthcoming, 2011). With the plant information, garden features and directional information it would be possible to reconstruct Darwin's garden. The well shaft must still exist under the old Full Street Police Station and the plaque is preserved in Derby Museum and Art Gallery.

14   E. Darwin, manuscript notebook.

15   E. Darwin, letter to R. Darwin (23 November 1799), King-Hele ed., *Collected Letters*, 535.

16   King-Hele, ed., *Collected Letters*, 579.

17   King-Hele, *Erasmus Darwin*; D. and S. Lysons, *Magda Britannia*, vol. 5, *Derbyshire* (London, 1817), 67–8; N. Redman, *Illustrated History of Breadsall Priory* (Derby, 1998); M. Craven and M. Stanley, *Derbyshire Country House*, 51–3; King-Hele ed., *Collected Letters*, 574, 578–9.

18   B. Trinder, *The Making of the Industrial Landscape* (Gloucester, 1982), 52–127; M. J. Daunton, *Progress and Poverty: An Economic and Social History of Britain, 1700–1850* (Oxford, 1995), 253–66.

19   C. W. J. Withers, *Geography, Science and National Identity: Scotland since 1520* (Cambridge, 2001).

20   D. Hudson and K. W. Luckhurst, *The Royal Society of Arts, 1754–1954* (London, 1954); G. Averley, 'English Scientific Societies of the Eighteenth and Early Nineeenth Centuries', unpublished PhD thesis, University of Teeside Polytechnic and Durham University (Durham, 1989), 329–81; S. Wilmot, *The Business of Improvement: Agriculture and Scientific Culture in Britain, c.1700–1870*, Historical Geography Research Series, 24 (1990).

21   H. C. Pawson, *Robert Bakewell: Pioneer Livestock Breeder* (London, 1957).

22   J. von Sachs, *History of Botany (1530–1860)*, translated by E. F. Garnsey and revised by I. B. Balfour (Cambridge, 1890).

23   J. S. Watson, *The Reign of George III* (London, 1960), 288.

24   King-Hele ed., *Collected Letters*, 304–7, which also reproduces the drawings.

25   E. Darwin, Commonplace Book, microfilm copy, Derby Local Studies Library, Derby, 107–9, 114; King-Hele, *Erasmus Darwin*, 187.

26   Elliott, *Derby Philosophers*; Schofield, *Lunar Society*, 76–82; Uglow, *Lunar Men*, 154–65; Jones, *Industrial Enlightenment*, 31–2.

27   Sachs, *History of Botany*; D. E. Allen, *The Naturalist in Britain: A Social History* (London, 1976); L. Brockway, *Science and Colonial Expansion: The Role of the British Botanic Gardens* (New York, 1979); P. Bowler, *History of the Environmental Sciences* (London, 1992); P. F. Stevens, *The Development of Biological Systematics: Antoine-Laurent de Jussieu, Nature and the Natural System* (New York, 1994); R. Desmond, *Kew: The History of the Royal Botanic Gardens* (London, 1995); N. Jardine, J. A. Secord and E. C. Spary eds., *Cultures of Natural History* (Cambridge, 1996); A. B. Shteir, *Cultivating Women, Cultivating Science: Flora's Daughters and Botany in England, 1760–1860* (Baltimore, 1996); D. P. McCracken, *Gardens of Empire: Botanical Institutions of the Victorian British Empire* (Leicester, 1997); R. Drayton, *Nature's Government: Science, Imperial Britain and the 'Improvement' of the World* (New Haven, 2000); C. Merchant, *Reinventing Eden: The Fate of Nature in Western Culture* (New York, 2004).

28   C. Linnaeus, *Miscellaneous Tracts Relating to Natural History, Husbandry and Physick*, translated by B. Stillingfleet (London, 1759); *A System of Vegetables ... Translated from the Thirteenth Edition of the' Systema Vegetabilium' of ... Lineus* [sic] *... by a Botanical Society at Lichfield*, 2 vols. (Lichfield, 1783); *The Families of Plants ... Translated from the Last Edition of the 'Genera Plantarum' ... by a Botanical Society at Lichfield* 2 vols. (Lichfield, 1787); J. E. Smith, *An Introduction to Physiological and Systematical Botany*, 6th edition (London, 1827); Sachs, *History of Botany*, 8–10, 79–108; E. Mayr, *The Growth of Biological Thought* (Cambridge MA, 1982), 177–80;

N. Jardine, 'Inner history; or, how to end Enlightenment', in W. Clark, J. Golinski and S. Schaffer eds., *The Sciences in Enlightened Europe* (Chicago, 1999), 477–94; Allen, *Naturalist in Britain*, 36–51; K. Thomas, *Man and the Natural World* (London, 1984), 65–9.

29   Mayr, *Growth of Biological Thought*, 196–202.

30   Sachs, *History of Botany (1530–1860)*, 108–54; Mayr, *Growth of Biological Thought*, 198–208; Stevens, *Development of Biological Systematics*.

31   Darwin, *The Loves of The Plants*, poem; 'Loves of the Triangles' *Anti-Jacobin; or, Weekly Examiner* (16 April, 23 April, and 7 May 1798); J. V. Logan, 'The Poetry and Aesthetics of Erasmus Darwin', *Princeton Studies in English*, 15 (1936), 46–92; D. King-Hele, *Erasmus Darwin and the Romantic Poets* (London, 1986); M. McNeil, 'The scientific muse: the poetry of Erasmus Darwin', in L. J. Jordanova ed., *Languages of Nature: Critical Essays on Science and Literature* (London, 1986); M. McNeil, *Under the Banner of Science: Erasmus Darwin and His Age* (Manchester, 1987); A. Bewell, '"Jacobin plants": botany as social theory in the 1790s', *Wordsworth Circle*, 20 (1989): 132–9; J. Browne, 'Botany for gentlemen: Erasmus Darwin and the Loves of the Plants', *Isis*, 80 (1989), 593–620; L. Schiebinger, 'The private life of plants: sexual politics in Carl Linnaeus and Erasmus Darwin', in M. Benjamin ed., *Science and Sensibility: Gender and Scientific Enquiry, 1780–1945* (Oxford, 1991); J. McGann, *The Poetics of Sensibility: A Revolution in Literary Style* (Oxford, 1996); Shteir, *Cultivating Women*; F. J. Teute, 'The loves of the plants; or, the cross-fertilization of science and desire at the end of the eighteenth century', *The Huntington Library Quarterly*, 63 (2000), 319–45; D. Coffey, 'Protecting the botanic garden: Seward, Darwin, and Coalbrookdale', *Women's Studies*, 31 (2002), 141–64; C. Packham, 'The science and poetry of animation: personification, analogy, and Erasmus Darwin's 'Loves of the Plants', *Romanticism*, 10 (2004), 191–208; M. Page, 'The Darwin before Darwin: Erasmus Darwin, visionary science, and romantic poetry', *Papers on Language and Literature*, 41 (2005), 146–69; T. Barnard, *Anna Seward: A Constructed Life: A Critical Biography* (Aldershot, 2009). I am grateful to Teresa Barnard of the University of Derby for allowing me to view her material on Seward and Darwin.

32   W. Withering, *A Botanical Arrangement of all the Vegetables Naturally growing in Great Britain, with descriptions of all the Genera and Species according to the celebrated system of the celebrated Linnaeus* (Birmingham, 1775); Schofield, *Lunar Society*, 122–7; King-Hele ed., *Collected Letters*, 130–7.

33   Schofield, *Lunar Society*, 306–13.

34   *A System of Vegetables … translated from the thirteenth edition of the Systema Vegetabilium*, 2 vols. (Lichfield, 1783), I, xi.

35   *A System of Vegetables*, I, 11–12.

36   C. W. J. Withers, *Placing the Enlightenment* (Chicago, 2007), 122–5.

37   Darwin, *Life of Erasmus Darwin*, 42.

38   Darwin, manuscript notebook.

39   Darwin, *Phytologia, Phytologia: or the Philosophy of Agriculture and Gardening* (London, 1800), 513.

40   Darwin, *Phytologia*, 513–14.

41   Darwin, *Phytologia*, 514–15.

42   E. Darwin, *Phytologia*, 194.

43   A. Bennet, *New Experiments on Electricity* (Derby and London, 1789); Darwin, *Phytologia*, 312.

44   Darwin, *Zoonomia*, 109–10; *Phytologia*, 39–40, 132–6, 471–2; P. Elliott, '"More subtle than the electric aura": Georgian medical electricity, the spirit of animation and the development of Erasmus Darwin's psychophysiology', *Medical History* (2008), 195–220.

45 Darwin, *Phytologia*, 39–40, 132–6; Elliott, 'More subtle than the electric aura'.
46 Darwin, *Zoonomia*, I, 33.
47 Darwin, *Temple of Nature*, 132–4.
48 King-Hele, *Erasmus Darwin*, 237–40, 264–6.
49 Darwin, *Life of Erasmus Darwin*, 31–4; King-Hele, *Erasmus Darwin*, 314–17.
50 *The Golden Age: A Poetical Epistle from Erasmus D---N to Thomas Beddoes, MD* (London, 1794), 7–8.
51 J. Gillray, *New Morality;-or-the Promis'd Installment of the High Priest of the Theophilanthropes, with the Homage of Leviathan and his Suite* (*Anti-Jacobin Review and Magazine*, 1798); V. Gatrell, *City of Laughter: Sex and Satire in Seventeenth-Century* (London (London, 2006), 464–6; *The Millenium; a Poem in Three Cantos*, 2 vols. (London, 1800–1), 99–102.
52 *Life and Works of Sir John Sinclair*, II, 85, quoted in King-Hele ed., *Collected Letters*, 517; Darwin, *Life of Erasmus Darwin*, 40–1.
53 Darwin, *Life of Erasmus Darwin*, 40–1.

CHAPTER 3

1 I. Parker *Dissenting Academies in England* (Cambridge, 1914); H. McLachlan, *English Education under the Test Acts* (Manchester, 1931); H. Robinson, 'Geography in the Dissenting academies', *Geography*, 36 (1951), 179–86; R. V. Holt, *The Unitarian Contribution to Social Progress in England* (London, 1952); J. W. A. Smith, *The Birth of Modern Education* (London, 1954); C. G. Bolam, J. Goring, H. L Short and R. Thomas, *The English Presbyterians* (London, 1968); A. Lincoln, *Some Political and Social Ideas of English Dissent, 1763–1800* (Cambridge, 1971); R. O'Day, *Education and Society, 1500–1800* (London, 1982); R. Watts, 'Joseph Priestley and education', *Enlightenment and Dissent*, 2 (1983), 83–97; R. Watts, 'Manchester College and Education 1786–1853', in B. Smith ed., *Truth, Liberty, Religion: Essays Celebrating Two Hundred Years of Manchester College* (Oxford, 1986); I. B. Cohen ed., *Puritanism and the Rise of Modern Science* (New Brunswick, 1990); D. L. Wykes, 'Religious Dissent and the penal laws: an explanation of business success?' *History*, 75 (1990), 39–62; A. P. F. Snell, 'Philosophy in the eighteenth-century Dissenting academies', *History of Universities*, 11 (1992), 75–122; R. Watts, 'Revolution and reaction: Unitarian academics, 1780–1800', *History of Education*, 20 (1991), 307–23; S. Skedd, 'Women teachers and the expansion of girl's schooling in England', in H. Barker and E. Chalus eds., *Gender in Eighteenth-Century England* (London, 1997), 101–25; A. N. Walters, 'Science and politeness in eighteenth-century England', *History of Science*, 35 (1997), 121–54; R. Watts, *The Unitarian Contribution to Education* (London, 1998); M. Mercer, 'Dissenting academies and the education of the laity, 1750–1850', *History of Education*, 30 (2001), 35–58; M. Bryant, *The London Experience of Secondary Education* (London, 1986), 97–119, p. 118; P. Wood, ed., *Science and Dissent in England, 1688–1945* (Aldershot, 2004); P. Elliott and S. Daniels, 'Pestalozzi, Fellenberg and British nineteenth-century geographical education', *Journal of Historical Geography*, 32 (2006), 752–74.
2 The expression 'modern' here reflects Georgian usage and refers to subjects not traditionally regarded as part of the classical curriculum such as modern languages, modern history and mathematics.
3 J. Phillips, *Electoral Behaviour in Unreformed England* (Princeton, 1982); F. O'Gorman, *Voters, Patrons and Parties: The Unreformed Electoral System of Hanoverian England* (Oxford, 1989); J. Bradley, 'Nonconformity and the electorate in eighteenth-century England', *Parliamentary History*, 6 (1987), 236–61; P. Langford, *Public Life and the Propertied Englishman, 1689–1798* (Oxford, 1991); H. T. Dickinson, *The Politics of the People in Eighteenth-Century Britain* (Basingstoke,

1994); M. Watts, *The Dissenters*, 2 vols. (Oxford, 1995); Bryant, *London Experience*, 98–9; P. Elliott, 'Provincial urban society, scientific culture and socio-political marginality in Britain in the eighteenth and nineteenth centuries', *Social History* 28 (2003), 361–87; J. Seed, 'History and narrative identity: Religious Dissent and the politics of memory in eighteenth-century England', *Journal of British Studies* 44 (2005), 46–63, pp. 48–9; Mercer, 'Dissenting academies', 56; Wood, *Science and Dissent*, 1–18; J. H. Brooke, 'Science and Dissent: some historiographical issues' and G. Cantor, 'Real disabilities?: Quaker schools as "nurseries" of science' in Wood ed., *Science and Dissent*, 19–38, 147–61.

4   D. N. Livingstone, *The Geographical Tradition* (Oxford, 1992); R. J. Mayhew, 'Geography in eighteenth-century British education', *Paedogogica Historica*, 34 (1998), 731–69; D. N. Livingstone and C. J. Withers eds., *Geography and Enlightenment* (Chicago, 1999); C. W. J. Withers and R. Mayhew, 'Rethinking "disciplinary" history: geography in British universities', *Transactions of the Institute of British Geographers*, 27 (2002), 11–29; Elliott and Daniels, 'Pestalozzi, Fellenberg'; P. Elliott and S. Daniels, '"No study so agreeable to the youthful mind": geographical education in the Georgian grammar school', *History of Education*, 39 (2010), 15–33.

5   L. Stewart, 'Seeing through the scholium: religion and reading Newton in the eighteenth century', *History of Science*, 34 (1996), 123–65; L. Thomas, *The Scourge: in Vindication of the Church of England* (London, 1720); S. Palmer, *A Defence of the Dissenters Education in their Private Academies* (London, 1703).

6   McLachlan, *English Education*, 52–3.

7   Smith, *Modern Education*, 83–7; McLachlan, *English Education*, 291–2; B. Simon, *The Two Nations and the Educational Structure, 1780–1870* (London, 1976), 17–71; Watts, 'Joseph Priestley and education' 83–97; J. Graham, 'Revolutionary philosopher: the political ideas of Joseph Priestley (1733–1804)', *Enlightenment and Dissent*, 8 (1989), 43–68; Bryant, *London Experience*, 112–13; D. Eshet, 'Re-reading Priestley: science at the intersection of theology and politics, *History of Science*, 39 (2001), 127–59.

8   Bryant, *London Experience*, 110–13; Smith, *Modern Education*, 152–9.

9   F. Watson, *The Beginnings of the Teaching of Modern Subjects in England* (London, 1909); N. Hans, *New Trends in Education in the Eighteenth Century* (London, 1951); R. S. Tompson, *Classics or Charity: The Dilemma of the Eighteenth-Century Grammar School* (Manchester, 1971); R. K. Merton, *Science, Technology and Society in Seventeenth Century England* (Atlantic Highlands, 1978); J. Money, *Experience and Identity: Birmingham and the West Midlands, 1760–1800* (London, 1978); J. Simon, 'Private classical schools in eighteenth-century England: a critique of Hans', *History of Education*, 8 (1979), 179–91; Bryant, *London Experience*, 120–44; Daniels and Elliott, 'No study so agreeable'.

10   J. Oldfield, *An Essay Towards the Improvement of Reason* (London, 1707), 18–25, 365; Smith, *Modern Education*, 121–6.

11   Parker, *Dissenting Academies*, 62, 138; McLachlan, *English Education*, 11, 49–52; Smith, *Modern Education*, 41–5, 87–90; Bryant, *London Experience*; S. Johnson, 'Life of Watts', in A. Murphy ed., *The Works of Samuel Johnson LLD*, 12 vols. (London, 1820) XI, 238–48.

12   I. Watts, *The Improvement of the Mind* (London, 1795), 3, 5–6.

13   Watts, *The Improvement of the Mind*, 256–7; Watts, *The Knowledge of the Heavens and Earth Made Easy*, fourth edition (London, 1745); D. Jennings, *Introduction to the Use of the Globes, and the Orrery: with the Application of Astronomy to Chronology* (London, 1747), iii.

14   Watts, *Heavens and Earth*, x–xii, 10, 20–1, 27–36, 211; Watts, *Improvement of the Mind*, 298–9.

15   McLachlan, *English Education*, 118–19; Smith, *Modern Education*, 95–6.

16   Watts, *Heavens and Earth*, xvi–xx; Watts, *Improvement of the Mind*, 199–217.

17   Watts, *Improvement of the Mind*, 295–9.

18   Watts, *Heavens and Earth*, xi–xii, 10, 20–1, 27, 29, 33.

19   Watts, *Heavens and Earth*, 35–6, 98,103, 104, 134–5.

20   E. Latham 'Exercitatis Physiologia', Derby Local Studies Library mss. 3368.

21   McLachlan, *English Education*, 131–34; J. Gregory, *A Manual of Modern Geography* first and third editions (London and Derby, 1739 and 1748); *Derby Mercury* (18 January 1754); J. Clegg, *The Diary of James Clegg* edited by V. S. Doe (Matlock, 1981), appendix II, 945–60.

22   Gregory, *Modern Geography* (1739), preface.

23   Smith, *Modern Education*, 134, 95–6; McLachlan, *English Education*, 6–62, 118–19, 123; J. C. Ryland, *An Easy and Pleasant Introduction to Sir Isaac Newton's Philosophy* (London, 1772).

24   Daventry Academy, the last Congregational College to admit lay students closed in 1789 and Baptist colleges had never admitted them, a position that did not begin to change until the 1840s.

25   D. Neal, *The History of the Puritans or Protestant Non-Conformists, from the Reformation in 1517 to the Revolution in 1688*, edited by J. Toulmin, 3 vols. (London, 1837); D. Bogue and J. Bennett, *History of Dissenters: from the Revolution in 1688, to the Year 1808*, 4 vols. (London, 1812); Watts, *The Dissenters*, II, 482–5; Phillips, *Electoral Behaviour*; O'Gorman, *Voters, Patrons and Parties*; Bradley, 'Nonconformity and the electorate'; Langford, *Public Life and the Propertied Englishman*; Dickinson, *Politics of the People*; P. Wood, Introduction, J. H. Brooke, 'Science and Dissent: some historiographical issues' and J. Money, 'Science, technology and Dissent in English provincial culture: from Newtonian transformation to agnostic incarnation', in P. Wood ed., *Science and Dissent*, 1–18, 19–38, 67–112.

26   Mercer, 'Dissenting academies', 54.

27   A. Thackray, 'Natural knowledge in cultural context: the Manchester Literary and Philosophical Society', *American Historical Review*, 79 (1974), 672–709; I. Inkster, Introduction, and D. Orange, 'Rational Dissent and provincial science: William Turner and the Newcastle Literary and Philosophical Society', in I. Inkster and J. B. Morell eds., *Metropolis and Provinces: Science in British Culture, 1780–1850* (London, 1983), 224; M. Fitzpatrick, 'Rational Dissent and the Enlightenment', *Faith and Freedom*, 38 (1985), 83–101; D. Spadafora, *The Idea of Progress in Eighteenth-Century Britain* (New Haven, 1990), 13; M. Fitzpatrick, 'Heretical religion and radical political ideas in late eighteenth-century England', in E. Helmuth ed., *The Transformation of Political Culture* (London, 1990), 339–72; L. Stewart, *The Rise of Public Science* (Cambridge, 1992); D. L. Wykes, 'The contribution of the Dissenting academy to the emergence of rational Dissent', in K. Haakonssen ed., *Enlightenment and Religion: Rational Dissent in Eighteenth-Century Britain* (Cambridge, 1996), 99–139; Elliott, 'Provincial urban society'; S. D. Snobelin, 'The Discourse of God: Isaac Newton's heterodox theology and his natural philosophy', in Wood ed., *Science and Dissent*, 39–66.

28   J. Strutt, 'On the relative advantages and disadvantages of the English and Scottish universities', Strutt papers, Fitzwilliam Museum, Cambridge, MS 48 – 1947.

29   *New Academical Institution*, 24 June 1788 (London, 1788); *List of Subscribers to the New Academical Institution* (London, 1788). Rees produced a new edition of Chambers' *Cyclopaedia* in 4 volumes (London, 1788) and was the editor of the 39 volume *Cyclopaedia: or, Universal Dictionary of Arts and Sciences* (London, 1819) which included extensive geographical entries and an atlas.

30   Minutes of the High Pavement Presbyterian Chapel, Nottingham, 26 February 1786, Nottingham University Library, HiM: 1, 1777–1812; A. Rees, *The Advantages*

of Knowledge Illustrated and Recommended (London, 1788); McLachlan, *English Education*, 175–88; Smith, *Modern Education*, 192–7; Bryant, *London Experience*, 108–12.

31 Strutt, 'On the relative advantages'.

32 Seed, 'History and narrative'; E. Gleason and R. Watts, 'Making women visible in the history of education: the late eighteenth century experience', *History of Education Society Bulletin*, 59 (1997), 37–46; Bryant, *London Experience*, 110–12.

33 J. Priestley, *Lectures on History and General Policy*, 2 vols. (London, 1793), 201–2; J. Priestley, 'Account of a course of lectures on the history of England, and a plan of the course', in J. T. Rutt ed., *The Theological and Miscellaneous Works of Joseph Priestley, LLD* (London, 1831), XXIV, 440–41; J. Graham, 'Revolutionary philosopher: the political ideas of Joseph Priestley (1733–1804)', *Enlightenment and Dissent*, 8 (1989), 43–68.

34 T. Barnes, *A Discourse Delivered at the Public Commencement of the Manchester Academy, September 14th 1786* (Manchester, 1786); J. F. Fulton, 'The Warrington Academy (1757–86) *Bulletin of the Institute of the History of Medicine*, 1 (1933); H. McLachlan, *Warrington Academy: Its History and Influence,* Chetham Society 107 (Manchester, 1943); G. M. Ditchfield, 'The early history of Manchester College', *Transactions of the Historical Society of Lancashire and Cheshire* 11 (1972); J. Seed, 'Manchester College, York: an early nineteenth-century Dissenting academy', *Journal of Educational Administration and History*, 14 (1982); D. Wykes, 'Sons and subscribers: lay support and the College 1786–1840', R. Watts, 'Manchester College and education, 1786–1853', J. Raymond and J. Pickstone, 'The natural sciences and the learning of the English Unitarians: an explanation of the roles of Manchester College' and C. Webster and J. Barry, 'The Manchester medical revolution', in B. Smith ed., *Truth, Liberty, Religion: Essays Celebrating Two Hundred Years of Manchester College* (Oxford, 1986), 31–78, 79–111, 127–38, 165–84; P. O'Brien, *Warrington Academy 1757–86: Its Predecessors and Successors* (Wigan, 1989); D. Wykes, 'Manchester College at York (1803–1840): its intellectual and cultural contribution', *Yorkshire Archaeological Journal*, 63 (1991).

35 S. Addington, *An Inquiry into the Reasons for and Against Inclosing Open-Fields* (Coventry, 1772), iii–iv, 8–11, 42–8.

36 S. Addington, *The Youth's Geographical Grammar* (London, 1770), 44–5, 47, 163; J. C. Davies, *Georgian Harborough* (Market Harborough, 1969), 85–9.

37 W. A. C. Stewart and W. P. McCann, *The Educational Innovators, 1750–1880* (Basingstoke, 1967), 35–52; L. Stewart, 'The public culture of radical philosophers in eighteenth-century London', in Wood ed., *Science and Dissent*, 113–29; T. H. Levere, 'Natural philosophers in a coffee house: Dissent, radical reform and pneumatic chemistry', in Wood ed., *Science and Dissent*, 131–46.

38 McLachlan, *English Education* (1931), 52–62; D. Williams, *Lectures on Education*, 3 vols. (London, 1789), III, 11–12, 22–27; D. Williams, *Lectures on the Universal Principles and Duties of Religion and Morality* 2 vols. (London, 1779); P. D. Lowe, 'Locals and Cosmopolitans: A Model for the Social Organisation of Provincial Science in the Nineteenth Century', unpublished MPhil thesis, University of Sussex (Brighton, 1978); C. Russell, *Science and Social Change, 1700–1900* (Basingstoke, 1983), 69–219; D. A. Hinton, 'Popular Science in England, 1830–79', unpublished PhD thesis, University of Bath (Bath, 1979); I. Inkster, *Scientific Culture and Urbanisation in Industrialising Britain* (Aldershot, 1997); G. Kitteringham, 'Studies in the Popularisation of Science in England, 1800–30', unpublished PhD thesis, University of Sussex (Brighton, 1981); J. E. McLellan, *Science Reorganised* (New York, 1985); G. Averley, 'English Scientific Societies of the Eighteenth and Early Nineteenth Centuries', unpublished PhD thesis, Teeside Polytechnic, University of Durham

(Durham, 1989); L. Pyenson and S. Sheets-Pyenson, *Servants of Nature* (London, 2000), 74–100, 319–49.

39   P. Elliott, '"Improvement always and everywhere": William George Spencer (1790–1866) and mathematical, geographical and scientific education in nine-teenth-century England', *History of Education*, 33 (2004), 391–417.

40   Goodacre papers, Bromley House Subscription Library, Angel Row, Nottingham; Willoughby, *Nottingham Directory*, 14; I. Inkster, 'Robert Goodacre's astronomy lectures (1823–1825), and the structure of scientific culture in Philadelphia', in *Scientific Culture and Urbanisation*, VIII.

41   R. Goodacre, *An Essay on the Education of Youth* (Nottingham, 1808), 19–40.

42   Goodacre, *Essay on Education*, 44–63; R. Goodacre, *A Brief Explanation of the Prin-cipal Terms Made use of in Astronomy*, second edition (Nottingham, 1822).

43   Goodacre, *Essay on Education*, 65–70.

44   Goodacre, *Essay on Education*, 60–63.

45   R. Watts, *Gender, Power and the Unitarians in England, 1760–1860* (London, 1998), 112–15; Elliott and Daniels, 'Pestalozzi, Fellenberg'.

CHAPTER 4

1   M. C. Jacob, *The Radical Enlightenment: Pantheists, Freemasons and Republicans* (London, 1981), 121–2.

2   On the parodying societies: W. J. Chetwode Crawley, 'Mock masonry in the eighteenth century', *Transactions of the Ars Quatuor Coronatorum* (subsequently *AQC*), 18 (1905), 129–46.

3   J. Money, 'The Masonic moment; or, ritual, replica, and credit: John Wilkes, the Macaroni parson, and the making of the middle-class mind', *Journal of British Studies*, 32 (1993), 359–95, p. 370; P. Clark, *British Clubs and Societies 1580–1800: the Origins of an Associational World* (Oxford, 2000), 336.

4   Rev. R. Green, 'On the Masonic duties', in G. Oliver ed., *The Golden Remains of the Early Masonic Writers*, 5 vols. (London, 1848), I, 229.

5   P. Elliott, *The Derby Philosophers: Science and Urban Culture in Britain, c.1700–1850* (Manchester, 2009).

6   J. M. Roberts, 'Freemasonry: possibilities of a neglected topic', *English Historical Review*, 84 (1969), 323–35; J. Money, *Experience and Identity: Birmingham and the West Midlands, 1760–1800* (Montreal, 1977), 136; Money, 'Masonic moment', 358–95; A. Prescott, 'Freemasonry and the problem of Britain', in *Collected Studies in the History of Freemasonry* (Sheffield, 2003), digital copy; R. Coombes, 'Subject for enquiry: sources for research and historical bibliography in the Library and Museum of Freemasonry, London', in R. W. Weisberger, W. Mcleod and S. Brent Morris eds., *Freemasonry on Both Sides of the Atlantic* (Boulder, 2002), 755–80; S. McVeigh, 'Freemasonry and musical life in London in the late eighteenth century', in D. W. Jones ed., *Music in Eighteenth-Century Britain* (Aldershot, 2000), 72–100; P. Fara, 'John Theophilus Desaguliers 1683–1744)', *Dictionary of National Biography* (Oxford, 2004), subsequently DNB.

7   R. W. Weisberger, *Speculative Freemasonry and the Enlightenment* (Boulder, 1993); S. C. Bullock, *Revolutionary Brotherhood: Freemasonry and the Transformation of the American Social Order, 1730–1840* (Chapel Hill, 1996); Clark, *British Clubs and Soci-eties*, 309–349; J. Van Horn Melton, *The Rise of the Public in Enlightenment Europe* (Cambridge, 2001), 252–72.

8   N. Hans, *New Trends in Education in the Eighteenth Century* (London, 1953); B. J. Teeter Dobbs and M. C. Jacob, *Newton and the Culture of Newtonianism* (New Jersey, 1995), 103; S. Shapin and S. Schaffer, *Leviathan and the Air Pump: Hobbes, Boyle and the Experimental Life* (Princeton, 1985); L. Stewart, *The Rise of Public*

Science: Rhetoric, Technology and Natural Philosophy (Cambridge, 1992); S. Shapin, *The Scientific Revolution* (Chicago, 1998).

9   Clark, *British Clubs and Societies*, 312.

10   Melton, *Rise of the Public*, 257.

11   M. Jacob, *Living the Enlightenment: Freemasonry in Eighteenth-Century Europe* (Oxford, 1992), 4–15.

12   Dobbs and Jacob, *Newton and Newtonianism*, 102; Jacob, *Living the Enlightenment*, 65.

13   Jacob, *Radical Enlightenment*, 112–13.

14   J. Anderson, *The Constitutions of the Freemasons* (London, 1723); W. F. Gould, *History of Freemasonry*, edited by H. Poole, 4 vols. (London, 1951); A. S. Frere ed., *Grand Lodge, 1717–1967* (Oxford, 1967); J. Money, 'Freemasonry and the fabric of loyalism in Hanoverian England', in E. Hellmuth ed., *The Transformation of Political Culture, England and Germany in the Late Eighteenth Century* (London, 1990), 235–70; Jacob, *Radical Enlightenment*; Jacob, *Living the Enlightenment*; A. Newman, 'Politics and freemasonry in the eighteenth century', *AQC*, 104 (1991), 32–50; Money, 'Masonic moment', 373–92; D. Stevenson, *The Origins of Freemasonry: Scotland's Century, 1590–1710* (Cambridge, 1998); Clark, *British Clubs and Societies*, 309–49.

15   Jacob, *Living the Enlightenment*, 68; Clark, *British Clubs and Societies*, 336–9; R. Burt, 'Freemasonry and socio-economic networking during the Victorian period', *Archives*, 27 (2002), 31–8.

16   Jacob, *Radical Enlightenment*, 142.

17   M. C. Jacob, *The Cultural Meaning of the Scientific Revolution* (New York, 1988), 126–7.

18   Jacob, *Scientific Revolution*, 126.

19   Gould, *History of Freemasonry*, IV, 348–54; J. Stokes, 'Life of John Theophilus Desaguliers', *AQC*, 38 (1925), 285–308; G. Knoop and G. P. Jones, *The Genesis of Freemasonry* (Manchester, 1947), 172–74; Hans, *New Trends*, 137–41; J. R. Wigelsworth, 'Competing to popularise Newtonian philosophy: John Theophilus Desaguliers and the preservation of reputation', *Isis*, 94 (2003), 435–55; Fara, 'John Theophilus Desaguliers', DNB; Jacob, *Radical Enlightenment*, 122–5.

20   Jacob, *Radical Enlightenment*, 123.

21   Gould, *History of Freemasonry*, IV, 400; W. F. Gould, 'The Rev. Wm. Stukeley, M.D.', *AQC*, 6 (1887), 134.

22   For all of these see DNB.

23   A. Guerrini, 'James Douglas, fourteenth earl of Morton (1702–1768)', DNB; There is a large literature on Maclaurin and the Scottish Enlightenment, for an entry see the references in P. Jones ed., *Philosophy and Science in the Scottish Enlightenment* (London, 1988).

24   B. Taylor, *Contemplatio Philosophica: A Posthumous Work ... To Which is Prefaced a Life of the Author* (London, 1793), preface; J. Nichols, *Literary Anecdotes of the Eighteenth Century* (London, 1812), I, 171–3; Jacob, *Radical Enlightenment*, 112–13.

25   Jacob, *Radical Enlightenment*, 133–6.

26   Jacob, *Radical Enlightenment*, 123–4.

27   Jacob, *Radical Enlightenment*, 133, 136–7.

28   H. Lyons, *The Royal Society, 1660–1940: A History of Its Administration under Its Charters* (Cambridge, 1944); J. R. Clarke, 'The Royal Society and early grand lodge freemasonry', *AQC*, 80 (1967), 111–14; R. Sorrenson, 'Towards a history of the Royal Society in the eighteenth century', *Notes and Records of the Royal Society of London*, 50 (1996), 29–46.

29   R. Sweet, *Antiquaries: The Discovery of the Past in Eighteenth-Century Britain* (London, 2007), 81–91.

30   J. Habermas, *The Structural Transformation of the Public Sphere*, translated by T. Burger and F. Lawrence (Cambridge MA, 1989); W. Outhwaite ed., *The Habermas Reader* (Cambridge MA, 1996), 23–66; C. Calhoun ed., *Habermas and the Public Sphere* (Cambridge MA, 1992).

31   D. Hume, 'Of refinement in the arts', *Essays and Treatises on Several Subjects*, 2 vols. (Edinburgh, 1825), I, 274, 268.

32   R. J. Morris, 'Voluntary societies and British urban elites, 1780–1850: an analysis', *The Historical Journal*, 26 (1983), 95–118; R. J. Morris, 'Clubs, societies and associations', in F. M. L. Thompson ed., *The Cambridge Social History of Britain, 1750–1950*, 3 vols. (Cambridge, 1990), III, 395–443; R. J. Morris, *Class, Sect and Party: The Making of the British Middle Class: Leeds, 1820–1850* (Manchester, 1990); R. J. Morris, 'Civil society and the nature of urbanism: Britain, 1750–1850', *Urban History*, 25 (1998), 289–301; J. Barry, 'Bourgeois collectivism? Urban association and the middling sort', in J. Barry and C. Brooks eds., *The Middling Sort of People* (London, 1994), 84–112; Clark, *British Clubs and Societies*, 309–49.

33   Morris, 'Voluntary societies'; Clark, *British Clubs and Societies*, 234–73.

34   Stevenson, *Origins of Freemasonry*, 227.

35   T. Munck, *The Enlightenment: A Comparative Social History, 1721–1794* (London, 2000), 70.

36   *The History of the Robin Hood Society* (London, 1764); Hans, *New Trends*, 165–71; T. Fawcett, '18th century debating societies', *British Journal for 18th century Studies*, 3 (1980); Clarke, *British Clubs and Societies*, 48, 119.

37   E. Robinson, 'R. E. Raspe, Franklin's Club of Thirteen, and the Lunar Society', *Annals of Science*, 11 (1955), 127, 144; A. Musson and E. Robinson, *Science and Technology in the Industrial Revolution* (Manchester, 1969), 127; J. P. Simpson, 'Some old London taverns and masonry', *AQC*, 20 (1907), 28–46. For Raspe, Nichols, *Literary Anecdotes*, III, 230–1.

38   Jacob, *Radical Enlightenment*, 125.

39   R. Porter, 'Science, provincial culture and public opinion in Enlightenment England', *British Journal for Eighteenth Century Studies*, 3 (1980), 20–46; S. Schaffer, 'Natural philosophy and public spectacle in the eighteenth century', *History of Science*, 21 (1983), 1–43; A. N. Walters, 'Science and politeness in eighteenth-century England', *History of Science*, 35 (1997), 121–54; J. Brewer, *The Pleasures of the Imagination: English Culture in the Eighteenth Century* (London, 1997); W. Clark, J. Golinski and S. Schaffer eds., *The Sciences in Enlightened Europe* (Chicago, 1999).

40   Jacob, *Living the Enlightenment*, 65; Rev. R. Green, 'On the Masonic duties', in Oliver ed., *Golden Remains*, I, 229; see also the references to the sciences and improvement in the various editions of William Preston's *Illustrations of Masonry* published between 1772 and 1812 produced by A. Prescott (Academy Electronic Publications, 2001).

41   Rev. J. Indwood, 'The Mason's lodge, a school of virtue and science', sermon preached at Ramsgate, 3 September 1798 at the consecration and constitution of the Jacob's Lodge, in Oliver ed., *Golden Remains*, IV, 300–5.

42   *Farley's Bristol Journal* (12 March 1789), cited in Clark, *British Clubs and Societies*, 335–6.

43   A. Q. Morton and J. A. Wess, *Public and Private Science: The King George III Collection* (London, 1993), 89–122.

44   Gould, *History of Freemasonry*, IV, 352.

45   J. Senex, *A Treatise of the Description and Use of Both Globes* (London, 1718); G. W. Daynes, 'John Senex and the Royal Society', *AQC*, 37 (1924), 102–3; L. Worms, 'John Senex', DNB; Jacob, *Radical Enlightenment*, 125–6.

46   Nichols, *Literary Anecdotes*, V, 659–661; Jacob, *Radical Enlightenment*, 125–6.

47   Thomas Wright of Durham manuscripts, Newcastle City Library.

48   J. Toland, *Pantheisticon: or, the Form of Celebrating the Socratic-Society* (London, 1751); Jacob, *Living the Enlightenment*, 66; R. E. Sullivan, *John Toland and the Deist Controversy; a Study in Adaptations* (London, 1982); S. H. Daniel, *John Toland: His Methods, Manners and Mind* (London, 1984); see however, D. B. Haycock, *William Stukeley: Science, Religion and Archaeology in Eighteenth-Century England* (London, 2002), 180–5.

49   A. F. Calvert, *History of the Old King's Arms Lodge No. 28* (London, 1899), 11. I am grateful to Professor Andrew Prescott for information concerning the old King's Arms lodge and the lodge minutes.

50   M. Clare, *Youth's Introduction to Trade and Business* (London, 1720); *The Rates for Learning, Boarding and Tuition, at the Academy in Soho Square, London* (London, c.1745), 23–4; *General Advertiser* (20 May 1751) quoted in Calvert, *Old King's Arms Lodge*, 63; M. Clare, 'An address made to the body of free and accepted masons' (1735), *The Pocket Companion and History of Free-Masons*, second edition (London, 1759), 307–16, also Oliver, *Golden Remains*, I, 74–88; Hans, *New Trends*, 87–90.

51   M. Clare, *On the Motion of Fluids, Natural and Artificial; in particular that of the Air and Water* (London, 1735); minutes of the Old Kings Arms lodge, Library and Museum of Freemasonry, Grand Lodge, London (6, 20 October 1735); Hans, *New Trends*, 87–91.

52   Jacob, *Living the Enlightenment*, 56–7.

53   Anderson, *Constitution of the Freemasons*; Sweet, *Antiquaries*, 81–118; Jacob, *Living the Enlightenment*, 34–8.

54   S. A. E. Mendyk, *'Speculum Britanniae': Regional Study, Antiquarianism and Science in Britain to 1700* (Toronto, 1989); V. Jankovic, *Reading the Skies: A Cultural History of English Weather, 1650–1820* (Chicago, 2000).

55   M. Myrone, and L. Peltz, *Producing the Past; Aspects of Antiquarian Culture and Practice, 1700–1850* (Aldershot, 1999); Sweet, *Antiquaries*.

56   R. E. Anderson, revised N. Guicciardini, 'Charles Hayes (1678–1760)', W. P. Courtney, revised P. Davis, 'Richard Richardson (1663–1741)', A. Ross, 'John Arbuthnot (1667–1735)', DNB; Nichols, *Literary Anecdotes*, II, 322–6, 578–93; J. Woodward, *An Essay Toward a Natural History of the Earth, and Terrestrial Bodies* (London, 1695); *An Account of Some Roman Urns and Other Antiquities* (London, 1713); J. M. Levine, *Dr Woodward's Shield: History, Science, and Satire in Augustan England* (London, 1977); J. Redwood, *Reason, Ridicule and Religion: The Age of Enlightenment in England, 1660–1750*, second edition (London, 1996), 116–132; R. Porter, *The Making of the Science of Geology* (Cambridge, 1977).

57   Letter from Henry Baker FRS to Philip Doddridge, 24 November, 1747, quoted in Musson and Robinson, *Science and Technology*, 380; for York freemasonry see Gould, *History of Freemasonry*, II, 233–76; B. Cole ed., *The Ancient Constitutions of Free and Accepted Masons* (London, 1731).

58   F. Drake, *Eboracum: or, the History and Antiquities of the City of York* (York, 1736); F. Drake, speech on the history of freemasonry, in Cole ed., *Ancient Constitutions*, 13–19; Nichols, *Literary Anecdotes*, II, 87; Gould, *History of Freemasonry*, II, 238–9.

59   Nichols, *Literary Anecdotes*, VI, 1–162; W. Pickering, *The Gentlemen's Society at Spalding: Its Origin and Progress* (Spalding, 1851); M. Perry, 'The origin, progress and present state of the Spalding Gentlemen's Society', *Journal of the British Archaeological Society*, 5 (1889), 39–50; S. H. Perry, 'The Gentlemen's Society at Spalding', *AQC*, 53 (1942), 335–9; D. M. Owen, *The Minute-Books of the Spalding Gentlemen's Society, 1712–1755* (Lincoln, 1981); C. Dack, 'The Peterborough Gentlemen's Society', *Journal of the British Archaeological Society*, 5 (1889), 141–60; H. J. J. Winter, 'Scientific notes from the minutes of the Peterborough Society', *Isis*, 31 (1939), 32–59; S. Piggott, *William Stukeley, an Eighteenth-Century Antiquary* (London, 1985); D. and M. Honeybone eds., *Friendship and Knowledge: The Cor-*

respondence of the Spalding Gentleman's Society, 1710–1761 (Lincoln, 2010). As a guide to Masonic membership I have used W. J. Williams, 'Masonic personalities 1723–39', *AQC*, 40 (1927), 30–42, 126–38, 230–40; J. H. Lepper, 'The Earl of Orrery', *AQC*, 35 (1922), 76–8; Stokes, 'Life of Johanne Theophilus Desaguliers', *AQC*, 38 (1925), 285–308, *AQC*, 40 (1927), 170. The career of the antiquarian, cartographer and draughtsman Alexander Gordon is also of interest in this context. He spent much time on Continental tours and became grand master of the English masons, serving as secretary to the Society for Encouragement of Learning, secretary to the Society of Antiquaries, and secretary to the Egyptian Club and made donations to the Spalding Society; Nichols, *Literary Anecdotes*, V, 329–37.

60   Quoted in Knoop and Jones, *Genesis of Freemasonry*, 134; Nichols, *Literary Anecdotes*, V, 499–510; J. and J. B. Nichols, *Illustrations of the Literary History of the Eighteenth Century*, 8 vols. (London, 1817–58); Gould, 'The Rev. Wm. Stukeley, M.D.', 134; Piggott, *William Stukeley*; Haycock uses a comprehensive examination of Stukeley's voluminous papers and publications to re-situate him in the milieu of Newtonian science, conceived in the broadest possible sense to include his antiquarian, medical and natural philosophical work (Haycock, *William Stukeley*).

61   Haycock, *William Stukeley*, 174–80.

62   'Biographical history of Mr Thomas Wright', *Gentlemen's Magazine* 63 (1793), 9–12, 120–1; T. Wright, *An Original Theory or New Hypothesis of the Universe* (London, 1750), reprinted with introduction by M. Hoskin (London, 1971), see especially plates XXV and XXXII and the 'Elements of Existence' composed when he was 22 in 1734, 1–15; T. Wright, manuscript volumes, Newcastle Central Library, VI, 70–9, 80–9, VIII, 139–42; E. Hughes, 'The early journal of Thomas Wright of Durham', *Annals of Science*, 7 (1951), 1–24; F. A. Paneth, *Chemistry and Beyond* (New York), 91–122.

63   Jacob, *Radical Enlightenment*, 110.

64   Stevenson, *Origins of Freemasonry*, 4, 227.

65   Stevenson, *Origins of Freemasonry*, 216–33; for Jacob's response see Jacob, *Living the Enlightenment*, 35–46.

66   Musson and Robinson, *Science and Technology*; A. Thackray, 'Natural knowledge in cultural context: the Manchester Literary and Philosophical Society', *American Historical Review*, 79 (1974), 672–709; C. Russell, *Science and Social Change, 1700–1830* (London, 1983); I. Inkster and J. B. Morrell eds., *Metropolis and Province: Science in British Culture, 1780–1850* (London, 1983); Jacob, *Cultural Meaning*; Stewart, *Rise of Public Science*; P. Borsay, 'The London connection; cultural diffusion and the eighteenth-century provincial town', *London Journal*, 19 (1994), 21–35; I. Inkster, *Scientific Culture and Urbanization in Industrialising Britain* (Ashgate, 1997); V. Jankovic, *Reading the Skies: A Cultural History of English Weather, 1650–1820* (Chicago, 2000); P. Elliott, 'Provincial urban society, scientific culture and socio-political marginality in Britain in the eighteenth and nineteenth centuries', *Social History*, 28 (2003), 361–87; D. N. Livingstone, *Putting Science in its Place: Geographies of Scientific Knowledge*, Chicago, 2003; essays in a special issue of BJHS on the historical geography of science, 38 (2005); J. Golinski, *British Weather and the Climate of Enlightenment* (Chicago, 2009); Elliott, *Derby Philosophers*.

67   Frere, ed., *Grand Lodge*, 92–7; Jacob, *Living the Enlightenment*, 60–5.

68   Musson and Robinson, *Science and Technology*; Porter, 'Science, provincial culture and public opinion', 20–46; Russell, *Science and Social Change*; T. L. Hankins, *Science and the Enlightenment* (Cambridge, 1985); J. McClellan III, *Science Reorganised: Scientific Societies in the Eighteenth Century* (New York, 1985); G. Averley, English Scientific Societies of the Eighteenth Early Nineteenth Centuries', unpublished PhD thesis University of Teeside Polytechnic (1989); Elliott, *Derby Philosophers*.

69   R. Porter, 'Herschel, Bath and the Philosophical Society' in G. Hunt ed., *Uranus and the Outer Planets* (Cambridge, 1982), 23–4; R. S. Watson, *The History of the Library and Philosophical Society of Newcastle-upon-Tyne, 1793–1896* (Newcastle-Upon-Tyne, 1896), 20.

70   For the Friars, list of members, minute and account books, ms. history, etc., see Norfolk County Record Office, Colman mss. 9/- NRO; C. B. Jewson, *The Jacobin City* (Glasgow, 1975), 145–9; T. Fawcett, 'Measuring the provincial Enlightenment: the case of Norwich', *Eighteenth-Century Life*, 18 (1982), 19–20; C. B. Jewson, *Simon Wilkin of Norwich* (Norwich, 1980), 44–8.

71   Gould, *History of Freemasonry*, IV, 242–5; on the important links between Masonry and the navy see C. D. Rotch, 'Thomas Dunckerley and the Lodge of Friendship', *AQC*, 56 (1945), 59–113.

72   *Journal Book of the Royal Society* (10 December 1761), 221–4; J. Nichols, *The History and Antiquities of the County of Leicester* (Leicester, 1804), III, part ii, 717.

73   W. L. Clowes, *The Royal Navy: A History* (London, 1898), III, 566.

74   J. Whitehurst, *Tracts; Philosophical and Mechanical by John Whitehurst FRS* (London, 1792); R. E. Schofield, *The Lunar Society of Birmingham* (Oxford, 1963); C. Hutton, 'Authentic memoirs of the life and writings of the late John Whitehurst FRS', *Universal Magazine* (November, 1788), 225–9, *European Magazine* (1788), ii, 316–20, *Gentleman's Magazine* (1788), i, 182–3; W. Hutton, *History of Derby* (Birmingham, 1791), appendix; M. Craven, *John Whitehurst; Clockmaker and Scientist* (Ashbourne, 1996); Elliott, *Derby Philosophers*, 54–68; *Derby Mercury* (24 June 1748, 22 September 1749). Membership records of the Virgin's Inn Lodge do not survive.

75   H. T. C. de la Fontaine, 'Benjamin Franklin', *AQC*, 41 (1929); *Derby Mercury* (30 July 1779); *Harrison's Derby Journal* (29 July 1779); O. Manton, *Early Freemasonry in Derbyshire* (Manchester, 1913–16), 6–10; J. Whitehurst, *Enquiry into the State and Formation of the Earth*, first edition (London, 1778), Preface, i, 16–17.

76   S. Daniels, 'Loutherbourg's chemical theatre: Coalbrookdale by night', in J. Barrell ed., *Painting and the Politics of Culture: New Essays on British Art, 1700–1850* (Oxford, 1992), 195–230, 221–2; *Gentleman's Magazine*, 58 (February 1788), 182; Hutton, 'Authentic memoirs', 226; Hutton, 'Memoir', in Whitehurst, *Tracts; Philosophical and Mechanical*, 9–10; E. Robinson, 'R. E. Raspe, Franklin's Club of Thirteen, and the Lunar Society', *Annals of Science*, 11 (1955); Musson and Robinson, *Science and Technology*, 58, 74, 126–7, 141, 391.

77   J. Throsby, *Select Views in Leicestershire from Original Drawings* (London, 1789), 137; Nichols, *Literary Anecdotes*, III, part ii, 717. Likewise, Wright's 'the Alchymist', exhibited in 1771, as Daniels has suggested, has Masonic associations in dignifying 'a low subject' and turning 'a technical operation into a noble pursuit … in Masonic terms it presents a "chemical theatre" of material and spiritual transformation'; S. Daniels, *Joseph Wright* (London, 1999), 26–30.

78   *Derby Mercury* (30 July 1779); *Derby Journal* (29 July 1779); Manton, *Early Freemasonry in Derbyshire*, 6–10, 37, 40; P. Sturges, 'The membership of the Derby Philosophical Society', *Midland History*, 4 (1978), 212–29.

79   Melton, *Rise of the Public*, 254–5.

80   Jacob, *Radical Enlightenment*, 206–8; Jacob, *Living the Enlightenment*, 120–42; M. Burke and M. C. Jacob, 'French freemasonry, women and feminist scholarship', *Journal of Modern History*, 68 (1996); Melton, *Rise of the Public*, 257–60.

CHAPTER 5

1   J. von Sachs, *History of Botany (1530–1860)* (London, 1890); D. E. Allen, *The Naturalist in Britain: A Social History* (London, 1976); L. Brockway, *Science and Colonial Expansion: The Role of the British Botanic Gardens* (New York, 1979);

C. Russell, *Science and Social Change, 1700–1900* (London, 1983); P. Bowler, *History of the Environmental Sciences* (London, 1992); P. F. Stevens, *The Development of Biological Systematics: Antoine-Laurent de Jussieu, Nature and the Natural System* (New York, 1994); R. Desmond, *Kew: The History of the Royal Botanic Gardens* (London, 1995); N. Jardine, J. A. Secord and E. C. Spary eds., *Cultures of Natural History* (Cambridge, 1996); A. B. Shtier, *Cultivating Women, Cultivating Science: Flora's Daughters and Botany in England, 1760–1860* (Baltimore, 1996); D. P. McCracken, *Gardens of Empire: Botanical Institutions of the Victorian British Empire* (Leicester, 1997); R. Drayton, *Nature's Government: Science, Imperial Britain and the 'Improvement' of the World* (New Haven, 2000); E. Spary, *Utopia's Garden; French Natural History from Old Regime to Revolution* (Chicago, 2000); C. Merchant, *Reinventing Eden: The Fate of Nature in Western Culture* (New York, 2004); B. Elliott, *The History of the Royal Horticultural Society, 1804–2004* (London, 2004).

2   W. Wroth, *The London Pleasure Garden of the Eighteenth Century* (London, 1898); W. Wroth, *Cremorne and other Pleasure Gardens of Nineteenth-Century London*; M. Girouard, *The English Town* (New Haven, 1990), 145–58; P. Borsay, 'The rise of the promenade: the social and cultural use of space in the English provincial town, *c.*1660–1800', *British Journal for Eighteenth-Century Studies*, 9 (1986), 125–40; T. Longstaffe-Gowan, *The London Town Garden, 1700–1830* (London, 1990); M. Galinou ed., *London's Pride: The Glorious History of the Capital's Gardens* (London, 1990); P. Borsay, *The English Urban Renaissance* (Oxford, 1991), 162–72; J. Brewer, *The Pleasures of the Imagination: English Culture in the Eighteenth Century* (London, 1997); M. Ogborn, *Spaces of Modernity: London's Geographies, 1680–1780* (London, 1998), 116–57; J. Conlin, 'Vauxhall revisited: the afterlife of a London pleasure garden 1770–1859', *Journal of British Studies*, 45 (2006), 718–43; J. Conlin, 'Vauxhall on the boulevard: pleasure gardens in London and Paris, 1764–1784', *Urban History*, 35 (2008), 24–47.

3   A plan of the Linnaean garden is contained in Carl Linnaeus and Samuel Nauclér, *Hortus Upsaliensis* (Uppsala, 1745).

4   T. Faulkner, *An Historical and Topographical Description of Chelsea and its Environs* (London, 1810), 18–27; D. Lysons, *The Environs of London*, second edition, 2 vols. (London, 1810), I, 102–4; H. Field, *Memoirs Historical and Illustrative of the Botanick Garden at Chelsea* (London, 1820); F. Dawtrey Drewitt, *The Romance of the Apothecaries' Garden at Chelsea*, third edition (Cambridge, 1928); W. S. C. Copeman, *The Worshipful Society of Apothecaries of London: A History* (London, 1967); Allen, *Naturalist in Britain*, 8–9; H. Le Rougetel, *The Chelsea Gardener: Philip Miller, 1691–1771* (London, 1990); S. Minter, *The Apothecaries' Garden: A History of the Chelsea Physic Garden* (Stroud, 2000); Drayton, *Nature's Government*, 35–7.

5   J. Haynes, *An Accurate Survey of the Botanic Garden at Chelsea with the Elevation and Ichnography of the Green House and Stoves* (London, 1751); Faulkner, *Historical and Topographical Description*, 18–27; Lysons, *Environs of London*, I, 102–4; Field, *Memoirs Historical and Illustrative*; Drewitt, *Romance of the Apothecaries' Garden*; Copeman, *Worshipful Society of Apothecaries*; Allen, *Naturalist in Britain*, 8–9; Rougetel, *The Chelsea Gardener*; Minter, *Apothecaries' Garden*; Drayton, *Nature's Government*, 35–7.

6   R. Stungo, 'North American plants at the Chelsea Physic Garden: some new first records', *Garden History*, 21 (1993), 247–9.

7   Faulkner, *Historical and Topographical Description*, 18–29; Lysons, *Environs of London*, I, 102–4; Allen, *Naturalist in Britain*, 8–9.

8   J. C. Loudon, *Encyclopaedia of Gardening*, second edition (London, 1824), 1063, 1103.

9   A. Wood, *The History and Antiquities of the University of Oxford*, translated by J. Gutch, 2 vols. (Oxford, 1796), 896–9; A. R. Woolley, *The Clarendon Guide to Ox-*

ford (London, 1963); M. Batey, *Oxford Gardens: the University's Influence on Garden History* (Amersham, 1982), 31–3, 51–60.

10   P. Boniface ed., *In Search of English Gardens: the Travels of John Claudius Loudon and his Wife Jane* (London, 1987), 129–31; Woolley, *Guide to Oxford*, 38, 40, 46; *Bae-decker's Great Britain: A Handbook for Travellers*, second edition (London, 1890), 234; Batey, *Oxford Gardens*, 91–103, 130; J. C. Loudon ed., *The Landscape Gardening and Landscape Architecture of the Late Humphry Repton Esquire* (London, 1840), 292–6.

11   *A Short Account of the Late Donation of a Botanic Garden to the University of Cambridge* (Cambridge, 1762); Cambridge University Library, Department of Manuscripts and Special Collections, minute book of the trustees, 22 December 1770, 4 July 1772.

12   *A Short Account of the Late Donation of a Botanic Garden*, 4–6; T. Wright and H. Longueville Jones, *Memorials of Cambridge*, 2 vols. (London, 1847), 131–5; J. J. Smith, *The Cambridge Portfolio* (Cambridge, 1840); S. M. Walters, *Cambridge Botany* (Cambridge, 1986); N. C. Johnson, 'Cultivating science and planting beauty: the spaces of display in Cambridge's botanical gardens', *Interdisciplinary Science Reviews*, 31 (2006), 42–57.

13   Minute book of the trustees, 11 January 1772, 13 July, 22 December 1779.

14   *A Description of the University, Town and County of Cambridge* (Cambridge, 1796), 26–8.

15   T. Martyn, *Heads of a Course of Lectures in Natural History read at the Botanic Garden* (London, 1782); G. C. Gorham, *Memoirs of John Martyn FRS and Thomas Martyn, BD, FRS, FLS, Professors of Botany in the University of Cambridge* (London, 1830), 108.

16   *A Proposal for an Annual Subscription for the Support of the Botanic Garden at Cambridge* (Cambridge, 1765), 3–4; Gorham, *Memoirs*, 132–3, 139–41; Minute book of the trustees of the Botanical Garden, 7 March 1767, 27 October 1768, 4 July 1772, 12 December 1774, 9 January, 5 July 1775, 18 December 1778, 18 December 1780.

17   Trustee minute book, 15 January 1796; P. Miller, *The Gardner's Dictionary* edited by T. Martyn, 2 vols. in four parts (London, 1807); Gorham, *Memoirs*, 170–2.

18   Minute book of the trustees, 17 November 1794, 15 January 1796; J. Donn, *Hortus Cantabrigiensis; or a Catalogue of Plants, Indigenous and Exotic Cultivated in the Botanic Garden, Cambridge*, first edition (Cambridge, 1796), second edition (1800); *Hortus Cantabrigiensis,* third edition (Cambridge, 1804); *Hortus Cantabrigiensis,* fourth edition (Cambridge, 1807).

19   T. Martyn, *Mantissa Plantarum Horti Botanici Cantabrigiensis* (Cambridge, 1772); W. Combe, *A History of the University of Cambridge, its Colleges, Halls and Public Buildings*, 2 vols. (London, 1815).

20   A. G. Morton, *John Hope (1725–1786): Scottish Botanist* (Edinburgh, 1986).

21   Minute book of the trustees, 27 October 1768, 22 December 1770; Gorham, *Memoirs*, 117–20.

22   Martyn, *Mantissa Plantarum*; Walters, *Cambridge Botany*, 36–46.

23   W. Hanbury, *Essay on Planting, and a Scheme for Making it Conducive to the Glory of God and the Advantage of Society* (Oxford, 1758); W. Hanbury, *History of the Rise and Progress of the Charitable Foundations at Church Langton* (London, 1767); W. Hanbury, *A Complete Body of Planting and Gardening*, 2 vols. (London, 1770), preface, i–iv; N. Aston, 'William Hanbury (1725–1778)', DNB.

24   Richard Drayton's stimulating *Nature's Government*, for instance, focuses largely on Kew and its imperial connections saying little about other metropolitan and provincial botanical gardens. Where they are mentioned they are treated as outposts of Kew (for instance on the Cambridge garden, p. 134).

25   Galinou ed., *London's Pride*, 66–149; R. Porter, *London: A Social History* (London, 1996), 93–184; J. Brewer, *The Pleasures of the Imagination: English Culture in the Eighteenth Century* (London, 1997), 28–55; S. Inwood, *A History of London* (London, 1998), 241–410; Ogborn, *Spaces of Modernity*, 28–36; R. Porter, *Enlightenment: Britain and the Creation of the Modern World* (London, 2000), 34–40.

26   'Kewensis', 'Biographical anecdotes of William Curtis the botanist', *Gentleman's Magazine* (1799), 274–80.

27   'Kewensis', 'Biographical anecdotes of William Curtis', 274–80.

28   W. Curtis, *Flora Londinensis; or Plates and Descriptions of Such Plants as Grow Wild in the Environs of London* (London, 1777); 'Kewensis', 'Biographical anecdotes of William Curtis', 274–80.

29   W. Curtis, *Proposals for Opening by Subscription a Botanic Garden to be called the London Botanic Garden* (London, 1778); W. Curtis, *A Catalogue of the British, Medicinal, Culinary and Agricultural Plants Cultivated in the London Botanic Garden* (London, 1783).

30   W. Curtis, *The Subscription Catalogue of the Brompton Botanic Garden for the year 1790* (London, 1790), 13–16; 'Kewensis', 'Biographical anecdotes of William Curtis', 274–80; Porter, *Enlightenment*, 145–7.

31   Curtis, *Subscription Catalogue of the Brompton Botanic Garden*, 9–12.

32   Curtis, *Flora Londinensis*; 'Kewensis', 'Biographical anecdotes of William Curtis', 274–80; W. Curtis, *Companion to the Botanical Magazine; or A Familiar Introduction to the Study of Botany* (London, 1788); S. Curtis, *Lectures on Botany as Delivered in the Botanical Garden at Lambeth by the Late William Curtis FLS*, 2 vols. (London, 1805); Lysons, *Environs of London*, I, 103; Loudon, *Encyclopaedia of Gardening*, second edition (London, 1824), 1109; Allen, *The Naturalist in Britain*, 8–9, 106–7.

33   J. Sowerby, 'William Curtis's botanic garden at Lambeth Marsh', before 1787, J. Paul Getty Jr. K.B.E., Hunt Institute for Botanical Documentation, Carnegie Mellon University (Pittsburgh, PA), reproduced in Galinou ed., *London's Pride*, 114–15; Curtis, *Catalogue of ... Plants Cultivated in the London Botanical Garden* (London, 1783); Curtis, *Subscription Catalogue of the Brompton Botanic Garden*, 17–38.

34   Curtis, *Subscription Catalogue of the Brompton Botanic Garden*, iv–v.

35   Curtis, *Catalogue of British Medical, Culinary and Agricultural Plants*; Faulkner, *Historical and Topographical Description of Chelsea*, 29–33; M. Tyler-Whittle and C. Cook, *Curtis's Flower Garden Displayed* (Leicester, 1991), 1–5.

36   W. Roscoe, *An Address delivered before the Proprietors of the Botanic Garden in Liverpool* (Liverpool, 1802); H. Roscoe, *Life of William Roscoe*, 2 vols. (Liverpool, 1834), I, 253–65; N. Hans, *New Trends in Education in the Eighteenth Century* (London, 1953), 99–116; M. Deacon, *Scientists and the Sea, 1650–1900* (Aldershot, 1971); I.Inkster, 'Scientific culture and scientific education in Liverpool prior to 1812 – a case study in the social history of education', in, M. D. Stephens and G. W. Roderick, eds., *Scientific and Technical Education in Early Industrial Britain* (Nottingham, 1981), 28–47; G.Kitteringham, 'Science in provincial society: the case of Liverpool in the early nineteenth century', *Annals of Science*, 39 (1982), 329–448; A. Wilson, 'The cultural identity of Liverpool, 1790–1850: the early learned societies', *Transactions of the Historic Society of Lancashire and Cheshire*, 147 (1997), 58–73; Drayton, *Nature's Government*; J. Stobart, 'Culture versus commerce: societies and spaces for elites in eighteenth-century Liverpool', *Journal of Historical Geography*, 28 (2002), 471–85.

37   Roscoe, *Address*.

38   Roscoe, *Address*, 9, 11–12, 15.

39   Roscoe, *Address*, 16–37; Roscoe, *Life of William Roscoe*, I, 253–65.

40   Richards, *Catalogue*, 1808; Loudon, *Encyclopaedia of Gardening*, 1080.

41 P. W. Watson, *Dendrologia Britannica*, 2 vols. (London, 1825), I, xv.

42 B. Silliman, *A Journal of Travels in England, Holland and Scotland*, second edition, 2 vols. (Boston, 1812), I, 56–7.

43 W. Roscoe, 'On artificial and natural arrangements of plants: and particularly on the systems of Linnaeus and Jussieu', *Transactions of the Linnaean Society*, 11 (1815), 50–78, p. 76.

44 W. Roscoe, *Monandrian Plants of the Order Scitamineae, Chiefly Drawn from Living Specimens in the Botanic Garden at Liverpool*, first edition (1828); J. Edmondson, *William Roscoe and Liverpool's first Botanical Garden* (Liverpool, 2005), 11.

45 *The Address of the President and Treasurer at the first General Meeting of the Subscribers to the Hull Botanic Garden* (Hull, 1812), 11–12, 17–18; *Gardener's Magazine*, 3 (1828), 495; E. Baines, *History, Directory and Gazetteer of the County of York*, 2 vols. (1823), II, 294; G. Jackson, *Hull in the Eighteenth Century: A Study in Economic and Social History* (Oxford, 1972), 66–9; H. Calvert, *A History of Kingston-Upon-Hull* (Chichester, 1978); K. J. Allison, ed., *Victoria History of the County of York: East Riding vol. I: The City of Kingston-Upon-Hull* (Oxford, 1969), 174–214.

46 *Address ... to the Hull Botanic Garden*, 6, 9–10.

47 *Address ... to the Hull Botanic Garden*, 8.

48 *Address ... to the Hull Botanic Garden*, 7.

49 Watson, *Dendrologia Britannica*, I, xii.

50 J. A. R. Bickford and M. E. Bickford, *The medical profession in Hull, 1400–1900: a biographical dictionary* (Hull, 1983); R. L. Luffingham, *John Alderson and his family* (Hull, 1995); J. Loudon, DNB.

51 W. Spence, 'Mr Kirby's acquaintance with Mr Spence: origin and progress of the "Introduction to entomology"', in J. Freeman, *Life of the Rev. William Kirby* (London, 1852), 265–327; J. F. M. Clark, DNB.

52 D. Elliston, G. S. Boulger, Rev. J. Gross, 'Adrian Hardy Haworth (1768–1833)', DNB; W. T. Stearn, 'Adrian Hardy Haworth, 1768–1833', in A. H. Haworth, *Complete Works on Succulent Plants*, 5 vols. (London, 1965), I, 9–57; Allen, *Naturalist in Britain*, 94–5.

53 Stearn, 'Adrian Hardy Haworth, 1768–1833', I, 9–57; Allen, *Naturalist in Britain*, 94–5.

54 Stearn, 'Adrian Hardy Haworth, 1768–1833', I, 9–57.

55 *Address ... to the Hull Botanic Garden*, 5–6, 11–12, 17–18; Baines, *History, Directory and Gazetteer*, II, 294; Watson, *Dendrologia Britannica*, I, xiii; Loudon, *Encyclopaedia of Gardening*, 1079; The botanic garden was moved to a much larger 49-acre site around 1880 near Spring Bank but was sold to the corporation for Hymen's College around 1890, E. Wrigglesworth, *Brown's Illustrated Guide to Hull* (Hull, 1891), 134; *Victoria County History ... Kingston-Upon-Hull*, 382.

56 C. H. Persoon, *Synopsis Plantarum, seu Enchiridium Botanicum*, 2 vols. (Paris, 1805–7); Watson, *Dendrologia Britannica*, I, xiii–xiv; R. de Zeeuw, 'Notes on the life of Persoon', *Mycologia*, 31 (1939), 369–70; R. H. Petersen, 'Some brief reflections on C. H. Persoon', *Kew Bulletin*, 31 (1977), 695–98.

57 Watson, *Dendrologia Britannica*, I, xiv–xvi; Gorham, *Memoirs*, 137.

58 G. Young, *A Catalogue of the Plants ... with the Rules of the Institution ... Whitby Botanic Garden* (R. Rodgers; Whitby, 1814); *Gentleman's Magazine*, 88 (1818), 432–3; G. Young, *A Picture of Whitby and its Environs* (Whitby, 1824), 249; *The Floricultural and Gardening Miscellany* (1851), 255.

59 A. B. Lambert: *A Description of the Genus Pinus*, second edition, 2 vols. (London, 1828), I, v.

60 J. C. Loudon, *Hints on the Formation of Gardens and Pleasure Grounds* (London, 1812); Loudon, *Encyclopaedia of Gardening*, 1028–31, 1068; M. L. Simo, *Loudon and the Landscape: From Country Seat to Metropolis* (New Haven, 1988), 105–6; Drayton,

Nature and Empire, 134–5; J. Black, *George III: America's Last King* (New Haven, 2006), 406–10.

CHAPTER 6

1   Much of this chapter was originally published in *Urban History*, 32 (2005) and I am grateful to the editors for permission to incorporate it here. P. Borsay, *The English Urban Renaissance: Culture and Society in the Provincial Town, 1660–1770* (Oxford, 1991); J. Brewer, *The Pleasures of the Imagination: English Culture in the Eighteenth Century* (1997); P. Clark, *British Clubs and Societies, 1580–1800: The Origins of an Associational World* (Oxford, 2000); V. Jankovic, 'The place of nature and the nature of place: the chorographic challenge to the history of British provincial science', *History of Science*, 38 (2000), 79–113. P. Clark ed., *The Cambridge Urban History*, vol. 2, *1549–1840* (Cambridge, 2000); M. Redd, 'The culture of small towns in England 1600–1800', in P. Clark ed., *Small Towns in Early-Modern Europe* (Cambridge, 2002), 121–47; I. Inkster and J. Morrell eds., *Metropolis and Province: Science in British Culture, 1780–1850* (London, 1983); P. Elliott, 'Provincial urban society, scientific culture and socio-political marginality in Britain in the eighteenth and nineteenth centuries', *Social History*, 28 (2003), 361–87.

2   D. Read, *The English Provinces, c.1760–1960: A Study in Influence* (London, 1964); R. Porter, 'Science, provincial culture and public opinion in Enlightenment England', *British Journal for Eighteenth-Century Studies*, 3 (1980), 20–46; P. Borsay, 'The London connection: cultural diffusion and the eighteenth-century provincial town', *London Journal*, 19 (1994), 21–35; D. Wahrman, 'National society, communal culture: an argument about the recent historiography of eighteenth-century Britain', *Social History*, 17 (1992), 43–72; J. Ellis, 'Regional and county centres', in P. Clark ed., *The Cambridge Urban History*, vol. 2, *1549–1840* (Cambridge, 2000); C. B. Estabrook, *Urbane and Rustic England: Cultural ties and social spheres in the provinces, 1660–1780* (Manchester, 1998); Jankovic, 'The place of nature and the nature of place'; C. Withers, *Geography, Science and National Identity: Scotland Since 1600* (Cambridge, 2001); C. W. J. Withers and M. Ogborn eds., *Georgian Geographies: Essays on Space, Place and Landscape in the Eighteenth Century* (Manchester, 2004).

3   R. Sweet, *The English Town, 1680–1840* (London, 1999), 197; Ellis, 'Regional and county centres', in Clark (ed.) *Cambridge Urban History* II, 673–704.

4   V. Jankovic, *Reading the Skies: A Cultural History of English Weather, 1650–1820* (Chicago, 2000), 123, 105–13; B. J. Shapiro, *Probability and Certainty in Seventeenth-Century England* (London, 1983); A. Goldgar, *Impolite Learning: Conduct and Community in the Republic of Letters, 1680–1750* (New Haven, 1995); P. Corfield, *Power and the Professions in Britain, 1700–1850* (London, 1995).

5   A. Thackray, 'Natural knowledge in cultural context: the Manchester Model', *American Historical Review*, 79 (1974), 672–709.

6   R. Emerson, 'The Enlightenment and social structures', in P. Fritz and D. Williams, eds., *City and Society in the Eighteenth Century* (Toronto, 1973), 99–124. I. Inkster, Introduction, in Inkster and Morrell eds., *Metropolis and Province*, 11–54.

7   D. Livingstone, *Putting Science in its Place: Geographies of Scientific Knowledge* (Chicago, 2003), 87–134.

8   A. Goldgar, *Impolite Learning: Conduct and Community in the Republic of Letters, 1680–1750* (New Haven, 1995); M. Francesca-Spada and N. Jardine eds., *Books and the Sciences in History* (Cambridge, 2000).

9   A. Everitt, 'Country, county and town: patterns of regional evolution in England', in Borsay ed., *The Eighteenth-Century Town, a Reader in English Urban History, 1688–1820* (London, 1990), 83–115; Ellis, 'Regional and county centres, 1700–1840'.

10　For example, F. J. Ruggiu, 'The urban gentry in England, 1660–1780: a French approach', *Historical Research*, 74 (2001), 249–70.

11　T. Gisborne, *An Enquiry into the Duties of Men in the Higher and Middle Classes of Society in Great Britain*, seventh edition (London, 1824); Thackray, 'Natural knowledge in cultural context'; M. C. Jacob, *The Newtonians and the English Revolution* (London, 1976); Inkster and Morrell eds., *Metropolis and Province*; R. J. Morris, 'Voluntary societies and British urban elites, 1780–1850: an analysis', *Historical Journal*, 26 (1983), 96–118; C. Russell, *Science and Social Change, 1700–1900* (London, 1983); J. E. McClellan III, *Science Reorganized: Scientific Societies in the Eighteenth Century* (New York, 1985); G. Averley, 'English Scientific Societies of the Eighteenth Early Nineteenth Centuries', unpublished PhD thesis, Teeside Polytechnic, University of Durham (Durham, 1989); L. Stewart, *The Rise of Public Science: Rhetoric, Technology and Natural Philosophy in Newtonian Britain, 1660–1750* (Cambridge, 1992); I. Inkster, *Scientific Culture and Urbanisation in Industrialising Britain* (Aldershot, 1997); L. Pyenson and S. Sheets-Pyenson, *Servants of Nature: A History of Scientific Institutions, Enterprises and Sensibilities* (London, 2000); V. Jankovic, *Reading the Skies: A Cultural History of English Weather, 1650–1820* (Chicago, 2000), 115–20; J. Golinski, *British Weather and the Climate of Enlightenment* (Chicago, 2007).

12　S. Schaffer and S. Shapin, *Leviathan and the Air-Pump: Hobbes, Boyle, and the Experimental Life* (Princeton, 1985); S. Shapin, *A Social History of Truth: Civility and Science in Seventeenth-Century England* (Chicago, 1995).

13　Jankovic, *Reading the Skies*, 123, 105–13.

14　B. J. Shapiro, *Probability and Certainty in Seventeenth-Century England* (Princeton, 1983); Goldgar, *Impolite Learning*.

15　J. H. Plumb, 'The new world of children in eighteenth-century England', *Past and Present*, 67 (1975), 64–95, 73; R. M. Wiles, 'Provincial culture in early-Georgian England', in P. Fritz and D. Williams eds., *The Triumph of Culture: 18th-Century Perspectives* (Toronto, 1972), 49–68.

16　G. A. Cranfield, *The Development of the Provincial Newspaper* (Oxford, 1962); Borsay, *English Urban Renaissance*, 133–4.

17　J. M. Levine, *Dr Woodward's Shield: History, Science and Satire in Augustan England* (Berkeley, 1977); A. Everitt, *Landscape and Community in England* (London, 1985); S. A. E. Mendyk, *'Speculum Britanniae': Regional Study, Antiquarianism, and Science in Britain to 1700* (Toronto, 1989); R. Helgerson, *Forms of Nationhood: The Elizabethan Writing of England* (Chicago, 1994), 105–48; Estabrook, *Urbane and Rustic England*, 56–9; L. Daston and K. Park, *Wonders and the Order of Nature, 1150–1750* (New York, 1998); Jankovic, 'The place of nature and the nature of place'; R. Sweet, *Antiquaries: The Discovery of the Past in Eighteenth-Century Britain* (London, 2004).

18　J. Gregory, *Manual of Modern Geography*, fourth edition (London, 1760).

19　J. Goldsmith (Richard Phillips), *A Grammar of British Geography*, fourth edition (London, 1816); R. Mayhew, 'The character of English geography c.1660–1800: a textual approach', *Journal of Historical Geography*, 24 (1998), 385–412, p. 397–9.

20　R. Sweet, *The Writing of Urban Histories in Eighteenth-Century England* (Oxford, 1997).

21　J. Ellis, '"For the honour of the town": comparison, competition and civic identity in eighteenth-century England', *Urban History*, 30 (2003), 325–37; R. Sweet, 'History and identity in eighteenth-century York: Francis Drake's *Eboracum* (1736)' in J. Rendall and M. Hallet eds., *Eighteenth-Century York: Culture, Space and Society* (York, 2003), 13–23.

22　N. Hans, *New Trends in Education in the Eighteenth Century* (London, 1966); M. E. Bryant, *The London Experience of Secondary Education* (London, 1986), 56–174.

23 J. Golinski, *Making Natural Knowledge: Constructivism and the History of Science* (Cambridge, 1998); Elliott, 'Provincial urban society'; D. N. Livingstone, *Putting Science in its Place: Geographies of Scientific Knowledge* (Chicago, 2003), 87–134; E. G. R. Taylor, *The Mathematical Practitioners of Hanoverian England, 1714–1840* (Cambridge, 1966).

24 Jankovic, *Reading the Skies*, 78–124, 143–64; Allen, *Naturalist in Britain*; L. Barber, *The Heyday of Natural History, 1820–1870* (London, 1980).

25 Elliott, 'Provincial urban society'; Livingstone, *Putting Science in its Place*, 17–86.

26 The collation of detailed biographical information concerning individual members, such as religious and political affiliations.

27 A. Everett, 'Country, county and town: patterns of regional evolution in England', *Transactions of the Royal Historical Society*, 29 (1979), 79–109; Ellis, 'Regional and county centres'.

28 E. Kitson Clark, *The History of 100 Years of Life of the Leeds Philosophical and Literary Society* (Leeds, 1924); A. Musson and E. Robinson, *Science and Technology* (Manchester, 1969), 155–9; J. Morrell, 'Wissenschaft in Worstedopolis: public science in Bradford, 1800–1850', *British Journal for the History of Science*, 18 (1985), 1–23; E. Baines, *Leeds Commercial Directory* (Leeds, 1817), quoted in Musson and Robinson, *Science and Technology*, 159.

29 I. Inkster, 'The development of a scientific community in Sheffield, 1790–1850: a network of people and interests', in *Scientific Culture and Urbanisation*, IV, 99–131.

30 Hans, *New Trends*, 99–116; M. Deacon, *Scientists and the Sea, 1650–1900* (Aldershot, 1971); R. Drayton, *Nature's Government: Science, Imperial Britain, and the 'Improvement' of the World* (New Haven, 2000); G. Kitteringham, 'Science in provincial society: the case of Liverpool in the early nineteenth century', *Annals of Science*, 39 (1982), 329–448; I. Inkster, 'Scientific culture and scientific education in Liverpool prior to 1812 – a case study in the social history of education', in Inkster, *Scientific Culture and Urbanisation*, X, 28–47; A. Wilson, 'The cultural identity of Liverpool, 1790–1850: the early learned societies', *Transactions of the Historic Society of Lancashire and Cheshire*, 147 (1997), 58–73; J. Stobart, 'Culture versus commerce: societies and spaces for elites in eighteenth-century Liverpool', *Journal of Historical Geography*, 28 (2002), 471–85.

31 Dissenters also tended to be concentrated in the north rather than the south.

32 J. Throsby, *The History and Antiquities of the Town of Leicester* (1791), 380, quoted in A. T. Patterson, *Radical Leicester: A History of Leicester, 1780–1850* (Leicester, 1954), 16; for Ludlam, Weston and Phillips see DNB.

33 C. Grewcock, 'Social and Intellectual Life in Leicester, 1763–1835', unpublished MA thesis, University of Leicester (Leicester, 1973), 91–4.

34 *Leicester Journal* (1 October 1790, 5 August 1791); 'Laws for the Regulation of the Literary Society, Leicester' (30 July 1790), Leicestershire County Record Office, misc. 79.

35 F. B. Lott, *The Centenary Book of the Leicester Literary and Philosophical Society* (Leicester, 1935).

36 J. D. Humphrey ed., *The Correspondence of Philip Doddridge* (1831), vol. 5, 28, quoted in Musson and Robinson, *Science and Technology*, 380.

37 F. Drake, *Eboracum: or, the History and Antiquities of the City of York* (London, 1736); P. M. Tillot ed., *The Victoria County History of Yorkshire, The City of York* (London, 1961), 207–53, 531–4; A. Armstrong, *Stability and Change in an English County Town: A Social Study of York 1801–51* (Cambridge, 1974); J. Looney, 'Cultural life in the provinces, Leeds and York, 1720–1820', in A. C. Beier, D. Cannadine and J. M. Rozenheim eds., *The First Modern Society: Essays in English history* (Cambridge, 1989), 483–512; Rendall and Hallett, *Eighteenth-Century York*.

38 J. W. Knowles, 'History of Stonegate', York Reference Library, ms. YO40, 188–90.

39   *York Herald*, 2 August 1794; O. S. Tomlinson, 'Libraries in York', in A. Stacpoole ed., *The Noble City of York* (York, 1972), 969–94; W. F. Gould, *History of Freemasonry*, third edition, 4 vols. (London, 1951), revised by Rev. H. Poole, II, 233–76.

40   A. D. Orange, *Philosophers and Provincials: the Yorkshire Philosophical Society from 1822 to 1844* (York, 1973); D. Orange, 'Science in early nineteenth-century York: the Yorkshire Philosophical Society and the British Association', J. Mathew, 'Science and technology in York, 1831–1981', P. Addyman, 'Archaeology in York, 1831–1981', in C. H. Feinstein ed., *York 1831–1981; 150 years of scientific endeavour and social change* (York, 1981), 1–29, 30–52, 53–87; J. Morrell and A. Thackray, *Gentlemen of Science: The Early Years of the British Association for the Advancement of Science* (Oxford, 1981), 63–76.

41   *Yorkshire Observer* (22 February 1823), quoted in Orange, *Philosophers and Provincials*, 5.

42   The population of the city having risen from 28,881 in 1693 to 40,051 in 1786, actually fell to 36,832 by 1801 (T. Peck, *The Norwich Directory* [Norwich], 1801); M. F. Lloyd Pritchard, 'The decline of Norwich', *Economic History Review*, second series, 3 (1951), 371–7; J. K. Edwards, 'The decline of the Norwich textiles industry', *Yorkshire Bulletin of Economic and Social Research*, 26 (1964), 31–41; J. K. Edwards, 'Communications and trade 1800–1900', 'Industrial Development of the city 1800–1900', in C. Barrinder ed., *Norwich in the Nineteenth Century* (Norwich, 1984), 119–59; P. Corfield, 'A provincial capital in the late seventeenth century: The case of Norwich', in P. Clark ed., *The Early Modern Town* (London, 1976), 233–72.

43   P. Borsay, *English Urban Renaissance*, 121–7, 332–3; T. Fawcett, *Music in Eighteenth-Century Norwich* (Norwich, 1979).

44   It was in a Norwich coffee house that Samuel Clarke first discussed Newtonian natural philosophy with William Whiston in 1697.

45   T. Kelly, 'Norwich, pioneer of public libraries', *Norfolk Archaeology*, 84 (1966–9), 215–22.

46   Collection of material on the Public Library, Norfolk County Record Office, Norwich SO 50/1/-; Minute Book of the Tusculum Society, Norfolk County Record Office, Norwich, Ms. NNAS/G2; for Norwich politics, D. S. O'Sullivan, 'Politics in Norwich, 1701–1835', unpublished MPhil Thesis, University of East Anglia (Norwich, 1975); C. B. Jewson, *The Jacobin City* (Glasgow, 1975); F. Meeres, *A History of Norwich* (Chichester, 1998).

47   M. Rajnai, *The Norwich School of Painters* (Norwich, 1978); M. Allthrope-Guyton, 'The artistic and literary life in Norwich during the century', in C. Barringer ed., *Norwich in the Nineteenth Century (Norwich, 1982)*, 1–46; S. Daniels, *Humphrey Repton: Landscape Gardening and the Geography of Georgian England* (New Haven, 1999).

48   *Pigot and Company's Commercial Directory of Norfolk and Suffolk* (London, 1830), 32.

49   Jewson, *Jacobin City*, 143–56; T. Fawcett, 'Popular science in eighteenth-century Norwich', *History Today*, 22 (1972), 590–5; T. Fawcett, 'Measuring the provincial Enlightenment: the case of Norwich', *Eighteenth-Century Life*, 8 (1982), 13–27.

50   *Pigot and Company's Commercial Directory of Norfolk and Suffolk* (London, 1830), 31–2; W.White, *History, Gazetteer and Directory of Norfolk* (Sheffield, 1845), 149–50; the minutes, account books, subscription lists and catalogues of both the Mechanics' Institute (mss. 4334-5, 4263-4) and the Literary Institution (SO 50/2/1-50) are preserved at the Norfolk County Record Office, Norwich.

51   C. Bryant, *Flora Diaetetica, or the History of Esculent Plants, both Domestic and Foreign* (London, 1783); C. Bryant, *A Dictionary of the Ornamental Trees, Shrubs and Plants* (Norwich, 1790); J. E. Smith, *Memoir and Correspondence of the late James Edward Smith*, 2 vols. (London, 1832).

317

52   Colman mss. 9/- Norfolk County Record Office, Norwich; Jewson, *Jacobin City*, 145–9; T. Fawcett, 'Measuring the provincial Enlightenment', 19–20; C. B. Jewson, *Simon Wilkin of Norwich* (Norwich, 1980), 44–8. Members of the labouring classes also took part in some of these scientific cultural activities, see, Fawcett, 'Measuring the provincial Enlightenment', 17; *Catalogue of the Norwich Library for Working People* (Norwich, 1824); Jewson, *Simon Wilkin of Norwich*, 49.

53   J. and E. Taylor, *History of the Octagon Chapel*, Norwich (London, 1848).

54   Taylor, *Octagon Chapel*, 46–51; G. C. Morgan, *Lectures on Electricity*, 2 vols. (Norwich, 1794); A. Brooke, *Miscellaneous Experiments and Remarks on Electricity, the Air-Pump and the Barometer* (London, 1789).

55   R. Watts, *Gender, Power and the Unitarians in England, 1760–1860* (London, 1998), 89.

56   B. Nicholson, *Joseph Wright of Derby, Painter of Light*, 2 vols. (New York, 1967); P. Sturges, 'The membership of the Derby Philosophical Society 1783–1802, *Midland History*, 4 (1978), 212–29; M. Craven, *John Whitehurst Clockmaker and Scientist, 1713–88* (Ashbourne, 1997); S. Daniels, *Joseph Wright* (London, 1998); P. Elliott, *The Derby Philosophers; Science and Culture in Urban Society, c.1700–1850* (Manchester, 2009).

57   W. Adam, *The Gem of the Peak; or Matlock Bath and its Vicinity*, second edition (London, 1840), 331–51; Elliott, *The Derby Philosophers*.

58   E. Darwin, 'Paving and lighting', broadside, 5 January 1791, Derby Local studies Library, Derby; D. King-Hele, *Erasmus Darwin: A Life of Unequalled Achievement* (London, 2000); Elliott, *Derby Philosophers*.

59   *Anti-Jacobin Review and Magazine* (16, 23 April, 7 May 1798).

60   B. Boothby, *A Letter to the Right Honourable Edmund Burke* (London, 1791), 68–70; T. Gisborne, 'On the benefits and duties resulting from the institution of societies for the advancement of literature and philosophy', *Manchester Memoirs*, 5 (1798–1802), 70–88.

61   Elliott, *Derby Philosophers*; P. Elliott, 'William George Spencer (1790–1866): mathematical, geographical and scientific education in nineteenth-century England', *History of Education*, 33 (2004), 391–417.

62   P. D. Lowe, 'Locals and Cosmopolitans: A Model for the Social Organisation of Provincial Science in the Nineteenth Century', Unpublished MPhil thesis, University of Sussex (Brighton, 1978); Russell, *Science and Society*, 69–219; D. A. Hinton, 'Popular Science in England, 1830–79', Unpublished PhD thesis, University of Bath (Bath, 1979); I. Inkster, 'Studies in the Social History of Science in England during the Industrial Revolution, circa 1790–1850', unpublished PhD thesis, University of Sheffield, 2 vols. (Sheffield, 1978); G. Kitteringham, 'Studies in the Popularisation of Science in England, 1800–30', Unpublished PhD thesis, University of Sussex (Brighton, 1981); J. E. McClellan III, *Science Reorganised: Scientific Societies in the Eighteenth Century* (New York, 1985); Averley, 'English Scientific Societies'; L. Pyenson and S. Sheets-Pyenson, *Servants of Nature: A History of Scientific Institutions, Enterprises and Sensibilities* (London, 2000), 74–100, 319–49; E. P. Thompson, 'Hunting the Jacobin fox', *Past and Present*, 142 (1994), 94–140, p. 122.

63   W. B. Stephens, 'Illiteracy and schooling in the provincial towns, 1640–1870; a comparative approach', in D. A. Reader ed., *Urban Education in the Nineteenth Century* (London, 1977), 27–47; W. B. Stephens, *Education, Literacy and Society, 1830–70* (Manchester, 1987); Estabrook, *Urbane and Rustic England*. County town status also emerges as significant with respect to religious observance figures which offer a useful comparison to other forms of public urban culture, K. S. Inglis, 'Patterns of worship in 1851', *Journal of Ecclesiastical History*, 11 (1960), 74–86. According to Inglis, of the 11 large towns whose index of church attend-

ance exceeded that of the whole of England and Wales, seven (64%) were county towns. Likewise, in their analysis of the 1851 data K. D. M. Snell and P. S. Ell find that large-scale urbanisation tended to reduce attendance for all denominations. They argue that medium-sized urban centres of between 8,000 and 20,000 had higher attendances and suggest that this was partly because they were more integrated with their rural surroundings which helps to explain the continuing prominence of county towns, see K. D. M. Snell and P. S. Ell, *Rival Jerusalems: The Geography of Victorian Religion* (Cambridge, 2000), 395–420.

CHAPTER 7

1   Some of this material appeared in P.Elliott, '"Food for the mind and rapture to the sense": scientific culture in Nottingham, 1740–1800', *Transactions of the Thoroton Society*, 109 (2005) and I am grateful to the editors for permission to reproduce it here.

2   *Universal British Directory of Trade, Commerce and Manufacture*, 6 vols. (London, 1791–7), IV, 44–56; E. Willoughby, *The Nottingham Directory* (Nottingham, 1799); J. Blackner, *The History of Nottingham* (Nottingham, 1815); F. C. Laird, *Topographical and Historical Description of the County of Nottingham* (London, 1820); W. H. Wylie, *Old and New Nottingham* (Nottingham, 1853); *The Nottingham Date Book, 850–1884* (Nottingham, 1884); J. D. Chambers, *Modern Nottingham in the Making* (Nottingham, 1945); J. D. Chambers, 'Population change in a provincial town, Nottingham 1700–1800', in L. S. Pressnell ed., *Studies in the Industrial Revolution* (Cambridge, 1960); J. D. Chambers, *Nottinghamshire in the Eighteenth Century*, second edition (London, 1966); M.I.Thomis, *Old Nottingham* (London, 1968); M. I. Thomis, *Politics and Society in Nottingham, 1785–1835* (London, 1969); J. Beckett, *The East Midlands since AD 1000* (London, 1988); J. V. Beckett *et al.* eds., *A Centenary History of Nottingham* (Manchester, 1997); S. Wallwork, 'Population estimates before the census: Nottingham, 1570–1801', *East Midland Historian*, 9 (1999), 35–42.

3   Willoughby, *Nottingham Directory*, 72; on the popularity of the Goose Fair, *Nottingham Journal* (7 October 1786).

4   C. Deering, *Nottinghamia Vetus et Nova: or an Historical Account of the Ancient and Present State of Nottingham* (Nottingham, 1751), 12–14; *Nottingham Journal* (20 November 1779); Chambers, *Nottinghamshire*, 82–7.

5   Examples of coffee houses: *Nottingham Journal* (30 June, 14 July 1770, 10 April 1773); assemblies, Deering, *Nottinghamia*, 75–6; circulating libraries, *Nottingham Journal* (24 October 1772, 4 September 1773); concerts and theatrical performances, *Nottingham Journal* (24 February, 30 June 1770, 30 March 1771); music clubs, *Nottingham Journal* (4 September 1773, 2 October 1779); florists society, *Nottingham Journal* (22 July 1775); Willoughby, *Nottingham Directory*, 71–2.

6   Willoughby, *Nottingham Directory*, 72–106; Chambers, *Nottinghamshire*, 79–136; Beckett *et al.* eds., *Centenary History*, 143–8, 189–97.

7   Deering, *Nottinghamia*, introduction, 1–13, 73; R. Sweet, *The Writing of Urban Histories in Eighteenth-Century England* (Cambridge, 1997).

8   Deering, *Nottinghamia*, introduction, 13; J. Ellis, '"For the honour of the town": comparison, competition and civic identity in eighteenth-century England', *Urban History*, 30 (2003), 325–37; R. Sweet, 'History and identity in eighteenth-century York: Francis Drake's *Eboracum* (1736)' in in J. Rendall and M. Hallet eds., *Eighteenth-Century York: Culture, Space and Society* (York, 2003), 13–23.

9   Deering, *Nottinghamia*, introduction, 12–13.

10   Deering, *Nottinghamia*, 2, 72,76–82

11   Deering, *Nottinghamia*, 6–7.

12 Deering, *Nottinghamia*, 70–90; Blackner, *History of Nottingham*, 340–2; A. C. Wood, 'Dr Charles Deering', *Transactions of the Thoroton Society*, 45 (1941); A. Everitt, *Landscape and Community in England* (Oxford, 1985); S. A. E. Mendyk, *'Speculum Britanniae': Regional Study, Antiquarianism, and Science in Britain to 1700* (Toronto, 1989); C. Estabrook, *Urbane and Rustic England* (London), 6–59; J. Beckett and C. Smith, 'Dr Charles Deering: Nottingham's first historian', *Nottinghamshire Historian*, 63 (1999), 14–16; S. Piggot, *Ancient Britons and the Antiquarian Imagination* (London, 1989); R. Sweet, *Antiquaries: The Discovery of the Past in Eighteenth-Century Britain* (London, 2004).

13 *Nottingham Journal* (19 May, 20 October 1770, 13 August, 24 September 1774, 18 March 1775, 31 July, 7 August 1779); R. Lowe, *General View of the Agriculture of the County of Nottingham* (London, 1798); Chambers, *Nottinghamshire*, 137–72.

14 P. Elliott, 'Provincial urban society, scientific culture and socio-political marginality in Britain in the eighteenth and nineteenth centuries', *Social History*, 28 (2003), 394–442; D. N. Livingstone, *Putting Science in its Place* (Chicago, 2003), 87–134.

15 For a summary of Nottingham government and institutions, Willoughby, *Nottingham Directory*, 53–9, 71–2.

16 Willoughby, *Nottingham Directory*, 68–9; F. H. Jacob, *History of the Nottingham General Hospital* (Nottingham, 1951); N. Scarfe, 'Nottingham and its general hospital in 1786: extracts from the travel diaries of a French nobleman', *Transactions of the Thoroton Society*, 103 (1999), 141–7; M. C. Jacob, *The Newtonians and the English Revolution* (London, 1976); I. Inkster and J. B. Morrell eds., *Metropolis and Province* (London, 1983); C. Russell, *Science and Social Change, 1700–1900* (Basingstoke, 1983); L. Stewart, *The Rise of Public Science: Rhetoric, Technology and Natural Philosophy in Newtonian Britain, 1660–1750* (Cambridge, 1992); I. Inkster, *Scientific Culture and Urbanisation in Industrialising Britain* (Aldershot, 1997); P. Elliott, *The Derby Philosophers: Science and Urban Culture in Britain, 1700–1850* (Manchester, 2009).

17 *Nottingham Weekly Courant* (15 March 1750); 'Rules to be observed in the lending library' (1744) in *Catalogue of the Standfast Library* (1817) appendix c, quoted in P. Hoare, 'Nottingham subscription library', in R. T. Coope and J. Y. Corbett eds., *Bromley House 1752–1991* (Nottingham, 1991), 23.

18 *Nottingham Journal* (18 June 1785).

19 *Nottingham Journal* (29 September, 13 October 1837).

20 Manuscript catalogue and borrowing records of the Derby Philosophical Society, Derby Local Studies Library, Derby Public Library

21 For entry into the literature on Newton and Newtonianism see the essays in the special edition of *Social Studies of Science* (2004).

22 C. Hutton, 'Some account of the life and writings of Mr Thomas Simpson', *Annual Register* (1764), 29–38; F. M. Clarke, *Thomas Simpson and His Times* (New York, 1929); K. Pearson, *The History of Statistics in the 17th and 18th Centuries*, edited by E. S. Pearson (New York, 1978), 166–85.

23 *Nottingham Journal* (12 April 1783).

24 *Nottingham Journal* (20 January 1770, 3 April 1779, 17 May 1783); R. Mellors, *Men of Nottingham and Nottinghamshire* (Nottingham, 1924), 290.

25 *Nottingham Journal* (7 October 1775, 22 July 1780, 8 June 1782, 18 August 1787).

26 *Nottingham Journal* (24 November 1774, 8 April 1775); A. N. Walters, 'Science and politeness in eighteenth-century England', *History of Science*, 35 (1997), 121–54.

27 *Nottingham Journal* (17 February 1781).

28 *Nottingham Journal* (19 September, 3 October 1778).

29 *Nottingham Journal* (24 January 1778, 18 October 1788).

30 *Nottingham Journal* (13, 20 October, 24 November, 15 December, 1787 and 12 January, 10, 31 May 1788).

31 *Nottingham Journal* (29 June 1782, 23, 30 August 1783).

32 *Nottingham Journal* (6 July 1776).

33 *Nottingham Journal* (29 April, 20 May 1786).

34 G. Henson, *The Civil, Political and Mechanical History of the Framework Knitters* (Nottingham, 1831), 256–370; W. Felkin, *History of Machine Wrought Hosiery and Lace Manufacturers* (London, 1867), 84–142; Chambers, *Nottinghamshire*, 35–44, 79–136; R. S. Fitton and A. P. Wadsworth, *The Strutts and the Arkwrights* (Manchester, 1958); S. D. Chapman, *The Early Factory Masters* (London, 1967).

35 N. Hans, *New Trends in Education in the Eighteenth Century* (London, 1966); M. E. Bryant, *The London Experience of Secondary Education* (London, 1986), 56–174.

36 D. Wardle, *Education and Society in Nineteenth-Century Nottingham* (Cambridge, 1971), 173.

37 The scholar and political pamphleteer Henry Shipley (1763–1808) received his education at the academy under Wilkinson from 1776 later serving as his assistant and opening his own school in Nottingham (Blackner, *History of Nottingham*, 357–9).

38 *Nottingham Journal* (10 January 1778, 3 June 1780).

39 *Nottingham Journal* (20, 27 June 1778, 3, 10 July 1779, 25 March 1780); Mellors, *Men of Nottingham*, 291.

40 R. V. Wallis and P. J. Wallis, *Biobibliography of British Mathematics and its Applications* (Newcastle, 1986), II, 266, 281–2, 155.

41 Mellors, *Men of Nottingham*, 290–1.

42 T. Peat and J. Badder, *A Plan of the Town of Nottingham from an Accurate Survey* (Nottingham, 1744); *Nottingham Journal* (10 January 1778, 2 January 1779, 26 February, 5, 12 August 1780); Blackner, *History of Nottingham*, 353–5; *Nottingham Date Book*, 132; Wylie, *Old and New Nottingham*, 158; Wallis, *Biobibliography*, 308–9; Beckett and Smith, 'Charles Deering', 14–16; A. F. Pollard rev. R. Wallis, 'Thomas Peat', DNB.

43 T. Peat, *A Short Account of a Course of Mechanical and Experimental Philosophy and Astronomy …* (*c.*1756[1744?]), British Library 8704.aa.10.

44 Peat, *Short Account of a Course of Mechanical and Experimental Philosophy and Astronomy*; J. L. Heilbron, *Electricity in the 17th and 18th Centuries* (New York, 1999), 312–23.

45 S. Schaffer, 'Natural philosophy and public spectacle in the eighteenth century', *History of Science*, 21 (1983), 1–43.

46 Elliott, *Derby Philosophers*; Peat's lectures were copied off those by Griffis.

47 *Nottingham Journal* (13 August 1774).

48 *Nottingham Chronicle* (3, 17, 24 October 1772).

49 *Nottingham Journal* (19 February 1774).

50 *Nottingham Journal* (28 January 1775).

51 *Nottingham Journal* (27 May 1775).

52 *Nottingham Journal* (3 February 1781).

53 *Nottingham Journal* (10, 17 April, 8, 15 May 1779).

54 *Nottingham Journal* (20 November 1784).

55 *Nottingham Journal* (23, 30 July, 6, 13, 27 August 1785).

56 *Nottingham Journal* (13 August 1785).

57 R. Hall, evidence at the trial of Charles Tennant, *Proceedings in a Suit of Chancery … in the name of Mr Charles Tennant …* (London, 1803), 41–4, 189–94; Mellors, *Men of Nottingham*, 313–14; Chambers, *Nottinghamshire*, xiii; A. E. Musson, and A. E. Robinson, *Science and Technology in the Industrial Revolution* (Manchester, 1969), 316–17, 324.

58   J. Aikin 'Memoir of the Rev. George Walker', in L. Aikin, *Memoir of John Aikin MD*, 2 vols. (London, 1823), II, 406–16, p. 410.

59   Records of High Pavement Presbyterian Chapel, 1576–1982, Department of Manuscripts and Special Collections, University of Nottingham (NUMD), especially Account Book, 1797–1826, HiA; material kindly supplied by Professor John Beckett, School of History, Nottingham University; A. E. Cockburn, Report on the Corporation of Nottingham, *First Report of the Commissioners on the Municipal Corporations of England and Wales* (London, 1835), appendix, part iii, 1985–2008; B. Carpenter, *Some Account of the Original Introduction of Presbyterianism in Nottingham* (Nottingham, 1862); A. R. Henderson, *History of Castle Gate Congregational Church, Nottingham, 1655–1905* (Nottingham, 1905); J. C. Warren, 'Nottingham Presbyterian records', *Transactions of the Thoroton Society*, 19 (1916), 56–77; A. B. Clarke, 'Notes on the mayors of Nottingham, 1660–1775', *Transactions of the Thoroton Society*, 42 (1938), 105–20; Thomis, *Old Nottingham*, 97–116; M. Thomis, 'The politics of Nottingham enclosure', *Transactions of the Thoroton Society*, 71 (1968), 90–6; M. Thomis, *Politics and Society*, 114–42; J. Beckett and B. H. Tolley, 'Church, chapel and school', in Beckett *et al.* eds., *Centenary History*, 351–64.

60   G. Walker junior, memoir of Rev. George Walker in G. Walker, *Essays on Various Subjects*, 2 vols. (London, 1809), lxvii, lxxi, lxxx; Aikin, 'Memoir of the Rev. George Walker'; J.Aikin, 'Memoir of Gilbert Wakefield, BA', in L.Aikin ed., *Memoir of John Aikin MD*, 2 vols. (London, 1823), II, 346–65.

61   G. Walker, *Sermons on Various Subjects*, 2 vols. (London, 1790).

62   Walker, memoir of Walker, *Essays*, I, cciii; Carpenter, *Presbyterianism*, 167.

63   Wakefield, *Memoirs*, I, 296.

64   Wakefield, *Memoirs*, I, 296–8.

65   Quoted in M. McLachlan, *English Education under the Test Acts* (Manchester, 1931), 121.

66   McLachlan, *English Education*; Wakefield, *Memoirs*, 210; *Monthly Repository*, 8 (1815), 625–9; Carpenter, *Presbyterianism*, 169–70.

67   Wakefield, *Memoirs*, I, 226, 555.

68   Walker, *Essays*, I; Thomis, *Old Nottingham*, 103; F. D. Cartwright ed., *The Life and Correspondence of Major Cartwright*, 2 vols (London, 1826), I, 172–3.

69   Wakefield, *Memoirs*, I, 544.

70   McLachlan, *English Education*, 223, 229; V. D. Davies, *A History of Manchester College* (Oxford, 1932), 67–70.

71   G. Wakefield, *A Reply to Some Parts of the Bishop of Llandaff's Address to the People of Great Britain* (London, 1798).

72   Walker, memoir, in *Essays*, I, clxix, clx–clxi.

73   Walker, memoir, in *Essays*, I, clxiv–clxxi; G. Wakefield, *Address to the Inhabitants of Nottingham on the Subject of the Test Laws* (Nottingham, 1789); G. Wakefield, *Cursory Reflections...on the Repeal of the Corporation and Test Acts* (London, 1790); G. Walker, *The Dissenter's Plea; or the appeal of the Dissenters to Justice, the Honour and the Religion of the Kingdom* (London, 1790), also in *Essays*, II, 261–336; *Nottingham Journal* (29 January, 5, 26 February 1785, 2 March 1793).

74   J. F. S utton, *The Date Book of Nottingham* (Nottingham, 1852) (28 October 1782, 9 March 1785); T. Bailey, *The Annals of Nottinghamshire*, IV, 97–8, 108–9; Walker, memoir, *Essays*, I, cxvii–cxxxix, cxli–clvi.

75   Walker, memoir, *Essays*, I, cxliii–cxliv.

76   Thomis, *Politics and Society*, 217–19; J. V. Beckett, 'Responses to war: Nottingham in the French Revolutionary and Napoleonic wars, 1793–1815', *Midland History*, 22 (1997), 71–84; Elliott, *Derby Philosophers*, 93–8.

77   McLachlan, *English Education*, 25.

78 *Nottingham Journal*, 30 October 1784; Blackner, *History of Nottingham*, 127–8; Wardle, *Education and Society*, 38–45, 173; R. C. Swift, *Lively People: Methodism in Nottingham, 1740–1979* (Nottingham, 1982), 99–115; On the adult school, Mellors, *Men of Nottingham*, 217–18; R. Watts, *Gender, Power and the Unitarians* (London, 1998); for Wilkinson, Blanchard and Goodacre, I. Inkster, 'Robert Goodacre's astronomy lectures (1823–1825), and the structure of scientific culture in Philadelphia', in Inkster, *Scientific Culture and Urbanisation*, article VIII.

79 G. Wakefield, *Silva Critica* (Cambridge, 1795); T. Paine, *The Age of Reason* (London, 1796); G. Wakefield, *An Enquiry into the Expediency and Propriety of Public or Social Worship* (London, 1791); G. Wakefield, *An Examination of the Age of Reason* (London, 1794).

80 J. Priestley ed., *The Theological Repository*, vols. 4–6 (1784–1788).

81 G. Wakefield, 'An essay on the origin of alphabetical characters', in *Memoirs*, II, 339–62.

82 G. Walker, *A Treatise on the Conic Sections in five books* (London, 1794); G. Walker, *Essays*, 2 vols.

83 E. Darwin, *The Zoonomia; or, The Laws of Organic Life*, second ed., 2 vols. (1794–6), I, 534, 10, 32; P. Elliott, '"More subtle than the electric aura": Georgian medical electricity, the spirit of animation and the development of Erasmus Darwin's psychophysiology', *Medical History*, 52 (2008), 195–220; G. Walker, 'Probable argument in favour of the immateriality of the soul', in G. Walker, *Essays*, II, 51, 71–2.

84 G. Walker, 'On imitation and fashion', in *Essays*, II, 218–20; E. Darwin, letter to Gilbert Wakefield in Wakefield, *Memoirs*, II, 19 August 1800, 229–32; D. King-Hele ed., *The Letters of Erasmus Darwin* (1981), 323. Associationism is the attempt to explain mental phenomena through relations that appear to exist between them, when things happen at the same time for instance (such as church bells and church services) or in close physical proximity to each other, so that when one of them occurs again the other becomes equally expected.

85 *Nottingham Journal* (12 June 1784); W. Cockin, *A Rational and Practical Treatise of Arithmetic* (London, 1766), *The Art of Delivering Written Language* (London, 1775), *The Theory of the Syphon* (London, 1781), *The Rural Sabbath* (London, 1805); D. A. Cross, 'William Cockin (1736–1802)', DNB.

86 W. Cockin, 'Account of an extraordinary appearance in a mist', *Philosophical Transactions* 70 (1780), 157–62.

87 W. Cockin, *The Fall of Scepticism and Infidelity Predicted; An Epistle to Dr Beattie* (London, 1785).

88 Cockin, *Fall of Scepticism and Infidelity*, vii, 9–10, 105, 114–19, 123–25, 132–42, 146, 148; W. Cockin, *The Freedom of Human Action Explained and Vindicated: in which the Opinions of Dr Priestley on the Subject are particularly considered* (London, 1791).

89 H. Kirke White, *Poems, Hymns and Prose Writings*, edited by R. T. Beckwith (1985), 54–5.

90 H. Kirk White, letters to Neville (October and November, 1799), Kirk White Manuscripts, N.U.M.D., KW C60a, KWC 61, KWC 5; C. V. Fletcher, 'The Poems and Letters of Henry Kirke White, unpublished PhD thesis, 3 vols., University of Nottingham, 1980, II, 354–60.

91 No account apparently survives in the Nottingham newspapers.

92 M. C. Pottle, 'Loyalty and patriotism in Nottingham, 1792–1816', unpublished PhD thesis, University of Oxford, 1988.

93 I. Inkster, 'Scientific culture and education in Nottingham, 1800–1843', *Transactions of the Thoroton Society*, 82 (1978), 45–50; J. Russell, *A History of the Nottingham Subscription Library* (Nottingham, 1916); R. T. Coope and J. Y. Corbett eds., *Bromley House, 1752–1991* (Nottingham, 1991).

CHAPTER 8

1   M. Ogborn, *Spaces of Modernity: London's Geographies, 1680–1780* (London, 1998); T. Munck, *The Enlightenment; A Comparative Social History, 1721–94* (London, 2000); K. H. Baker and R. M. Reill eds., *Whats Left of Enlightenment?: A Postmodern Question* (Stanford, 2001); J. Van Horn Melton, *The Rise of the Public in Enlightenment Europe* (Cambridge, 2001); J. I. Israel, *Enlightenment Contested: Philosophy, Modernity and the Emancipation of Man* (Oxford, 2008).

2   Ogborn, *Spaces of Modernity*; Munck, *The Enlightenment*; Melton, *Rise of the Public*.

3   J. A. Yelling, *Common Field and Enclosure in England, 1450–1850* (London, 1977); J. M. Neeson, *Commons: Common Right, Enclosure and Social Change in England, 1700–1820* (Cambridge, 1993).

4   W. G. Hoskins, *The Making of the English Landscape* (London, 1970), 279–89; R. Sweet, 'Freemen and independence in English borough politics, *c.*1770–1830', *Past and Present*, 161 (1998), 84–115.

5   J. B. Bury, *The Idea of Progress: an Inquiry into its Origin and Growth* (Macmillan, 1932), 78–97; S. Pollard, *The Idea of Progress: History and Society* (London, 1968); R. Nisbet, *History of the Idea of Progress* (New York, 1980); J. H. Plumb, 'The acceptance of modernity', in N. McKendrich, J. Brewer and J. H. Plumb, *The Birth of Consumer Society: the Commercialisation of Eighteenth-century England* (London, 1982), 316–34; D. Spadafora, *The Idea of Progress in Eighteenth-Century Britain* (Yale, New Haven, 1990), 21–84.

6   Spadafora, *Idea of Progress*; S. and B. Webb, *English Local Government* IV, Statutory Bodies for Special Purposes (London, 1922); E. L. Jones and M. E. Falkus, 'Urban improvement and the English economy in the seventeenth and eighteenth centuries', in P. Borsay ed., *The Eighteenth-Century Town, 1688–1800* (London, 1990), 116–58; P. Borsay, *The English Urban Renaissance: Culture and Society in the Provincial Town, 1660–1770* (Oxford, 1989); R. Sweet, *The English Town, 1680–1840* (London, 1999); P. Clark, *British Clubs and Societies: the Origins of an Associational World* (Oxford, 2000), 166.

7   R. L. Emerson, 'Science and the origins and concerns of the Scottish Enlightenment', *History of Science*, 26 (1988), 333–66; A. J. Youngson, *The Making of Classical Edinburgh, 1750–1840* (Edinburgh, 1966); P. Reed, 'Form and content: a study of Georgian Edinburgh', in T. A. Markus ed., *Order and Space in Society: Architectural Form and its Context in the Scottish Enlightenment* (Edinburgh, 1982), 115–45; C. Philo, 'Edinburgh, Enlightenment, and the geographies of unreason', in D. Livingstone and C. Withers eds., *Geography and Enlightenment* (Chicago, 1999), 372–98.

8   R. F. Jones, *Ancients and Moderns* (St Louis, 1936); R. K. Faulkner, *Francis Bacon and the Project of Progress* (Lanham; 1993); B. H. G. Wormald, *Francis Bacon: History, Politics and Science, 1561–1626* (Cambridge, 1993); J. E. Leary, *Francis Bacon and the Politics of Science* (Ames, 1994); M. Peltonen ed., *The Cambridge Companion to Bacon* (Cambridge, 1996).

9   F. Bacon, *The Advancement of Learning and New Atlantis* (Cambridge, 1983), ix.

10   Bacon, *Advancement of Learning and New Atlantis*, 215–47.

11   W. Gilpin, *Three Essays on Picturesque Beauty*, second edition (London, 1794), 32.

12   S. Johnson, *Dictionary of the English Language* (London, 1797), unpaginated; J. Martyn ed., *The Georgics of Virgil* (London, 1741).

13   P. Borsay, 'The London connection: cultural diffusion and the eighteenth-century provincial town', *London Journal*, 19 (1994), 21–35.

14   E. Gibbon, *The History of the Decline and Fall of the Roman Empire*, edited by D. Womersley 3 vols. (London, 1994), I, 184–5.

15 D. Hume, 'Of refinement in the arts', in T. H. Green and T. H. Grose eds., *Essays, Moral, Political, and Literary*, 2 vols. (London, 1889), I, 299–309, p. 301.

16 H. Repton, *Sketches and Hints on landscape Gardening* (1795) in J. C. Loudon ed., *The Landscape Gardening and Landscape Architecture of the Late Humphry Repton Esquire* (London, 1840), 92–4.

17 Borsay, *English Urban Renaissance*; Sweet, *Eighteenth-Century Town*; J. Langton, 'Urban growth and economic change: from the late seventeenth century to 1841', P. Clark and R. A. Houston, 'Culture and leisure, 1700–1840', M. Reed, 'The transformation of urban space, 1700–1840', in Clark ed., *The Cambridge Urban History*, vol. 2 (Cambridge, 2000), 453–90, 575–613, 615–40.

18 Jones and Falkus, 'Urban improvement and the English economy'.

19 Borsay, *English Urban Renaissance*.

20 H. F. Hutchinson, *Sir Christopher Wren: A Biography* (London, 1976); J. A. Bennett, *The Mathematical Science of Christopher Wren* (Cambridge, 1982); J. Hanson, 'Order and structure in urban design: the plans for rebuilding of London after the Great Fire of 1666', *Ekistics*, 56 (1989); A. Tinniswood, *His Invention So Fertile: A Life of Christopher Wren* (London, 2001); L. Jardine, *On a Grander Scale: The Outstanding Career of Sir Christopher Wren* (London, 2002); C. C.Gillespie ed., *Dictionary of Scientific Biography* (New York, 1976), XIV, 509–11.

21 T. F. Reddaway, *The Rebuilding of London after the Great Fire* (London, 1951).

22 Ogborn, *Spaces of Modernity*.

23 Youngson, *Making of Classical Edinburgh*; R. Emerson, 'The Enlightenment and social structures', in P. Fritz and D. Williams eds., *City and Society in the Eighteenth Century* (Toronto, 1973), 122; A. Chitnis, *The Scottish Enlightenment: a Social History* (London, 1976); Reed, 'Form and content', 115–45; P. Wood, 'Science and the Aberdeen Enlightenment', in P.Jones ed., *Philosophy and Science in the Scottish Enlightenment* (Edinburgh, 1988), 57; Emerson, 'Origins and concerns of the Scottish Enlightenment'; Philo, 'Edinburgh, Enlightenment, and the geographies of unreason', 372–98.

24 M. Barfoot, 'Hume and the culture of science in the early eighteenth century', in M. A. Stewart ed., *Oxford Studies in the History of Philosophy* (Oxford, 1990).

25 'Proposals for carrying on certain public works in the City of Edinburgh', reproduced in Youngson, *Making of Classical Edinburgh*, 4–12.

26 R. Porter, 'Enlightenment London and urbanity', in, T. D. Hemming, E. Freeman and D. Meakin eds., *The Secular City* (London, 1994), 27–41.

27 Plan reproduced in Youngson, *Making of Classical Edinburgh*, 72–3, discussed, 70–110; town council minutes, 14 October 1767, quoted in Youngson, 80–1.

28 D. Daiches, *The Scottish Enlightenment: an Introduction* (1986), 21, quoted in Philo, 'Edinburgh, Enlightenment, and the geographies of unreason', 381; A. Walker, 'The Glasgow grid', in Markus ed., *Order in Space and Society*, 155–99; A. Hook and R. B. Sher eds., *The Glasgow Enlightenment* (East Linton, 1997).

29 M. Foucault, *Discipline and Punish: The Birth of the Prison* (London, 1979); M. Ignatieff, *A Just Measure of Pain: The Penitentiary in the Industrial Revolution, 1750–1850* (London, 1978); Markus, *Order in Space and Society*; T. A. Markus, 'Buildings and the ordering of minds and bodies', in Jones ed., *Philosophy and Science*, 169–224; P. Elliott, *The Derby Philosophers, Science and Urban Culture in Britain, 1700–1850* (Manchester, 2009), 163–89.

30 W. Chase, *The Norwich Directory; or, Gentlemen and Tradesmen's Assistant* (Norwich, 1783), iii–vi.

31 E. Fox Keller, *Reflections on Gender and Science* (New Haven, 1985); L. Schiebinger, *The Mind Has no Sex?: Women and the Origins of Modern Science* (Cambridge MA, 1991); E. P. Thompson, 'Time, work and industrial capitalism', in *Customs in Common* (London, 1994), 352–403.

32   Markus, 'Buildings and the ordering of minds and bodies', 169–224; J. C. D. Clarke, *English Society, 1660–1832: Religion, Ideology and Politics during the Ancien Regime*, second edition (Cambridge, 2000).

33   D. Hume, *Treatise of Human Nature*, edited by L. A. Selby-Bigge, revised by P. H. Nidditch (Oxford, 1978), 567–76, 586–7; D. Hume, *Enquiry Concerning the Principles of Morals* in *Essays, Moral, Political and Literary*, edited by T. H. Green and T. H. Grose (London, 1889), 202.

34   H. Home, Lord Kames, *Elements of Criticism*, seventh edition, 2 vols. (Edinburgh, 1788), II, 432–3, 455.

35   W. Hogarth, *The Analysis of Beauty* (London, 1753), edited by R. Paulson (New Haven, 1997), 25–48.

36   A. Smith, *Essays on Philosophical Subjects*, quoted in W. Watson, 'People, prejudice and place', in F. W. Boal and D. N. Livingstone eds., *The Behavioural Environment: Essays in Reflection, Application, and Re-evaluation* (London, 1989), 93–110, p. 100.

37   E. Burke, *A Philosophical Enquiry into the Origin of our Ideas of the Sublime and Beautiful*, in *The Works of Edmund Burke*, 2 vols. (London, 1834), I.

38   Home, *Elements of Criticism*, II, 433–4.

39   A. Alison, *Essays on the Nature and Principles of Taste* (1790), fifth edition (Edinburgh, 1817), I, 13ff, 317, II, 20ff, 33ff; R. Payne Knight, *Analytical Inquiry into the Principles of Taste* (London, 1805), 169, both quoted in R. Wittkower, *Palladio and English Palladianism* (London, 1974), 153.

40   H. Repton, *Observations on the Theory and Practice of Landscape Gardening* (1803) in J. C. Loudon ed., *The Landscape Gardening and Landscape Architecture of the Late Humphry Repton Esq.* (London, 1840), 262–5, 312–19; S. Daniels, *Humphry Repton: Landscape Gardening and the Geography of Georgian England* (New Haven, 1999).

41   Borsay, *English Urban Renaissance*, 100–1; R. S. Neale, *Bath: A Social History, 1680–1850* (London, 1981); P. Borsay, *The Image of Georgian Bath, 1700–2000* (Oxford, 2000).

42   J. Wood, *Choir Gaure, Vulgarly Called Stonehenge* (Oxford, 1747); J. Wood, *A Description of the Exchange at Bristol* (Bath, 1745); *The Origin of Building; or, The Plagiarism of the Heathens Detected* (Bath, 1741); *A Dissertation Upon the Orders of Columns, and their Appendages* (London, 1750); T. Mowl and B. Earnshaw, *John Wood: Architect of Obsession* (Bath, 1988).

43   Borsay, *English Urban Renaissance*; Ogborn, *Spaces of Modernity*; Sweet, *English Town*; J. Ellis, *The Georgian Town, 1680–1840* (London, 2001), 97–105; Elliott, *Derby Philosophers*, 113–31.

44   Youngson, *Classical Edinburgh*, 4–12; Ogborn, *Spaces of Modernity*, 98–104.

45   D. King-Hele ed., *The Collected Letters of Erasmus Darwin* (Cambridge, 2007), 395–6; Elliott, *Derby Philosophers*, 118–21.

46   *Universal British Directory of Trade, Commerce and Manufacture*, 6 vols. (London, 1791–7), IV, 44–56; E. Willoughby, *The Nottingham Directory* (Nottingham, 1799); J. Blackner, *The History of Nottingham* (Nottingham, 1815); F. C. Laird, *Topographical and Historical Description of the County of Nottingham* (London, 1820); W. H. Wylie, *Old and New Nottingham* (Nottingham, 1853); *The Nottingham Date Book, 850–1884* (Nottingham, 1884); J. D. Chambers, *Modern Nottingham in the Making* (Nottingham, 1945); J. D. Chambers, 'Population change in a provincial town, Nottingham 1700–1800', in L. S. Pressnell ed., *Studies in the Industrial Revolution* (Cambridge, 1960); J. D. Chambers, *Nottinghamshire in the Eighteenth Century*, second edition (London, 1966); M. I. Thomis, *Old Nottingham* (London, 1968); M. Thomis, 'The politics of Nottingham enclosure', *Transactions of the Thoroton Society*, 71 (1968), 90–6; M. I. Thomis, *Politics and Society in Nottingham, 1785–1835* (London, 1969); J. Beckett, *The East Midlands since AD 1000* (London, 1988); E. Biagini, 'The complex pattern of opposition to the Nottingham enclosure', *East*

Midland Geographer, 19 (1996), 20–37; J. V. Beckett *et al.* eds., *A Centenary History of Nottingham* (Manchester, 1997); S. Wallwork, 'Population estimates before the census: Nottingham, 1570–1801', *East Midland Historian*, 9 (1999), 35–42; P. Elliott, 'The politics of urban improvement in Georgian Nottingham: the enclosure dispute of the 1780s', *Transactions of the Thoroton Society*, 110 (2006), 87–102.

47   Beckett, *East Midlands since AD 1000*, 151–64; J. V. Beckett and J. E. Heath, 'When was the Industrial Revolution in the East Midlands?', *Midland History*, 13 (1988), 77–94.

48   Chambers, *Nottinghamshire*, 83.

49   M. E. Falkus, 'Lighting in the dark ages of English economic history: town streets before the industrial revolution', in D. C. Coleman and A. H. John eds., *Trade, Government and Economy in Pre-Industrial England* (London, 1976), 248–73.

50   S. D. Chapman, 'Working-class housing in Nottingham during the industrial revolution', *Transactions of the Thoroton Society*, 67 (1963), 69–77; Church, *Economic and Social Change*, 5–10.

51   *Nottingham Journal* (1 September 1787).

52   Records of High Pavement Presbyterian Chapel, 1576–1982, Department of Manuscripts and Special Collections, University of Nottingham (NUMD), especially Account Book, 1797–1826, HiA; B. Carpenter, *Some Account of the Original Introduction of Presbyterianism in Nottingham* (Nottingham, 1862); A. R. Henderson, *History of Castle Gate Congregational Church, Nottingham, 1655–1905* (Nottingham, 1905); J. C. Warren, 'Nottingham Presbyterian records', *Transactions of the Thoroton Society*, 19 (1916), 56–77; A. B. Clarke, 'Notes on the mayors of Nottingham, 1660–1775', *Transactions of the Thoroton Society*, 42 (1938), 105–20; Thomis, *Old Nottingham*, 97–116; M. Thomis, *Politics and Society*, 114–42; J. Beckett and B. H. Tolley, 'Church, chapel and school', in Beckett *et al.* eds., *Centenary History*, 351–64; High Pavement records, chapel minutes, vestry meetings (7 May 1780 to 22 June 1794); Elliott, 'The politics of urban improvement', 96–7, 101–2.

53   Elliott, 'Politics of urban improvement', 97.

54   Elliott, 'Politics of urban improvement'.

55   High Pavement records, chapel minutes, vestry meeting (26 February 1786).

56   High Pavement records, minute books, vestry meetings (4 September 1779, 24 June 1781, 26 February 1786, 4 March, 16 December 1787).

57   *Nottingham Date Book* (28 October 1782, 9 March 1785); Wakefield, *Memoirs*, I, 297–9.

CHAPTER 9

1   Parts of this chapter are based upon P. Elliott, 'Abraham Bennet FRS (1750–1749): a provincial electrician in eighteenth-century England', *Notes and Records of the Royal Society of London*, 53 (1999), 59–78. I am grateful to the editors for permission to employ this material here.

2   A. Bennet, 'Description of a new electrometer, in a letter from the Rev. Abraham Bennet, MA to the Rev. Joseph Priestley, LLP, FRS', *Philosophical Transactions*, 76 (1786), 26–34; 'An account of a doubler of electricity, or a machine by which the least conceivable quantity of positive or negative electricity may be continually doubled, till it becomes perceptible by common electrometer, or visible sparks ... communicated by the Rev. Richard Kaye, LLD, FRS', *Philosophical Transactions*, 77 (1787), 288–96; 'Letter on attraction and repulsion: communicated by DrPercival, October 11, 1786', *Manchester Memoirs*, 3 (1788), 116–23; *New Experiments on Electricity* (Derby and London, 1789); 'A new suspension of the magnetic needle, invented for the discovery of minute quantities of magnetic attraction: also an air

vane of great sensibility; with new experiments on the magnetism of iron filings and brass ... communicated by the Rev. Sir Richard Kaye, Bart, FRS', *Philosophical. Transactions*, 82 (1792), 81–98.

3   G. Pancaldi, *Volta: Science and Culture in the Age of Enlightenment* (Princeton, 2003).

4   V. Jankovic, *Reading the Skies: A Cultural History of English Weather, 1650–1820* (Chicago, 2000); J. Golinski, *British Weather and the Climate of Enlightenment* (Chicago, 2007).

5   Jankovic, *Reading the Skies*, 111.

6   P. L. Mottelay, *A Bibliographical History of Electricity and Magnetism* (London, 1922), 289–91; H. Ostwald, *Electrochemistry: History and Theory* (Leipzig 1896), translated by N. P. Date (Washington, 1980), I, chap. IV; W. C. Cameron Walker, 'The detection and estimation of electric charges in the eighteenth-century', *Annals of Science* (1936), 66–100; R. Schofield, *The Lunar Society of Birmingham: A Social History of Provincial Science and Industry in the Eighteenth Century* (Oxford, 1966), 166, 275;
J. Heilbron, *Electricity in the 17th and 18th Centuries: A Study of Early Modern Physics*, revised edition (New York, 1999), 450–1, 457–8.

7   J. Gascoigine, *Joseph Banks and the English Enlightenment* (Cambridge, 1994); H. Carter, *Sir Joseph Banks, 1743–1820* (London, 1988); R. Drayton, *Nature's Government: Science, Imperial Britain and the 'Improvement' of the World* (New Haven, 2000), 94–106; S. Wilmot, *The Business of Improvement: Agriculture and Scientific Culture in Britain c.1700–1870* (London, 1990).

8   *Derby* Mercury (23 May 1799, 5 July 1826); *Derby* Reporter (6 July, 1826); *European Magazine* and *London Review*, 35 (1799), 429; *Gentleman's Magazine*, 69 (1799), i, 443; *A New Catalogue of Living English Authors* (London, 1799), I, 225–6. Bennet died in 1799 after a 'severe illness' and was buried on 9 May. His notebook has survived and a portrait of him hangs in the vestry of Wirksworth Church which is half-length, done in oils and about 10.5 by 9 inches in size. The picture shows a profiled figure in clerical dress with grey hair, a large forehead and a straight, aquiline nose and dates to between 1789 and 1799 because Bennet is shown holding a copy of his *New Experiments* with another book and a roll of parchment, probably representing his scientific papers. Bennet's notebook (A. Bennet, memoranda miscellania, mss., Derby Local Studies Library) contains ideas, extracts from books, periodicals and newspapers.

9   G. Wilson, *Life and Works of the Hon. Henry Cavendish* (London, 1851); J. Clerk Maxwell, *Electrical Researches* (Cambridge, 1879); Mottelay, *Bibliographical History*, 238–9; J. Pearson, *Stags and Serpents: The Story of the House of Cavendish* (London, 1983), 86–115; T. Brighton, *The Discovery of the Peak District* (Chichester, 2004), 74–5.

10   'Francis Russell, Duke of Bedford (1765–1802)', DNB; S. Daniels, S. Seymour and C. Watkins, 'Enlightenment, improvement and the geographies of horticulture in later Georgian England', in D. N. Livingstone and C. Withers eds., *Geography and Enlightenment* (Chicago, 1999), 345–71, p. 349.

11   A. Volta, 'On the electricity excited by the mere contact of conducting substances of different kinds', *Philosophical Transactions*, 90 (1800), 403–31, English translation from Ostwald, *Electrochemistry: History and Theory*, 115–41.

12   J. Farey, *A General View of the Agriculture and Minerals of Derbyshire*, 3 vols. (London, 1811–17); K. C. Edwards, *The Peak District* (London, 1973), 63–70.

13   J. Pilkington, *A View Of the Present State of Derbyshire*, 2 vols. (Derby, 1789), II, 299–301; J. Aikin, *A Description of the County from Thirty to Forty Miles round Manchester* (London, 1795), 504–5; N. Pevsner, *The Buildings of England: Derbyshire* (London, 1953), 246–9; Edwards, *The Peak District*, 183–91; A. Menuge, 'The cotton mills of the Derbyshire Derwent and its tributaries', *Industrial Archaeology Re-*

view 16 (1993), 38–61, pp. 39, 41; G. Turbutt, *A History of Derbyshire*, 4 vols. (Cardiff, 1999), IV, 1431–6, 1498; J. Barnatt and K. Smith, *The Peak District: Landscapes Through Time* (Macclesfield, 2004); D. Hey, *Derbyshire: A History* (Lancaster, 2008).

14 Farey, *General View of the Agriculture and Minerals of Derbyshire*; W. Adam, *Gem of the Peak*, second edition (Derby, 1840); Edwards, *Peak District*, 183–91; T. D. Ford ed., *Limestone and Caves of the Peak District* (Norwich, 1977); T. D. Ford and J. H. Rieuwerts eds., *Lead Mining in the Peak District*, fourth edition (Ashbourne, 2000).

15 Heilbron, *Electricity in the 17th and 18th Centuries*, 312–34; Walker, 'The detection and estimation of electric charges', 70–72.

16 J. Priestley, *The History and Present State of Electricity, with Original Experiments* (London, 1767), 455–79; Heilbron, *Electricity in the 17th and 18th Centuries*, 337–9, 384–7.

17 Bennet, 'Letter on attraction and repulsion', 116–17, 119–22.

18 T. Cavallo, 'Some new experiments in electricity with the description and use of two new electrical instruments', *Philosophical Transactions*, 70 (1780), 15–29; Bennet, 'Description of a new electrometer', 26–34.

19 Bennet, *New Experiments*, 112–14 and plate III.

20 Heilbron, *Electricity in the 17th and 18th Centuries*, 412–17, 457, Bennet, 'Description of a new electrometer', 32.

21 Bennet, 'An account of a doubler', 289–91; Bennet, *New Experiments*, 75–9.

22 Bennet, *New Experiments*, 78.

23 T. Cavallo, 'Of the methods of manifesting the presence, and ascertaining the quality of small quantities of natural or artificial electricity', *Philosophical Transactions*, 78 (1788), 2.

24 W. Nicholson, 'A description of an instrument which by the turning of a winch, produces the two states of electricity without friction or communication with the earth', *Philosophical Transactions*, 78 (1788), 403–7; Bennet, *New Experiments*, 83.

25 Bennet, *New Experiments*, 83.

26 Bennet, *New Experiments*, 83.

27 Bennet, *New Experiments*, 83–9.

28 Bennet, *New Experiments*, 91.

29 Bennet, *New Experiments*, 97–100.

30 A. Volta, 'Of the method of rendering very sensible the weakest natural or artificial electricity by Mr Alexander Volta, Professor of Experimental Philosophy in Como ... communicated by the Right Hon. George Earl Cowper, FRS', *Philosophical Transactions*, 122 (1782), appendix, vii–xxxxiii.

31 L. Galvani, *De Viribus electricitatis in motu musculari commentarius cum Joannis Aldini dissertatione et notis* (Modena, 1792); M. Pera, *The Ambiguous Frog: the Galvani-Volta Controversy on Animal Electricity*, translated by J. Mandelbaum (Princeton, 1992), 111.

32 Ostwald, *Electrochemistry: History and Theory*, 126.

33 *Analytical Review*, 2 (1788), 387–8.

34 G. Adams, *An Essay on Electricity*, fifth edition corrected and edited by Jones (London, 1799), 406–11.

35 G. J. Singer, *Elements of Electricity and Electro-Chemistry* (London, 1814), 277–8, 128–9.

36 C. H. Wilkinson, *Elements of Galvanism in Theory and Practice*, 2 vols. (London, 1804), II, 23.

37 E. Darwin, *The Temple of Nature: or the origin of society, additional note, 50, 62* (London, 1803); H. Davy, 'Bakerian lecture, on some chemical agencies of electricity', *Philosophical Transactions*, 97 (1807), 297.

38 Bennet, *New Experiments*, 91, 100.

39   Pera, *Ambiguous Frog*, 111.
40   A. Volta, 'On the electricity excited by the mere contact of conducting substances of different kinds', 431.
41   Schofield, *Lunar Society*; M. McNeil, *Under the Banner of Science, Erasmus Darwin and His Age* (Manchester, 1987), 14–24; J. Uglow, *The Lunar Men: the Friends who Made the Future* (London, 2002); P. M. Jones, *Industrial Enlightenment: Science, Technology and Culture in Birmingham and the West Midlands, 1760–1820* (Manchester, 2008); P. Elliott, *The Derby Philosophers: Science and Culture in British Urban Society, 1700–1850* (Manchester, 2009).
42   Bennet, *New Experiments*, 21.
43   Bennet, *New Experiments*, 50.
44   Schofield, *Lunar Society*, 252–3; D.King-Hele ed., *The Collected Letters of Erasmus Darwin* (Cambridge, 2007), 248–60, 269–71, 281–2, 292–4.
45   Schofield, *Lunar Society*, 252–3; King-Hele ed., *Collected Letters*, 254–5.
46   McNeil, *Under the Banner of Science*; D. King-Hele, *Erasmus Darwin and the Romantic Poets* (London, 1986); D. King-Hele, *Erasmus Darwin: a Life of Unequalled Achievement* (London, 1998).
47   King-Hele ed., *Collected Letters*, 285–7.
48   E. Darwin, *Zoonomia; or the Laws of Organic Life* (London, 1794) I, 120.
49   E. Darwin, 'Remarks on the opinion of Henry Eeles Esq., concerning the ascent of vapour', *Philosophical Transactions*, 50 (1757), 240–54.
50   Darwin, *Temple of Nature, additional note XII*, 46–79; E. Darwin, 'Frigorific experiments on the mechanical expansion of air', *Philosophical Transactions*, 78 (1788); King-Hele, *Erasmus Darwin*, 226–8.
51   E. Darwin, *The Economy of Vegetation* (London, 1791); Bennet, 'An account of a doubler of electricity', 289.
52   Bennet, *memoranda miscellania*, 136; Bennet, 'Letter on attraction and repulsion', 121–2; Darwin, *Economy of Vegetation*, note XIII, 25; Darwin, *Temple of Nature, additional note XII*, 52.
53   In 1775 Gregorio Fontana had suggested that spider's thread be used as a substitute for wires, Mottelay, *Bibliographical History*, 290.
54   Bennet, 'A new suspension of the magnetic needle', 92–6; King-Hele, ed., *Collected Letters*, 296–7.
55   Bennet, 'A new suspension of the magnetic needle', 88; H. Lloyd, *Elementary Treatise on the Wave Theory of Light* (London, 1857), 10–11 quoted in G. N. Cantor, *Optics after Newton: Theories of Light in Britain and Ireland, 1704–1840* (Manchester, 1984), 57–8.
56   E. W. Gray, 'Account of an earthquake felt in various parts of England, November 18, 1795: with some observations thereon', *Philosophical Transactions*, 86 (1796), 361.
57   S. Glover, *History and Gazetteer of the Town of Derby*, 2 vols. (Derby, 1833), II, 601; *Derby Mercury* (2 October 1832); Elliott, *Derby Philosophers*, 241–2.
58   T. D. Ford, 'White Watson (1760–1835) and his geological tablets', *Mercian Geologist*, 13 (1995), 157–64.
59   Darwin, *Economy of Vegetation*, canto I, 1.137–42, additional note VI.
60   J. Whitehurst, *Inquiry into the Original State and Formation of the Earth*, first edition (London, 1778); M. Craven, *John Whitehurst of Derby; Clockmaker and Scientist, 1731–88* (Ashbourne, 1996).
61   Gray, 'Account of an earthquake', 353–81; Darwin, *Temple of Nature, additional note XII*, 68–72; Bennet, 'A new suspension of the magnetic needle', 92; P. Bowler, *The Fontana History of the Environmental Sciences* (London, 1992), 413–16.

62   G. J. Singer, Elements of Electricity and Electro-Chemistry (London, 1814), 357–8.

FINAL CONCLUSIONS

1   H. Spencer, 'Progress; its law and cause', in *Essays: Scientific, Political and Speculative*, 3 vols. (London, 1891), I, 56–8.

2   D. E. Allen, *The Naturalist in Britain: A Social History* (London, 1976); L. Brockway, *Science and Colonial Expansion: The Role of the British Botanic Gardens* (New York, 1979); C. Russell, *Science and Social Change, 1700–1900* (London, 1983); I. Inkster and J. Morell eds., *Metropolis and Provinces: Science in British Culture, 1780–1850* (London, 1983); R. M. Young, *Darwin's Metaphor: Nature's Place in Victorian Culture* (London, 1985); L. L. Merrill, *The Romance of Victorian Natural History* (Oxford, 1989); P. Bowler, *History of the Environmental Sciences* (London, 1992); P. F. Stevens, *The Development of Biological Systematics: Antoine-Laurent de Jussieu, Nature and the Natural System* (New York, 1994); N. Jardine, J. A. Secord and E. C. Spary eds., *Cultures of Natural History* (Cambridge, 1996); A. B. Shteir, *Cultivating Women, Cultivating Science: Flora's Daughters and Botany in England, 1760–1860* (Baltimore, 1996); B. Lightman ed., *Victorian Science in Context* (Chicago,1997); D. P. McCracken, *Gardens of Empire: Botanical Institutions of the Victorian British Empire* (Leicester, 1997); R. Drayton, *Nature's Government: Science, Imperial Britain and the 'Improvement' of the World* (New Haven, 2000); C. Merchant, *Reinventing Eden: The Fate of Nature in Western Culture* (New York, 2004); C. Yanni, *Nature's Museums: Victorian Science and the Architecture of Display* (New York, 2005); M. Daunton ed., *The Organisation of Knowledge in Victorian Britain* (Oxford, 2005).

3   J. C. Loudon, *Encyclopaedia of Gardening*, second edition (London, 1824), 1072–3.

4   M. Batey, *Oxford Gardens: The University's Influence on Garden History* (Amersham, 1982), 139–42.

5   N. Johnson, 'Cultivating science and planting beauty: the spaces of display in Cambridge's botanical gardens', *Interdisciplinary Science Reviews*, 31 (2006),42–57, pp. 49–53;

6   J. C. Loudon, *Arboretum et Fruticetum Britannicum*, 8 vols. (London, 1838); P. Elliott, 'The Derby Arboretum (1840): the first specially designed municipal public park in Britain', *Midland History* (2001); P. Elliott, C. Watkins and S. Daniels eds., 'Cultural and historical geographies of the arboretum', special issue of *Garden History* (2007); P. Elliott, C. Watkins and S. Daniels, 'The Nottingham Arboretum (1852): natural history, leisure and public culture in a Victorian regional centre', *Urban History* (2008), 48–71; P. Elliott, C. Watkins and S. Daniels, *The British Arboretum: Trees, Science and Culture in the Nineteenth Century* (London, 2011).

7   J. Endersby 'Classifying sciences: systematics and status in mid-Victorian natural history', in Daunton ed., *Organisation of Knowledge*, 61–85; H. Ritvo, 'Zoological nomenclature and the empire of Victorian science', in B. Lightman, ed., *Victorian Science in Context*, 334–53; N. Johnson, 'Names, labels and planting regimes: regulating trees at Glasnevin Botanical Gardens, Dublin, 1795–1850', in Elliott, Watkins and Daniels, eds., *Cultural and Historical Geographies of the Arboretum*, 53–70, pp. 60–1.

8   Batey, *Oxford Gardens*, 142–4.

9   T. Wright and H. Longueville Jones, *Memorials of Cambridge*, 2 vols. (London, 1847), II, 134–5; Johnson, 'Cultivating science', 49–53.

10   Elliott, 'Derby Arboretum'; R. Desmond, *Kew: The History of the Royal Botanic Gardens* (London, 1995); Shteir, *Cultivating Women*, 149–65; Drayton, *Nature's Government*, 153–268; B. Elliott, *The History of the Royal Horticultural Society, 1804–2004* (London, 2004); P. Elliott, *The Derby Philosophers: Science and Culture in British Urban Society, 1700–1850* (Manchester, 2009), 235–59.

Bibliography

MANUSCRIPT MATERIAL

Birmingham Central Library, Archives and Heritage

Galton papers, Birmingham Central Library, Archives and Heritage (MS3101)

British Library

Peat, T., *A Short Account of a Course of Mechanical and Experimental Philosophy and Astronomy* (*c.*1756[1744?]), (8704.aa.10)

Cambridge University Library, Department of Manuscripts and Special Collections

Cambridge University Botanical Garden, minute books of the trustees and other papers
Darwin Papers (DAR) (material relating to Erasmus Darwin)
Darwin, E., manuscript notebook, Cambridge University Library, DAR 227.2.11

Derby Local Studies Library, Derby

Bennet, A., 'Memoranda Miscellania', commonplace book
Derby Friargate Chapel minutes and accounts, 1697–1819 (BA 288, DER A.22207)
Derby Philosophical Society, manuscript catalogue and charging ledger, 1785–9 and cash ledger, 1813–45 (BA 106, 9229–30)
Latham, E., 'Exercitatis Physiologia' (mss. 3368)
Nun's Green box of broadsides and pamphlets (box 27)
Paving and Lighting Commissioners, Minute Book of the Paving and Lighting Commissioners, 1789–1825 (BA 625/8, 16048)

Strutt papers and correspondence on microfilm (D125/-). Originals moved to Derbyshire County Records Office

Derbyshire County Records Office, Matlock

Friargate Chapel, Derby, birth and baptismal registers (M690 RG 4/5, M695 RG 4/2033, M695 RG 4/2034, M691 RG 4/499), and minutes and accounts from the period (1312 D/A1, A2), c.1700–1860
Strutt Family papers, including deeds, estate and family papers and correspondence (D1564, D2912, D2943M, D3772, D5303)

Durham University Library, Palace Green, Durham

Thomas Wright Manuscripts (GB-0033-WRM)

Fitzwilliam Museum, Cambridge

Strutt papers (MS 48 – 1947)

Harris Manchester College, Oxford

Martineau papers

Keele University Library

Wedgwood Manuscripts

Leicestershire County Record Office

'Laws for the Regulation of the Literary Society, Leicester', 30 July 1790 (misc. 79)

Library and Museum of Freemasonry, Grand Lodge, London

Minutes of the Old Kings Arms Lodge, London

Newcastle Central Library

Wright, Thomas (of Durham), bound manuscript papers

Norfolk County Record Office, Norwich

Colman manuscripts (9/-)
Collection of material concerning the Public Library, Norwich (SO 50/1/-)
Minute Book of the Tusculum Society, Norwich (mss. NNAS/G2)
Minutes, account books, subscription lists and catalogues of the Norfolk
    Literary Institution (SO 50/2/1–50)
Minutes, account books, subscription lists and catalogues of the Norwich
    Mechanics' Institute (mss. 4334–5, 4263–4)

Nottingham Subscription Library, Bromley House, Nottingham

Goodacre papers

Royal Society of London, Library and Archives

Journal Book of the Royal Society

University of Nottingham, Department of Manuscripts and Special Collections
Kirk White, H., Kirk White mss., (KW C60a, KWC 61, KWC 5)
Records of High Pavement Presbyterian Chapel, Nottingham 1676–1982: ac-
    count books, 1798–1826 (HiA), minute books, 1777–1812 (HiM1),
    baptismal registers, 1691–1837 (HiR), A–V

MAIN PERIODICALS USED, 1680–1850

Analytical Review
Derby Mercury
European Magazine and London Review
Gardener's Magazine
Gentleman's Magazine
Harrison's Derby and Nottingham Journal
Leicester Journal
Nicholson's Journal
Nottingham Journal
Nottingham Review
Philosophical Magazine
Philosophical Transactions

WORKS PUBLISHED BEFORE 1850

Adam, W., *Gem of the Peak*, second edition (Derby, 1840)

Adams, G., *An Essay on Electricity*, fifth edition (London, 1799)

Addington, S., *An Inquiry into the Reasons for and Against Inclosing Open-Fields* (Coventry, 1772)

——, *The Youth's Geographical Grammar* (London, 1770)

Aikin, J., *A Description of the County from Thirty to Forty Miles round Manchester*, (London, 1795)

Alison, A., *Essays on the Nature and Principles of Taste* (1790), fifth edition (Edinburgh, 1817) ·

Anderson, J., *The Constitutions of the Freemasons* (London, 1723)

Anonymous, 'Biographical history of Mr Thomas Wright', *Gentlemen's Magazine*, 63 (1793), 9–12, 120–1

Anonymous, *The Golden Age: A Poetical Epistle from Erasmus D---N to Thomas Beddoes, MD* (London, 1794)

Anonymous, *The Millenium; a Poem in Three Cantos*, 2 vols. (London, 1800–1)

Arbuthnot, J., *An Examination of Dr Woodward's Account of the Deluge* (London, 1697)

Baines, E., *History, Directory and Gazetteer of the County of York*, 2 vols. (1823)

Bennet, A., 'A new suspension of the magnetic needle, invented for the discovery of minute quantities of magnetic attraction ... communicated by the Rev. Sir Richard Kaye, Bart, FRS', *Philosophical Transactions*, 82 (1792), 81–98

——, 'An account of a doubler of electricity, or a machine by which the least conceivable quantity of positive or negative electricity may be continually doubled, till it becomes perceptible by common electrometer, or visible sparks ... communicated by the Rev. Richard Kaye, LLD, FRS', *Philosophical Transactions*, 77 (1787), 288–96

——, 'Letter on attraction and repulsion: communicated by Dr Percival, October 11, 1786', *Manchester Memoirs*, 3 (1788), 116–23

——, *New Experiments on Electricity* (Derby and London, 1789)

——, 'Description of a new electrometer, in a letter from the Rev. Abraham Bennet, MA to the Rev. Joseph Priestley, LLP, FRS', *Philosophical Transactions*, 76 (1786), 26–34

Blackner, T., *The History of Nottingham* (Nottingham, 1815)

Bogue, D. and Bennett, J., *History of Dissenters: From the Revolution in 1688, to the Year 1808*, 4 vols. (London, 1812)

Brooke, A., *Miscellaneous Experiments and Remarks on Electricity, the Air-Pump and the Barometer* (London, 1789)

Bryant, M., *A Compendious System of Astronomy* (London, 1797)

Burke, E., A Philosophical Enquiry into the Origin of our Ideas of the Sublime and Beautiful in E. Burke, *The Works of Edmund Burke*, 2 vols. (London, 1834)

——, *The Works of Edmund Burke*, 2 vols. (London, 1834)

Cambridge Botanical Garden, *A Short Account of the Late Donation of a Botanic Garden to the University of Cambridge* (Cambridge, 1762)

——, *A Proposal for an Annual Subscription for the Support of the Botanic Garden at Cambridge* (Cambridge, 1765)

Cavallo, T., 'Of the methods of manifesting the presence, and ascertaining the quality of small quantities of natural or artificial electricity', *Philosophical Transactions*, 78 (1788)

——, 'Some new experiments in electricity with the description and use of two new electrical instruments', *Philosophical Transactions*, 70 (1780), 15–29

Chase, W., *The Norwich Directory; or, Gentlemen and Tradesmen's Assistant* (Norwich, 1783)

Clare, M., 'An address made to the body of free and accepted masons' (1735), *The Pocket Companion and History of Free-Masons*, second edition (London, 1759), 307–16

——, *On the Motion of Fluids, Natural and Artificial* (London, 1735)

Cole B., ed., *The Ancient Constitutions of Free and Accepted Masons* (London, 1731)

Curtis, S., *Lectures on Botany as Delivered in the Botanical Garden at Lambeth by the Late William Curtis FLS*, 2 vols. (London, 1805)

Curtis, W., *A Catalogue of the British, Medicinal, Culinary and Agricultural Plants Cultivated in the London Botanic Garden* (London, 1783)

——, *Companion to the Botanical Magazine; or a Familiar Introduction to the Study of Botany* (London, 1788)

——, *Flora Londinensis; or Plates and Descriptions of Such Plants as Grow Wild in the Environs of London* (London, 1777)

——, *Proposals for Opening by Subscription a Botanic Garden to be called the London Botanic Garden* (London, 1778)

——, *The Subscription Catalogue of the Brompton Botanic Garden for the year 1790* (London, 1790)

Darwin, E., *The Botanic Garden*, second edition, 2 vols. (London, 1791)

——, 'Frigorific experiments on the mechanical expansion of air', *Philosophical Transactions*, 78 (1788)

——, *Phytologia; or the Philosophy of Agriculture and Gardening* (London, 1800)

——, 'Remarks on the opinion of Henry Eeles Esq., concerning the ascent of vapour', *Philosophical Transactions*, 50 (1757), 240–54

——, *The Temple of Nature; or the Origins of Society* (London, 1803)

——, *Zoonomia; or the Laws of Organic Life*, second edition (London, 1796)

Deering, C., *Nottinghamia Vetus et Nova: or, An Historical Account of the Ancient and Present State of Nottingham* (Nottingham, 1751)

Donn, J., *Hortus Cantabrigiensis; or a Catalogue of Plants, Indigenous and Exotic Cultivated in the Botanic Garden, Cambridge*, first edition (Cambridge, 1796); second edition (1800); *Hortus Cantabrigiensis* third edition (Cambridge, 1804); fourth edition (Cambridge, 1807)

Drake, F., *Eboracum: or, the History and Antiquities of the City of York* (York, 1736)

Edgeworth, M., *Letters for Literary Ladies* (1795)

Edgeworth, R. L. and Edgeworth, M., *Practical Education*, 2 vols. (London, 1798)

Farey, J., *A General View of the Agriculture and Minerals of Derbyshire*, 3 vols. (London, 1811–17)

Faulkner, T., *An Historical and Topographical Description of Chelsea and its Environs* (London, 1810)

Ferguson, J., *Analysis of a Course of Lectures on Mechanics, Pneumatics, Hydrostatics and Astronomy* (London, 1761)

Field, H., *Memoirs Historical and Illustrative of the Botanick Garden at Chelsea* (London, 1820)

Gilpin, W., *Three Essays on Picturesque Beauty,* second edition (London, 1794)

Gisborne, T., 'On the benefits and duties resulting from the institution of societies for the advancement of literature and philosophy', *Manchester Memoirs*, 5 (1798–1802), 70–88

——, *An Enquiry into the Duties of Men in the Higher and Middle Classes of Society in Great Britain,* seventh edition (London, 1824)

Glover, S., *History and Gazeteer of the Town of Derby*, 2 vols. (Derby, 1833)

Goldsmith, J., (Richard Phillips), *A Grammar of British Geography*, fourth edition (London, 1816)

Gorham, G. C., *Memoirs of John Martyn FRS and Thomas Martyn, BD, FRS, FLS, Professors of Botany in the University of Cambridge* (London, 1830)

Gray, E. W., 'Account of an earthquake felt in various parts of England, November 18, 1795: with some observations thereon', *Philosophical Transactions*, 86 (1796)

Gregory, J., *A Manual of Modern Geography,* first and third editions (London and Derby, 1739 and 1748)

Hanbury, W., *A Complete Body of Planting and Gardening*, 2 vols. (London, 1770)

——, *Essay on Planting, and a Scheme for Making It Conducive to the Glory of God and the Advantage of Society* (Oxford,1758)

Haynes, J., *An Accurate Survey of the Botanic Garden at Chelsea with the Elevation and Ichnography of the Green House and Stoves* (London, 1751)

Hogarth, W., *The Analysis of Beauty* (London, 1753), edited by R. Paulson, (New Haven, 1997)

Home, H., Lord Kames, *Elements of Criticism*, seventh edition, 2 vols. (Edinburgh, 1788)

Hull Botanic Garden, *The Address of the President and Treasurer at the First General Meeting of the Subscribers to the Hull Botanic Garden* (Hull, 1812)

Hume, D., *Essays and Treatises on Several Subjects,* 2 vols. (Edinburgh, 1825)

——, *Essays, Moral, Political, and Literary,* edited by T. H. Green and T. H. Grose, 2 vols. (London, 1889)

Hutton, C., 'Authentic memoirs of the life and writings of the late John Whitehurst FRS', *Universal Magazine* (November, 1788), 225–9

Hutton, W., *History of Derby* (Birmingham, 1791)

Jennings, D., *Introduction to the Use of the Globes, and the Orrery: With the Application of Astronomy to Chronology* (London, 1747)

Johnson, S., 'Life of Watts', in A. Murphy ed., *The Works of Samuel Johnson LLD*, 12 vols. (London, 1820) XI, 238–48

——, *Dictionary of the English Language* (London, 1797)

Jones, S., *Rudiments of Reason: or the Young Experimental Philosopher*, 3 vols. (London, 1793)

'Kewensis', 'Biographical anecdotes of William Curtis the botanist', *Gentleman's Magazine*, (1799), 274–80

Lambert, A. B., *A Description of the Genus Pinus*, second edition, 2 vols. (London, 1828)

Linnaeus, C., *Miscellaneous Tracts Relating to Natural History, Husbandry and Physick*, translated by B. Stillingfleet (London, 1759)

——, *A System of Vegetables ... Translated from the Thirteenth Edition of the 'Systema Vegetabilium' ... by a Botanical Society at Lichfield*, 2 vols. (Lichfield, 1783)

——, *The Families of Plants ... Translated from the Last Edition of the 'Genera Plantarum' ... by a Botanical Society at Lichfield* 2 vols. (Lichfield, 1787)

Loudon, J. C., *The Encyclopaedia of Gardening*, second edition (London, 1824)

——, *Hints on the Formation of Gardens and Pleasure Grounds* (London, 1812)

——, ed., *The Landscape Gardening and Landscape Architecture of the Late Humphry Repton Esquire* (London, 1840)

Lowe, R., *General View of the Agriculture of the County of Nottingham* (London, 1798)

Lysons, D., *The Environs of London*, second edition, 2 vols. (London, 1810), I, 102–4

Lysons, D. and S., *Magda Britannia*, vol. 5, *Derbyshire* (London, 1817)

Magna Britannia et Hibernia, Antiqua; Or, A New Survey of Great Britain 6 vols. (London, 1720–31)

Martyn, T., *Catalogus Horti Botanici Cantabrigiensis* (Cambridge, 1771)

——, *Heads of a Course of Lectures in Natural History read at the Botanic Garden* (London, 1782)

——, *Mantissa Plantarum Horti Botanici Cantabrigiensis* (Cambridge, 1772)

Miller, P., *The Gardener's Dictionary* edited by T. Martyn, 2 vols. in four parts (London, 1807)

Morgan, G. C., *Lectures on Electricity*, 2 vols. (Norwich, 1794)

Murphy, A., ed., *The Works of Samuel Johnson LLD*, 12 vols. (London, 1820)

Neal, D., *The History of the Puritans or Protestant Non-Conformists, from the Reformation in 1517 to the Revolution in 1688*, edited by J. Toulmin, 3 vols. (London, 1837)

Nichols J., *The History and Antiquities of the County of Leicester* (Leicester, 1804)

——, *Literary Anecdotes of the Eighteenth Century*, (London, 1812)

Nichols, J. and J. B., *Illustrations of the Literary History of the Eighteenth Century*, 8 vols. (London, 1817–58)

Nicholson, W., 'A description of an instrument which by the turning of a winch, produces the two states of electricity without friction or communication with the earth', *Philosophical Transactions*, 78 (1788), 403–7

Oldfield, J., *An Essay Towards the Improvement of Reason* (London, 1707)

Oliver, G., ed., *The Golden Remains of the Early Masonic Writers*, 5 vols. (London, 1848)

Peat, T. and Badder, J., *A Plan of the Town of Nottingham from an Accurate Survey* (Nottingham, 1744)

Pilkington, J., *A View Of the Present State of Derbyshire*, 2 vols. (Derby, 1789)

Priestley, J., *A Description of a Chart of Biography*, fourth edition (London, 1770)

——, *A Description of a New Chart of History* (London, 1786)

——, *History and Present State of Electricity*, first edition (London, 1767)

——, *History and Present State of Electricity*, third edition, 2 vols. (London, 1775)

——, *Lectures on History and General Policy*, 2 vols. (London, 1793)

Rees, A., *Chambers' Cyclopaedia* in 4 volumes (London, 1788)

——, *Cyclopaedia: or, Universal Dictionary of Arts and Sciences*, 39 vols. (London, 1819)

Roscoe, W., An Address delivered before the Proprietors of the Botanic Garden in Liverpool (Liverpool, 1802)

——, *Life of William Roscoe*, 2 vols. (Liverpool, 1834)

——, *Monandrian Plants of the Order Scitamineae, Chiefly Drawn from Living Specimens in the Botanic Garden at Liverpool*, first edition (1828)

——, 'On artificial and natural arrangements of plants: and particularly on the systems of Linnaeus and Jussieu', *Transactions of the Linnaean Society*, 11 (1815), 50–78

Rutt, J. T. ed., *The Theological and Miscellaneous Works of Joseph Priestley, LLD*, 25 vols. (London, 1831)

Ryland, J. C., *An Easy and Pleasant Introduction to Sir Isaac Newton's Philosophy* (London, 1772).

Senex, J., *A Treatise of the Description and Use of Both Globes* (London, 1718)

Seward, A., *Memoirs of the Life of Dr Darwin* (London, 1804)

Shenstone, W., *Works in Verse and Prose*, 2 vols. (London, 1764)

Singer, G. J., *Elements of Electricity and Electro-Chemistry* (London, 1814)

Smith, J. E., *An Introduction to Physiological and Systematical Botany*, sixth edition (London, 1827)

——, *Memoir and Correspondence of the late James Edward Smith*, 2 vols. (London, 1832)

Smith, J. J. ed., *The Cambridge Portfolio* (London, 1840)

Throsby, J., *Select Views in Leicestershire from Original Drawings* (London, 1789)

Toland, J., *Pantheisticon: or, the Form of Celebrating the Socratic Society* (London, 1751)

Universal British Directory of Trade, Commerce and Manufacture (London, 1791–7)

Volta, A., 'Of the method of rendering very sensible the weakest natural or artificial electricity ... communicated by the Right Hon. George Earl Cowper, FRS', *Philosophical Transactions*, 122 (1782), appendix, vii–xxxxiii

——, 'On the electricity excited by the mere contact of conducting substances of different kinds', *Philosophical Transactions*, 90 (1800), 403–31, English translation from Ostwald, *Electrochemistry; History and Theory*, 115–41

Wakefield, G., *Address to the Inhabitants of Nottingham on the Subject of the Test Laws* (Nottingham, 1789)

——, *Cursory Reflections ... on the Repeal of the Corporation and Test Acts* (London, 1790)

Walker, G., *Essays on Various Subjects*, 2 vols. (London, 1809)

——, *Sermons on Various Subjects*, 2 vols. (London, 1790)

Watson, P. W., *Dendrologia Britannica*, 2 vols. (London, 1825)

Watts, I., *The Improvement of the Mind* (London, 1795)

——, *The Knowledge of the Heavens and Earth Made Easy*, fourth edition (London, 1745)

Wesley, J., *The Desideratum, or Electricity Made Plain and Useful* (London, 1760)

Whitehurst, J., *Enquiry into the Original State and Formation of the Earth*, first edition (London, 1778)

——, *Tracts; Philosophical and Mechanical by John Whitehurst FRS* (London, 1792)

Wilkinson, C. H., *Elements of Galvanism in Theory and Practice*, 2 vols. (London, 1804)

Williams, D., *Lectures on Education*, 3 vols. (London, 1789)

——, *Lectures on the Universal Principles and Duties of Religion and Morality* 2 vols., (London, 1779)

Withering, W., *A Botanical Arrangement of all the Vegetables Naturally growing in Great Britain ...* (Birmingham, 1775)

Wood, A., *The History and Antiquities of the University of Oxford*, translated by J. Gutch, 2 vols. (Oxford, 1796)

——, *The Origin of Building; or, The Plagiarism of the Heathens Detected* (Bath, 1741)

Woodward, J., *An Essay Toward a Natural History of the Earth* (London, 1695)

Wright, T., *An Original Theory or New Hypothesis of the Universe* (London, 1750), reprinted with introduction by M. Hoskin (London, 1971)

Wright, T. and Longueville, Jones H., *Memorials of Cambridge*, 2 vols. (London, 1847)

Young, G., *A Catalogue of the Plants ... with the Rules of the Institution ... Whitby Botanic Garden* (Whitby, 1814)

WORKS PUBLISHED AFTER 1850

Agar, J. and Smith, C. eds., *Making Space for Science: Territorial Themes in the Shaping of Knowledge* (London, 1998)

Alberti, S., 'Placing nature: natural history collections and their owners in nineteenth-century provincial England', *British Journal for the History of Science*, 35 (2002)

Allen, D. E., *The Naturalist in Britain: A Social History* (London, 1976)

Allison, K. J. ed., *Victoria History of the County of York: East Riding* vol. I: *The City of Kingston-Upon-Hull* (Oxford, 1969)

Allthrope-Guyton, M., 'The artistic and literary life in Norwich during the century', in C. Barringer ed., *Norwich in the Nineteenth Century* (Norwich, 1984), 1–46

Armstrong, A., *Stability and Change in an English County Town: A Social Study of York 1801–51* (Cambridge, 1974)

Averley, G., 'English Scientific Societies of the Eighteenth and Early Nineteenth Centuries', unpublished PhD thesis, University of Teeside Polytechnic and Durham University (Durham, 1989)

Bacon, F., *The Advancement of Learning and the New Atlantis* (Cambridge, 1983)

Baedecker's Great Britain: A Handbook for Travellers, second edition (London, 1890)

Baker, K. H. and Reill, R. M. eds., *Whats Left of Enlightenment?: A Postmodern Question* (Stanford, 2001)

Barfoot, M., 'Hume and the culture of science in the early eighteenth century', in M. A. Stewart ed., *Oxford Studies in the History of Philosophy* (Oxford, 1990)

Barnard, T., *Anna Seward: A Constructed Life: A Critical Biography* (Aldershot, 2009)

Barnatt, J. and Smith, K., *The Peak District: Landscapes Through Time* (Macclesfield, 2004)

Barnett, A., 'In with the new: novel goods in domestic provincial England, c.1700–1790', in B. Blondé, N. Coquery, J. Stobart and I. Van Damme eds., *Fashioning Old and New: Changing Consumer Patterns in Western Europe (1650–1900)*, (Turnhout, 2009), 81–94

Barrell. J., *The Dark Side of the Landscape: the Rural Poor in English Paintings, 1730–1840* (Cambridge, 1980)

Barry, J., 'Bourgeois collectivism? Urban association and the middling sort', in J. Barry and C. Brooks eds., *The Middling Sort of People* (London, 1994), 84–112

Batey, M., *Oxford Gardens: the University's Influence on Garden History* (Amersham, 1982)

Beckett, J. V., *The East Midlands since AD 1000* (London, 1988)

——, 'Responses to war: Nottingham in the French Revolutionary and Napoleonic wars, 1793–1815', *Midland History*, 22 (1997), 71–84

——, and Heath J. E., 'When was the industrial revolution in the East Midlands?', *Midland History*, 13 (1988), 77–94

——, *et al.* eds., *A Centenary History of Nottingham* (Manchester, 1997)

Benjamin, M., ed., *Science and Sensibility: Gender and Scientific Enquiry, 1780–1945* (Oxford, 1994)

Berg, M., *Luxury and Pleasure in Eighteenth-Century Britain*, second edition (Oxford, 2007)

Bermingham, A., *Landscape and Ideology: the English Rustic Tradition, 1740–1860* (Berkeley, 1986)

Bertucci, P., 'Revealing sparks: John Wesley and the religious utility of electrical healing', *British Journal for the History of Science*, 39 (2006), 341–62

Bewell, A., '"Jacobin plants": botany as social theory in the 1790s', *Wordsworth Circle*, 20 (1989): 132–9

Biagini, E., 'The complex pattern of opposition to the Nottingham enclosure', *East Midland Geographer*, 19, (1996), 20–37

Billinge, M., 'Hegemony, class and power in late-Georgian and early-Victorian England', in R. H. Baker and D. Gregory eds., *Explorations in Historical Geography* (Cambridge, 1984), 28–67

Black, J., *George III: America's Last King* (New Haven, 2006).

Blunt, A., 'Cultural geography: cultural geographies of home', *Progress in Human Geography*, 29 (2005), 505–15

———, and Dowling, R., *Home* (London, 2006)

Bolam, C. G., Goring, J., Short, H. L. and Thomas, R., *The English Presbyterians* (London, 1968)

Boniface, P., ed., *In Search of English Gardens: the Travels of John Claudius Loudon and his Wife Jane* (London, 1987)

Borsay, P., 'The London connection; cultural diffusion and the eighteenth-century provincial town', *London Journal*, 19 (1994), 21–35

———, *The English Urban Renaissance, 1660–1770* (Oxford, 1991)

———, 'The rise of the promenade: the social and cultural use of space in the English provincial town, c.1660–1800', *British Journal for Eighteenth-Century Studies*, 9 (1986), 125–40

———, *The Image of Georgian Bath, 1700–2000* (Oxford, 2000)

Bowler, P., *History of the Environmental Sciences* (London, 1992)

Bradley, J., 'Nonconformity and the electorate in eighteenth-century England', *Parliamentary History*, 6 (1987), 236–61

Brewer, J., *The Pleasures of the Imagination: English Culture in the Eighteenth Century*, (London, 1997)

Brighton, T., *The Discovery of the Peak District* (Chichester, 2004)

Brockway, L., *Science and Colonial Expansion: The Role of the British Botanic Gardens* (New York, 1979)

Brooke, J. H., 'Science and Dissent: some historiographical issues' in P. Wood ed., *Science and Dissent in England, 1688–1945* (Aldershot, 2004), 19–38

Browne, J., 'Botany for gentlemen: Erasmus Darwin and the Loves of the Plants', *Isis*, 80 (1989), 593–620

Bryant, M., *The London Experience of Secondary Education* (London, 1986)

Buckle, H. T., *History of Civilisation in England*, 3 vols. (London, 1869)

Bullock, S. C., *Revolutionary Brotherhood: Freemasonry and the Transformation of the American Social Order, 1730–1840* (Chapel Hill, 1996)

Burke, M. and Jacob M. C., 'French freemasonry, women and feminist scholarship', *Journal of Modern History*, 68 (1996), 513–49

Burton, E., *The Georgians at Home* (London, 1973)

Bury, J. B., *The Idea of Progress: An Inquiry into its Origin and Growth* (London, 1932)

Calhoun, C. ed., *Habermas and the Public Sphere* (Cambridge MA, 1992)

Calvert, H., *A History of Kingston-Upon-Hull* (Chichester, 1978)

Cameron Walker, W. C., 'The detection and estimation of electric charges in the eighteenth-century', *Annals of Science* (1936), 66–100

Cantor, G. N., 'Real disabilities?: Quaker schools as "nurseries" of science' in P. Wood ed., *Science and Dissent in England, 1688–1945* (Aldershot, 2004), 147–61

———, *Optics After Newton: Theories of Light in Britain and Ireland, 1704–1840* (Manchester, 1984)

Carter, H., *Sir Joseph Banks, 1743–1820* (London, 1988)

Chambers, J., *The English House* (London, 1985)

Chambers, J. D., *Nottinghamshire in the Eighteenth Century* (London, 1966)

Chapman, S. D., 'Working-class housing in Nottingham during the industrial revolution', *Thoroton Transactions*, 67 (1963), 69–77

Clarke, J. C. D., *English Society, 1660–1832: Religion, Ideology and Politics during the Ancien Regime*, second edition (Cambridge, 2000)

Clark P., *British Clubs and Societies 1580–1800: the origins of an associational world* (Oxford, 2000)

Clark, P. and Houston, R. A., 'Culture and leisure, 1700–1840', in P. Clark ed., *The Cambridge Urban History*, vol. 2 (Cambridge, 2000), 575–613

Clark, W., Golinski, J. and Schaffer, S. eds., *The Sciences in Enlightened Europe* (Chicago, 1999)

Clarke, J. R., 'The Royal Society and early grand lodge freemasonry', *AQC*, 80 (1967), 111–14

Clegg, J., *The Diary of James Clegg*, edited by V. S. Doe, (Matlock, 1981)

Clow, A. and Clow, N., *The Chemical Revolution: A Contribution to Social Technology* (London, 1952)

Coffey, D., 'Protecting the botanic garden: Seward, Darwin, and Coalbrookdale', *Women's Studies*, 31 (2002), 141–64

Cohen, I. B. ed., *Puritanism and the Rise of Modern Science* (New Brunswick, 1990)

Cohen, M., '"To think, to compare, to combine, to methodize": notes towards rethinking girls' education in the eighteenth century', in S. Knott and B. Taylor eds., *Women, Gender and Education* (London, 2005)

——, 'Gender and method in eighteenth-century English education', *History of Education*, 33 (2004), 585–95

Conlin, J., 'Vauxhall on the boulevard: pleasure gardens in London and Paris, 1764–1784', *Urban History*, 35 (2008), 24–47

——, 'Vauxhall revisited: the afterlife of a London pleasure garden 1770–1859', *Journal of British Studies*, 45 (2006), 718–43

Copeman, W. S. C., *The Worshipful Society of Apothecaries of London: A History* (London, 1967)

Corfield, P. 'A provincial capital in the late seventeenth century: The case of Norwich', in P. Clark ed., *The Early Modern Town* (London, 1976), 233–72

——, *Power and the Professions in Britain, 1700–1850* (London, 1995)

Cosgrove, D. and Daniels, S. eds., *The Iconography of Landscape* (Cambridge, 1988)

Cranfield, G. A., *The Development of the Provincial Newspaper* (Oxford, 1962)

Craven, M., *John Whitehurst; Clockmaker and Scientist* (Ashbourne, 1996)

Daniel, S. H., *John Toland: His Methods, Manners and Mind* (London, 1984)

Daniels, S., *Humphry Repton: Landscape Gardening and the Geography of Georgian England* (New Haven, 1999)

——, *Joseph Wright* (London, 1999)

——, 'Loutherbourg's chemical theatre: Coalbrookdale by night', in J. Barrell ed., *Painting and the Politics of Culture: New Essays on British Art, 1700–1850*, (Oxford, 1992), 195–230

——, 'The political iconography of woodland in later Georgian England', in D. Cosgrove and S. Daniels eds., *The Iconography of Landscape* (Cambridge 1988), 43–82

——, Seymour, S. and Watkins, C., 'Enlightenment, improvement and the geographies of horticulture in later Georgian England', in D. N. Livingstone and C. Withers eds., *Geography and Enlightenment* (Chicago, 1999), 345–71

Darwin, C., *Life of Erasmus Darwin*, edited by D. King-Hele (Cambridge, 2003)

Daston, L. and Park, K., *Wonders and the Order of Nature, 1150–1750* (New York, 1998)

Daunton, M. J., *Progress and Poverty: An Economic and Social History of Britain, 1700–1850* (Oxford, 1995)

Davies, J. C., *Georgian Harborough* (Market Harborough, 1969)

Deacon, M., *Scientists and the Sea, 1650–1900* (Aldershot, 1971)

Desmond, R., *Kew: The History of the Royal Botanic Gardens* (London, 1995)

Dickinson, H. T., *The Politics of the People in Eighteenth-Century Britain* (Basingstoke, 1994)

Ditchfield, G. M., 'The early history of Manchester College', *Transactions of the Historical Society of Lancashire and Cheshire*, 11 (1972)

Dixon Hunt, J., *The Picturesque Garden in Europe* (London, 2002)

Drayton, R., *Nature's Government: Science, Imperial Britain and the 'Improvement' of the World* (New Haven, 2000)

Drewitt, F. D., *The Romance of the Apothecaries' Garden at Chelsea*, third edition (Cambridge, 1928)

Duncan, J. S., and Lambert, D., 'Landscapes of home', in J. S. Duncan, N. C. Johnson and R.H. Schein eds., *A Companion to Cultural Geography* (Oxford, 2003), 382–403

Earle, P., *The Making of the English Middle Class* (London, 1991)

Edmondson, J., *William Roscoe and Liverpool's first Botanical Garden*, (Liverpool, 2005)

Edwards, J. K., 'Communications and trade 1800–1900', 'Industrial development of the city 1800–1900', in C. Barrinder ed., *Norwich in the Nineteenth Century* (Norwich, 1984), 119–59

Edwards, K. C., *The Peak District* (London, 1973)

Elliott, B., *The History of the Royal Horticultural Society, 1804–2004* (London, 2004)

Elliott, P., 'Erasmus Darwin's Trees', in L. Auricchio, E. H. Cook and G. Pacini, *Arboreal Values in Europe and North America, 1660–1830*, (forthcoming, 2011)

——, 'Provincial urban society, scientific culture and socio-political marginality in Britain in the eighteenth and nineteenth centuries', *Social History*, 28 (2003), 361–87

——, '"Improvement always and everywhere": William George Spencer (1790–1866) and mathematical, geographical and scientific education in nineteenth-century England', *History of Education*, 33 (2004), 391–417

——, 'The politics of urban improvement in Georgian Nottingham: the enclosure dispute of the 1780s', *Transactions of the Thoroton Society*, 110 (2006), 87–102

——, '"More subtle than the electric aura": Georgian medical electricity, the spirit and animation and the development of Erasmus Darwin's psychophysiology', *Medical History*, 52 (2008), 195–220.

——, *The Derby Philosophers: Science and Culture in English Urban Society, 1700–1850* (Manchester, 2009)

——, and Daniels, S., '"No study so agreeable to the youthful mind": geographical education in the Georgian grammar school', *History of Education* 39 (2010), 15–33

——, and Daniels, S., 'Pestalozzi, Fellenberg and British nineteenth-century geographical education', *Journal of Historical Geography*, 32 (2006), 752–74

Ellis, J., 'Regional and county centres', in P. Clark ed., *The Cambridge Urban History*, vol. 2, *1549–1840* (Cambridge, 2000), 673–704

——, '"For the honour of the town": comparison, competition and civic identity in eighteenth-century England', *Urban History*, 30 (2003), 325–37

Emerson, R. L., 'Science and the origins and concerns of the Scottish Enlightenment', *History of Science*, 26 (1988), 333–66

——, 'The Enlightenment and social structures', in, P. Fritz and D. Williams eds., *City and Society in the Eighteenth Century* (Toronto, 1973)

Eshet, D., 'Re-reading Priestley: science at the intersection of theology and politics', *History of Science*, 39 (2001), 127–59

Estabrook, C., *Urbane and Rustic England* (Manchester, 1998)

Everitt, A., *Landscape and Community in England* (Oxford, 1985)

——, 'Country, county and town: patterns of regional evolution in England', in P. Borsay ed., *The Eighteenth-Century Town* (London, 1990), 83–115

Falkus, M. E., 'Lighting in the dark ages of English economic history: town streets before the industrial revolution', in D. C. Coleman and A. H. John eds., *Trade, Government and Economy in Pre-Industrial England* (London, 1976), 248–73

Fawcett, T., '18th century debating societies', *British Journal for 18th century Studies*, 3 (1980)

——, 'Measuring the provincial Enlightenment: the case of Norwich', *Eighteenth-Century Life*, 8 (1982), 13–27

——, 'Popular science in eighteenth-century Norwich', *History Today*, 22 (1972), 590–95

Feinstein, C. H. ed., *York 1831–1981; 150 Years of Scientific Endeavour and Social Change* (York, 1981)

Fitzpatrick, M., 'Heretical religion and radical political ideas in late eighteenth-century England', in E. Helmuth ed., *The Transformation of Political Culture* (London, 1990), 339–72

Fitton, R. S. and Wadsworth, A. P., *The Strutts and the Arkwrights* (Manchester, 1958)

Fitzpatrick, M., 'Rational Dissent and the Enlightenment', *Faith and Freedom*, 38 (1985), 83–101

Ford, T. D., 'White Watson (1760–1835) and his geological tablets', *Mercian Geologist*, 13 (1995), 157–64

——, and Rieuwerts J. H. eds., *Lead Mining in the Peak District*, fourth edition (Ashbourne, 2000)

Foucault, M., *Discipline and Punish: The Birth of the Prison* (London, 1979)

Frere, A. S. ed., *Grand Lodge, 1717–1967* (Oxford, 1967)

Fulton, J. F., 'The Warrington Academy (1757–86)', *Bulletin of the Institute of the History of Medicine*, 1 (1933)

Galinou, M., ed., *London's Pride: The Glorious History of the Capital's Gardens* (London, 1990)

Gascoigine, J., *Joseph Banks and the English Enlightenment* (Cambridge, 1994)

George, D. M., *London Life in the Eighteenth Century* (London, 1965)

Gibbon, E., *The History of the Decline and Fall of the Roman Empire*, edited by D. Womersley, 3 vols. (London, 1994)

Girouard, M., *The English Town* (New Haven, 1990)

Gleason, E. and Watts, R., 'Making women visible in the history of education: the late eighteenth century experience', *History of Education Society Bulletin*, 59 (1997), 37–46

Goldgar, A., *Impolite Learning: Conduct and Community in the Republic of Letters, 1680–1750* (New Haven, 1995)

Golinksi, J., *Making Natural Knowledge: Constructivism and the History of Science* (Cambridge, 1998)

——, *British Weather and the Climate of Enlightenment* (Chicago, 2007)

——, *Science as Public Culture: Chemistry and Enlightenment in Britain, 1760–1820* (Cambridge, 1992)

Goodman, J., 'Troubling histories and theories: gender and the history of education' *History of Education*, 32 (2003), 157–74

Gould, W. F., *History of Freemasonry*, edited by H. Poole, 4 vols. (London, 1951)

Graham, J., 'Revolutionary philosopher: the political ideas of Joseph Priestley (1733–1804)', *Enlightenment and Dissent*, 8 (1989), 43–68

Gregory, D., 'Interventions in the historical geography of modernity: social theory, spatiality and the politics of representation', *Geografiska Annaler*, 73 (1991), 17–44

Gross, P. R., *Higher Superstition: The Academic Left and its Quarrels with Science* (Baltimore, 1998)

——, Levitt, N. and Lewis, M. eds., *The Flight from Science and Reason* (New York, 1996)

Gruber, H. E., *Darwin on Man: A Psychological Study of Scientific Creativity* (London, 1974)

Habermas, J., *The Structural Transformation of the Public Sphere*, translated by T. Burger and F. Lawrence (Cambridge MA, 1989)

Hackmann, W. D., *Electricity from Glass: The History of the Frictional Electrical Machine 1600–1850* (Netherlands, 1978)

Hamblyn, R., *The Invention of Clouds* (London, 2001)

Hankins, T. L., *Science and the Enlightenment* (Cambridge, 1985)

Hans, N., *New Trends in Education in the Eighteenth Century*, (London, 1951)

Haycock, D. B., *William Stukeley: Science, Religion and Archaeology in Eighteenth-Century England* (London, 2002)

Haworth, A. H., *Complete Works on Succulent Plants*, 5 vols. (London, 1965)

Heffernan, M., 'Edme Mentelle's geographies and the French Revolution', in D. N. Livingstone and C. W. J. Withers eds., *Geography and Revolution* (Chicago, 2005), 273–303

Heilbron, J. L., *Electricity in the 17th and 18th Centuries: A Study in Early Modern Physics*, revised edition (New York, 1999)

Helgerson, R., *Forms of Nationhood: The Elizabethan Writing of England* (Chicago, 1994)

Hey, D., *Derbyshire: A History* (Lancaster, 2008)

Herschel, J., *Memoir and Correspondence of Caroline Herschel* (London, 1876)

Hinton, D. A., 'Popular Science in England, 1830–79', unpublished PhD thesis, University of Bath (Bath, 1979)

Holt, R. V., *The Unitarian Contribution to Social Progress in England* (London, 1952)

Hook, A. and Sher, R. B. eds., *The Glasgow Enlightenment* (East Linton, 1997)

Hooper-Greenhill, E., *Museums and the Shaping of Knowledge* (London, 1992)

Hoskins, W. G., *The Making of the English Landscape* (London, 1970)

Howell, P., 'Public space and the public sphere: political theory and the historical geography of modernity', *Environment and Planning D: Society and Space*, 11 (1993), 303–22

Hudson, D. and Luckhurst, K. W., *The Royal Society of Arts, 1754–1954* (London, 1954)

Hughes, E., 'The early journal of Thomas Wright of Durham', *Annals of Science*, 7 (1951), 1–24

Hume, D., *Treatise of Human Nature*, edited by L. A. Selby-Biggge, revised by P. H. Nidditch (Oxford, 1978)

——, *Enquiry Concerning the Principles of Morals*, in *Essays, Moral, Political and Literary*, edited by T. H. Green and T. H. Grose (London, 1889)

Ignatieff, M., *A Just Measure of Pain: The Penitentiary in the Industrial Revolution, 1750–1850* (London, 1978)

Inwood, S., *A History of London* (London, 1998)

Inkster, I., *Scientific Culture and Urbanisation in Industrialising Britain* (Aldershot, 1997)

——, 'Scientific culture and education in Nottingham, 1800–1843', *Transactions of the Thoroton Society*, 82 (1978), 45–50

——, 'Scientific culture and scientific education in Liverpool prior to 1812 – a case study in the social history of education', in M. D. Stephens and G. W. Roderick eds., *Scientific and Technical Education in Early Industrial Britain* (Nottingham, 1981), 28–47

——, and Morrell, J. B. eds., *Metropolis and Province: Science in British Culture* (London, 1983)

Israel, J. I., *Enlightenment Contested: Philosophy, Modernity and the Emancipation of Man* (Oxford, 2008)

Jackson, G., *Hull in the Eighteenth Century: A Study in Economic and Social History* (Oxford, 1972)

Jacob, M., *Living the Enlightenment: Freemasonry in Eighteenth-Century Europe* (Oxford, 1992)

Jacob, M. C., *The Cultural Meaning of the Scientific Revolution* (New York, 1988)

——, *The Radical Enlightenment: Pantheists, Freemasons and Republicans*, (London, 1981)

Jankovic V., 'The place of nature and the nature of place: the chorographic challenge to the history of British provincial science', *History of Science*, 38 (2000), 80–113

——, *Reading the Skies; A Cultural History of English Weather, 1650–1820* (Chicago, 2000)

Jardine, L., *On a Grander Scale: The Outstanding Career of Sir Christopher Wren* (London, 2002)

Jardine, N., 'Inner history; or, how to end Enlightenment', in J. Clark, J. Golinski and S. Schaffer eds., *The Sciences in Enlightened Europe* (Chicago, 1999), 477–94

——, Secord, J. A. and Spary, C. eds., *Cultures of Natural History* (Cambridge, 1996)

Jewson, C. B., *The Jacobin City* (Glasgow, 1975)

Johnson, N. C., 'Cultivating science and planting beauty: the spaces of display in Cambridge's botanical gardens', *Interdisciplinary Science Reviews*, 31 (2006), 42–57

Jones, E. L. and Falkus M. E., 'Urban improvement and the English economy in the seventeenth and eighteenth centuries', in P. Borsay, ed., *The Eighteenth-Century Town, 1688–1800* (London, 1990), 116–58

Jones, P. ed., *Philosophy and Science in the Scottish Enlightenment* (London, 1988)

Jones, P. M., *Industrial Enlightenment: Science, Technology and Culture in Birmingham and the West Midlands, 1760–1820* (Manchester, 2009)

King-Hele, D., *The Collected Letters of Erasmus Darwin* (Cambridge, 2007)

——, *Erasmus Darwin and the Romantic Poets* (London, 1986)

——, *Erasmus Darwin: A Life of Unequalled Achievement* (London, 1999)

Kitteringham, G., 'Studies in the Popularisation of Science in England, 1800–30', unpublished PhD thesis, University of Sussex (Brighton, 1981)

——, 'Science in provincial society: the case of Liverpool in the early nineteenth century', *Annals of Science*, 39 (1982), 329–448

Knoop, G. and Jones, G. P., *The Genesis of Freemasonry* (Manchester, 1947)

Knorr-Cetina, K., *The Manufacture of Knowledge: An essay on the constructivist and contextual nature of science* (New York, 1981)

Langford, P., *A Polite and Commercial People: England, 1727–1783*, second edition (Oxford, 1998)

——, *Public Life and the Propertied Englishman, 1689–1798* (Oxford, 1991)

Langton, J., 'Urban growth and economic change: from the late seventeenth century to 1841', in P. Clark ed., *The Cambridge Urban History*, vol. 2 (Cambridge, 2000), 453–90

Latour, B. and Woolgar, S., *Laboratory Life: The Construction of Scientific Facts* (Princeton, 1981)

Leary, J. E., *Francis Bacon and the Politics of Science* (Ames, 1994)

Le Rougetel, H., *The Chelsea Gardener: Philip Miller, 1691–1771* (London, 1990)

Levere, T. H., 'Natural philosophers in a coffee house: Dissent, radical reform and pneumatic chemistry', in P. Wood ed., *Science and Dissent in England, 1688–1945* (Aldershot, 2004), 131–46

Levine, J. M., *Dr Woodward's Shield: History, Science, and Satire in Augustan England*, (London, 1977)

Lincoln, A., *Some Political and Social Ideas of English Dissent, 1763–1800* (Cambridge, 1971)

Livingstone, D. N., *Putting Science in Its Place: Geographies of Scientific Knowledge* (Chicago, 2003)

——, The spaces of knowledge: contributions towards a historical geography of science', *Environment and Planning D: Society and Space*, 13 (1995), 5–34

——, and Withers, C. W. J. eds., *Geography and Enlightenment* (Chicago, 1999)

Logan, J. V., 'The poetry and aesthetics of Erasmus Darwin', *Princeton Studies in English*, 15 (1936), 46–92

Longstaffe-Gowan, T., *The London Town Garden, 1700–1830* (London, 1990)

Looney, J., 'Cultural life in the provinces, Leeds and York, 1720–1820', in A.C. Beier, D. Cannadine and J. M. Rozenheim eds., *The First Modern Society: Essays in English History* (Cambridge, 1989), 483–512

Lowe, P. D., 'Locals and Cosmopolitans: A Model for the Social Organisation of Provincial Science in the Nineteenth Century', unpublished MPhil thesis, University of Sussex (Brighton, 1978)

Mallett, S., 'Understanding home: a critical review of the literature', *The Sociological Review*, 52 (2004), 62–89

Markus, T. A., 'Buildings and the ordering of minds and bodies', in P. Jones ed., *Philosophy and Science in the Scottish Enlightenment* (London, 1988), 169–224

Mayhew, R. J., 'Geography in eighteenth-century British education', *Paedogogica Historica*, 34 (1998), 731–69

——, 'The character of English geography *c.*1660–1800: a textual approach', *Journal of Historical Geography*, 24 (1998), 385–412

——, 'Mapping science's imagined community: geography as a Republic of Letters, 1600–1800', in *British Journal for the History of Science*, 38 (2005), 73–92

Mayr, E., *The Growth of Biological Thought* (Cambridge MA, 1982)

McClellan III, J. E., *Science Reorganised: Scientific Societies in the Eighteenth Century*, (New York, 1985)

McCracken, D. P., *Gardens of Empire: Botanical Institutions of the Victorian British Empire* (Leicester, 1997)

McGann, J., *The Poetics of Sensibility: A Revolution in Literary Style* (Oxford, 1996)

McLachlan, H., *English Education under the Test Acts* (Manchester, 1931)

——, *Warrington Academy: Its History and Influence* Chetham Society 107 (Manchester, 1943)

McNeil, M., 'The scientific muse: the poetry of Erasmus Darwin', *Languages of Nature: Critical Essays on Science and Literature*, edited by L. J. Jordanova (London, 1986)

——, *Under the Banner of Science: Erasmus Darwin and His Age* (Manchester, 1987)

McVeigh, S., 'Freemasonry and musical life in London in the late eighteenth century' in D. W. Jones ed., *Music in Eighteenth-Century Britain* (Aldershot, 2000), 72–100

Meeres, F., *A History of Norwich* (Chichester, 1998).

Melton, J. Van Horn, *The Rise of the Public in Enlightenment Europe* (Cambridge, 2001)

Mendyk, S. A. E., *'Speculum Britanniae': Regional Study, Antiquarianism and Science in Britain to 1700* (Toronto, 1989)

Menuge, A., 'The cotton mills of the Derbyshire Derwent and its tributaries', *Industrial Archaeology Review,* 16 (1993), 38–61

Mercer, E., *English Vernacular Houses* (London, 1975)

Mercer, M., 'Dissenting academies and the education of the laity, 1750–1850', *History of Education*, 30 (2001), 35–58

Merchant, C., *Reinventing Eden: The Fate of Nature in Western Culture* (New York, 2004)

Merton, R. K., *Science, Technology and Society in Seventeenth Century England* (Atlantic Highlands, 1978)

Millburn, H. C. and King, J. R., *Geared to the Stars: The Evolution of Planetariums, Orreries and Astronomical Clocks* (Toronto, 1978)

Minter, S., *The Apothecaries' Garden: A History of the Chelsea Physic Garden* (Stroud, 2000)

Money, J., *Experience and Identity: Birmingham and the West Midlands, 1760–1800* (Montreal, 1977)

——, 'Freemasonry and the fabric of loyalism in Hanoverian England', in E. Hellmuth ed., *The Transformation of Political Culture, England and Germany in the Late Eighteenth Century* (London, 1990), 235–70

——, 'The Masonic moment; or, ritual, replica, and credit: John Wilkes, the Macaroni parson, and the making of the middle-class mind', *Journal of British Studies*, 32 (1993), 359–95

——, 'Science, technology and Dissent in English provincial culture: from Newtonian transformation to agnostic incarnation', in P. Wood ed., *Science and Dissent in England, 1688–1945* (Aldershot, 2004), 67–112

Morrell, J. and Thackray, A., *Gentlemen of Science: The Early Years of the British Association for the Advancement of Science* (Oxford, 1981)

Morris, R. J., 'Civil society and the nature of urbanism: Britain, 1750–1850', *Urban History*, 25 (1998), 289–301

——, 'Clubs, Societies and Associations', in F. M. L. Thompson ed., *The Cambridge Social History of Britain, 1750–1950*, 3 vols. (Cambridge, 1990), III, 395–443

——, 'Voluntary societies and British urban elites, 1780–1850: an analysis', *The Historical Journal*, 26 (1983), 95–118

——, and Gunn S. eds., *Identities in Space: Contested Terrains in the Western City since 1850* (Aldershot, 2001)

Morton, A. Q. and Wess, J. A., *Public and Private Science: The King George III Collection* (London, 1993)

Mottelay, P. F., *Bibliographical History of Electricity and Magnetism* (London, 1922)

Mowl, T. and Earnshaw, B., *John Wood: Architect of Obsession* (Bath, 1988)

Munck, T., *The Enlightenment: A Comparative Social History, 1721–1794* (London, 2000)

Musson, A. E., and Robinson, A. E., *Science and Technology in the Industrial Revolution* (Manchester, 1969)

Myrone, M. and Peltz, L., *Producing the Past; Aspects of Antiquarian Culture and Practice, 1700–1850* (Aldershot, 1999)

Naylor, S., 'Historical geographies of science: places, contexts, cartographies', *British Journal for the History of Science*, 38 (2005), 1–12.

——, 'The field, the museum, and the lecture hall: the places of natural history in Victorian Cornwall', *Transactions of the Institute of British Geographers*, 27 (2002)

——, 'Nationalising provincial weather: Meteorology in nineteenth-century Cornwall', *British Journal for the History of Science*, 39 (2006), 1–27

Neale, R. S., *Bath: A Social History, 1680–1850* (London, 1981)

Neeson, J. M., *Commons: Common Right, Enclosure and Social Change in England, 1700–1820* (Cambridge, 1993)

Newman, A., 'Politics and freemasonry in the eighteenth century', *AQC*, 104 (1991), 32–50

Nicholson, B., *Joseph Wright of Derby, Painter of Light*, 2 vols. (New York, 1967)

Nisbet, R., *History of the Idea of Progress* (New York, 1980)

Noble, D. F., *A World without Women: The Christian Clerical Culture of Western Science* (New York, 1992)

O'Brien, P., *Warrington Academy 1757–86: Its Predecessors and Successors* (Wigan, 1989)

O'Day, R., *Education and Society, 1500–1800* (London, 1982)

O'Gorman, F., *Voters, Patrons and Parties: The Unreformed Electoral System of Hanoverian England* (Oxford, 1989)

Ogborn, M., *Spaces of Modernity: London's Geographies, 1680–1780* (London, 1998)

Ophir, A. and Shapin S., 'The place of knowledge: the spatial setting and its relation to the production of knowledge', *Science in Context*, 4 (1991), special issue

Orange, D., 'Rational Dissent and provincial science: William Turner and the Newcastle Literary and Philosophical Society', in I. Inkster and J. B. Morell eds., *Metropolis and Provinces: Science in British Culture, 1780–1850* (London, 1983)

Orange, A. D., *Philosophers and Provincials: the Yorkshire Philosophical Society from 1822 to 1844* (York, 1973)

Ostwald, H., *Electrochemistry: History and Theory* (Leipzig, 1896), translated by N.P. Date, 2 vols. (Washington, 1980)

Outhwaite, W. ed., *The Habermas Reader* (Cambridge MA, 1996)

Owen, D. M. ed., *The Minute-Books of the Spalding Gentlemen's Society, 1712–1755* (Lincoln, 1981)

Packham, C., 'The science and poetry of animation: personification, analogy, and Erasmus Darwin's *Loves of the Plants*', *Romanticism*, 10 (2004), 191–208

Page, M., 'The Darwin before Darwin: Erasmus Darwin, visionary science, and romantic poetry', *Papers on Language and Literature*, 41 (2005), 146–69

Pancaldi, G., *Volta: Science and Culture in the Age of Enlightenment* (Princeton, 2003)

Paneth, F. A., *Chemistry and Beyond*, edited by H. Dingle (New York, 1964)

Parker, I., *Dissenting Academies in England* (Cambridge, 1914)

Patterson, A. T., *Radical Leicester: A History of Leicester, 1780–1850* (Leicester, 1954)

Pawson, H. C., *Robert Bakewell: Pioneer Livestock Breeder* (London, 1957)

Pearson, J., *Stags and Serpents: The Story of the House of Cavendish* (London, 1983)

Peltonen, M. ed., *The Cambridge Companion to Bacon* (Cambridge, 1996)

Pera, M., *The Ambiguous Frog: the Galvani-Volta Controversy*, translated by J. Mandelbaum (Princeton, 1992)

Pevsner, N., *The Buildings of England: Derbyshire* (London, 1953)

Phillips, J., *Electoral Behaviour in Unreformed England* (Princeton, 1982)

Philo, C., 'Edinburgh, Enlightenment, and the geographies of unreason', in D. Livingstone and C. Withers eds., *Geography and Enlightenment* (Chicago, 1999), 372–98

Piggott, S., *Ancient Britons and the Antiquarian Imagination* (London, 1989)

Piggott, S., *William Stukeley, an Eighteenth-Century Antiquary*, second edition (London, 1985)

Plumb, J. H., 'The acceptance of modernity', in N. McKendrich, J. Brewer and J. H. Plumb eds., *The Birth of Consumer Society: the Commercialisation of Eighteenth-century England* (London, 1982), 316–34

——, 'The new world of children in eighteenth-century England', *Past and Present*, 67 (1975), 64–95

Pollard, S., *The Idea of Progress: History and Society* (London, 1968)

Porter, R., 'Enlightenment London and urbanity', in T. D. Hemming, E. Freeman and D. Meakin eds., *The Secular City* (London, 1994), 27–41

——, *Enlightenment: Britain and the Creation of the Modern World* (London, 2000)

——, *London: A Social History* (London, 1996)

——, *The Making of the Science of Geology* (Cambridge, 1977)

——, 'Science, provincial culture and public opinion in Enlightenment England', *British Journal for Eighteenth Century Studies*, 3 (1980), 20–46

——, ed., *The Cambridge History of Science, vol. 4: Eighteenth-Century Science* (Cambridge, 2003)

——, and Teich, M. eds., *The Enlightenment in National Context* (Cambridge, 1991)

Prescott, A., *Collected Studies in the History of Freemasonry* (Sheffield, 2003)

Pyenson, L. and Sheets-Pyenson, S., *Servants of Nature: A History of Scientific Institutions, Enterprises and Sensibilities* (London, 2000)

Raymond, J. and Pickstone, J., 'The natural sciences and the learning of the English Unitarians: an explanation of the roles of Manchester College', in B. Smith ed., *Truth, Liberty, Religion: Essays Celebrating Two Hundred Years of Manchester College* (Oxford, 1986), 127–38

Read, D., *The English Provinces, c. 1760–1960: A Study in Influence* (London, 1964)

Records of the Borough of Nottingham, VII, 1760–1800 (Nottingham, 1947)

Redd, M., 'The culture of small towns in England 1600–1800', in P. Clark ed., *Small Towns in Early-Modern Europe* (Cambridge, 2002), 121–47

Reddaway, F. T., *The Rebuilding of London after the Great Fire* (London, 1951)

Redwood, J., *Reason, Ridicule and Religion: The Age of Enlightenment in England, 1660–1750*, second edition (London, 1996)

Reed, M., 'The transformation of urban space, 1700–1840', in P. Clark ed., *The Cambridge Urban History*, vol. 2 (Cambridge, 2000), 615–40

Reed, P., 'Form and content: a study of Georgian Edinburgh', in T. A. Markus ed., *Order and Space in Society: Architectural Form and Its Context in the Scottish Enlightenment* (Edinburgh, 1982), 115–45

Relph, E., *Place and Placelessness* (London, 1976)

Roberts, J. M., 'Freemasonry: possibilities of a neglected topic', *English Historical Review*, 84 (1969), 323–35

Robinson, E., 'R. E. Raspe, Franklin's Club of Thirteen, and the Lunar Society', *Annals of Science*, 11 (1955)

Robinson, H., 'Geography in the Dissenting academies', *Geography*, 36 (1951), 179–86

Rousseau, G. S. and Porter, R. eds., *The Ferment of Knowledge* (Cambridge, 1980)

Ruggiu, F. J., 'The urban gentry in England, 1660–1780: a French approach', *Historical Research*, 74 (2001), 249–70

Russell, C., *Science and Social Change, 1700–1900* (Basingstoke, 1983)

Sachs, J. von, *History of Botany (1530–1860)*, translated by E. F. Garnsey and revised by I. B. Balfour (Cambridge, 1890)

Schaffer, S., 'Natural philosophy and public spectacle in the eighteenth century', *History of Science*, 21 (1983), 1–43

——, and Shapin S., *Leviathan and the Air-Pump: Hobbes, Boyle, and the Experimental Life* (Princeton, 1985)

Schiebinger, L., *The Mind Has no Sex?: Women and the Origins of Modern Science* (Cambridge MA, 1989)

——, 'The private life of plants: sexual politics in Carl Linnaeus and Erasmus Darwin', in M. Benjamin ed., *Science and Sensibility: Gender and Scientific Enquiry, 1780–1945* (Oxford, 1991)

Schofield, R. E., *The Lunar Society of Birmingham* (Oxford, 1963)

Secord, J., 'Newton in the nursery: Tom Telescope and the philosophy of tops and balls', *History of Science*, 23 (1985), 127–51

Seed, J., 'History and narrative identity: Religious Dissent and the politics of memory in eighteenth-century England', *Journal of British Studies*, 44 (2005), 46–63

——, 'Manchester College, York: an early nineteenth-century Dissenting academy', *Journal of Educational Administration and History*, 14 (1982)

Seymour, S. and Calcovoressi, R., 'Landscape parks and the memorialisation of empire: the Pierreponts' "naval seascape" in Thoresby Park, Nottinghamshire during the French Wars' 1793–1815, *Rural History* 18 (2007), 95–118

——, Watkins, S. and Daniels, S., 'Estate and Empire: Sir George Cornewall's management of Moccas, Herefordshire and La Taste, Grenada, 1771–1819', *Journal of Historical Geography*, 24 (1998), 313–51

Shapin, S., 'The house of experiment in seventeenth-century England', *Isis*, 79 (1988), 373–404

——, 'Placing the view from nowhere: Historical and sociological problems in the location of science, *Transactions of the Institute of British Geographers*, 23 (1998), 5–12

——, *A Social History of Truth: Science and Civility in Seventeenth-Century England* (Chicago, 1995)

——, *The Scientific Revolution* (Chicago, 1998)

——, and Ophir A., 'The place of knowledge: a methodological survey', *Science in Context*, 4 (1991), 3–21

——, and Schaffer S., *Leviathan and the Air Pump: Hobbes, Boyle and the Experimental Life*, (Princeton, 1985)

Shapiro, B. J., *Probability and Certainty in Seventeenth-Century England* (London, 1983)

Shteir, A. B., *Cultivating Women, Cultivating Science: Flora's Daughters and Botany in England, 1760–1860* (Baltimore, 1996)

Simo, M. L., *Loudon and the Landscape: From Country Seat to Metropolis* (New Haven, 1988)

Simon, B., *The Two Nations and the Educational Structure, 1780–1870* (London, 1976)

Skedd, S., 'The Education of Women in Hanoverian Britain, c.1760–1820', DPhil. thesis, University of Oxford (1997)

——, 'Women teachers and the expansion of girl's schooling in England in H. Barker and E. Chalus eds., *Gender in Eighteenth-Century England* (London, 1997), 101–25

Smith, B. ed., *Truth, Liberty, Religion: Essays Celebrating Two Hundred Years of Manchester College* (Oxford, 1986)

Smith, J. W. A., *The Birth of Modern Education* (London, 1954)

Smith, C. U. M., and Arnott, R. eds., *The Genius of Erasmus Darwin* (Aldershot, 2005)

Snell, A. P. F., 'Philosophy in the eighteenth-century Dissenting academies', *History of Universities*, 11 (1992), 75–122

Snell, K. D. M. and Ell, P. S., *Rival Jerusalems: The Geography of Victorian Religion* (Cambridge, 2000)

Snobelin, S. D., 'The discourse of God: Isaac Newton's heterodox theology and his natural philosophy', in P. Wood ed., *Science and Dissent in England, 1688–1945* (Aldershot, 2004), 39–66

Sorrenson, R., 'Towards a history of the Royal Society in the eighteenth century', *Notes and Records of the Royal Society of London*, 50 (1996), 29–46

Spadafora, D., *The Idea of Progress in Eighteenth-Century Britain* (New Haven, 1990)

Spary, E., *Utopia's Garden; French Natural History from Old Regime to Revolution* (Chicago, 2000)

Stacpoole, A. ed., *The Noble City of York* (York, 1972)

Stafford, B. M., *Body Criticism: Imaging the Unseen in Enlightenment Art and Medicine* (Cambridge MA, 1991)

Stephens, W. B., 'Illiteracy and schooling in the provincial towns, 1640–1870; a comparative approach', in D. A. Reader ed., *Urban Education in the Nineteenth Century* (London, 1977), 27–47

Stevens, P. F., *The Development of Biological Systematics: Antoine-Laurent de Jussieu, Nature and the Natural System* (New York, 1994)

Stevenson, D., *The Origins of Freemasonry: Scotland's Century, 1590–1710* (Cambridge, 1998)

Stewart, L., 'The public culture of radical philosophers in eighteenth-century London', in P. Wood ed., *Science and Dissent in England, 1688–1945* (Aldershot, 2004), 113–29

——, *The Rise of Public Science: Rhetoric, Technology and Natural Philosophy in Newtonian Britain, 1660–1750* (Cambridge, 1992)

——, 'Seeing through the scholium: religion and reading Newton in the eighteenth century', *History of Science*, 34 (1996), 123–65

Stewart, W. A. C. and McCann, W. P., *The Educational Innovators, 1750–1880* (Basingstoke, 1967)

Stobart, J., 'Culture versus commerce: societies and spaces for elites in eighteenth-century Liverpool', *Journal of Historical Geography*, 28 (2002), 471–85

——, *Spaces of Consumption: Shopping and Leisure in the English Town, c. 1680–1830* (London, 2007)

Sullivan, R. E., *John Toland and the Deist Controversy; a Study in Adaptations* (London, 1982)

Summerson, J., *Architecture in Britain, 1530–1830*, eighth edition (London, 1991)

——, *Georgian London* (London, 1962)

Sweet, R., *Antiquaries: The Discovery of the Past in Eighteenth-Century Britain* (London, 2004)

——, *The Writing of Urban Histories in Eighteenth-Century England* (Oxford, 1997)

——, *The English Town, 1680–1840* (London, 1999)

Taylor, E. G. R., *The Mathematical Practitioners of Hanoverian England, 1714–1840* (Cambridge, 1966)

Teeter Dobbs, B. J. and Jacob, M. C., *Newton and the Culture of Newtonianism* (New Jersey, 1995)

Teute, F. J., 'The Loves of the Plants; or, the cross-fertilization of science and desire at the end of the eighteenth century', *The Huntington Library Quarterly*, 63 (2000), 319–45

Thacker, C., *The Genius of Gardening* (London, 1994)

Thackray, A., 'Natural knowledge in cultural context: the Manchester model', *American Historical Review*, 79 (1974), 672–709

Thomas, K., *Man and the Natural World* (London, 1984)

Thomis, M., *Politics and Society in Nottingham, 1785–1835* (London, 1969)

Thompson, E. P., 'Time, work and industrial capitalism', in *Customs in Common* (London, 1994)

——, 'Hunting the Jacobin fox', *Past and Present*, 142 (1994), 94–140.

Tillot, P. M. ed., *The Victoria County History of Yorkshire, The City of York* (London, 1961)

Tinniswood, A., *His Invention So Fertile: A Life of Christopher Wren* (London, 2001)

Tompson, R. S., *Classics or Charity: The Dilemma of the Eighteenth-Century Grammar School* (Manchester, 1971)

Trinder, B., *The Making of the Industrial Landscape* (Gloucester, 1982)

Tuan, Y. F., *Topophllia: A Study of Environmental Perception, Attitudes, and Values* (Englewood Cliffs NJ, 1974)

Turbutt, G., *A History of Derbyshire*, 4 vols. (Cardiff, 1999)

Tyler-Whittle, M. and Cook, C., *Curtis's Flower Garden Displayed* (Leicester, 1991)

Uglow, J., 'But what about the women? The Lunar Society's attitude to women and science, and to the education of girls', in C. U. M. Smith and R. Arnott eds., *The Genius of Erasmus Darwin* (Aldershot, 2005), 179–94

——, *The Lunar Men: the Friends who Made the Future* (London, 2002)

Vickery, A., *The Gentlemen's Daughter: Women's Lives in Georgian England* (New Haven, 1998)

Wahrman, D., 'National society, communal culture: an argument about the recent historiography of eighteenth-century Britain', *Social History*, 17 (1992), 43–72

Walker, A., 'The Glasgow grid', in T. Markus ed., *Order in Space and Society*, 155–99

Wallis, R. V. and Wallis, P. J., *Biobibliography of British Mathematics and Its Applications* (Newcastle, 1986)

Walters, A. N., 'Science and politeness in eighteenth-century England', *History of Science*, 35 (1997), 121–54

Walters, S. M., *The Shaping of Cambridge Botany* (Cambridge, 1981)

Watson, F., *The Beginnings of the Teaching of Modern Subjects in England* (London, 1909)

Watson, J. S., *The Reign of George III* (London, 1960)

Watson, W., 'People, prejudice and place', in, F. W. Boal and D. N. Livingstone eds., *The Behavioural Environment: Essays in Reflection, Application, and Re-evaluation* (London, 1989), 93–110

Watts, M., *The Dissenters*, 2 vols. (Oxford, 1995)

Watts, R., *Gender, Power and the Unitarians in England, 1760–1860* (London, 1998)

——, *Women in Science: A Social and Cultural History* (Abingdon, 2007)

Weatherill, L., *Consumer Behaviour and Material Culture in Britain, 1660–1760*, second edition (London, 1996)

Webb, S. and B., *English Local Government* IV, Statutory Bodies for Special Purposes (London, 1922)

Weisberger, R. W., *Speculative Freemasonry and the Enlightenment* (Boulder, 1993)

Wigelsworth, J. R., 'Competing to popularise Newtonian philosophy: John Theophilus Desaguliers and the preservation of reputation', *Isis*, 94 (2003), 435–55

Wiles, R. M., 'Provincial culture in early-Georgian England', in P. Fritz and D. Williams eds., *The Triumph of Culture: 18th-Century Perspectives* (Toronto, 1972), 49–68

Wilmot, S., *The Business of Improvement: Agriculture and Scientific Culture in Britain, c. 1700–1870*, Historical Geography Research Series 24 (Cheltenham, 1990)

Wilson, A., 'The cultural identity of Liverpool, 1790–1850: the early learned societies', *Transactions of the Historic Society of Lancashire and Cheshire*, 147 (1997), 58–73

Withers, C., *Geography, Science and National Identity: Scotland since 1520* (Cambridge, 2002)

———, *Placing the Enlightenment: Thinking Geographically about the Age of Reason* (Chicago, 2007)

Withers, C. W. J. and Mayhew, R., 'Rethinking "disciplinary" history: geography in British universities', *Transactions of the Institute of British Geographers*, 27 (2002), 11–29

———, and Ogborn, M. eds., *Georgian Geographies: Essays on Space, Place and Landscape in the Eighteenth Century* (Manchester, 2004)

Wittkower, R., *Palladio and English Palladianism* (London, 1974)

Wolf, A., *A History of Science, Technology and Philosophy in the 18th Century*, revised edition, 2 vols. (New York, 1961)

Wood, P., 'Science and the Aberdeen Enlightenment', in P. Jones ed., *Philosophy and Science in the Scottish Enlightenment* (Edinburgh, 1988)

———, ed., *Science and Dissent in England, 1688–1945* (Aldershot, 2004)

Woolley, A. R., *The Clarendon Guide to Oxford* (London, 1963)

Wormald, B. H. G., *Francis Bacon: History, Politics and Science, 1561–1626* (Cambridge, 1993)

Wroth, W. W., *Cremorne and the later London Gardens* (London, 1907)

———, *The London Pleasure Garden of the Eighteenth Century* (London, 1898)

Wykes D. L., 'The contribution of the Dissenting academy to the emergence of rational Dissent', K. Haakonssen ed., *Enlightenment and Religion: Rational Dissent in Eighteenth-Century Britain* (Cambridge, 1996), 99–139

———, 'Religious Dissent and the penal laws: an explanation of business success?' *History*, 75 (1990), 39–62

Yelling, J. A., *Common Field and Enclosure in England, 1450–1850* (London, 1977)

Youngson, A. J., *The Making of Classical Edinburgh, 1750–1840* (Edinburgh, 1966)

Index